R00069 59972

CHICAGO PUBLIC LIBRARY
HAROLD WASHINGTON LIBRARY CENTER
R0006959972

D1786098

REF QH IMCO/FAO/UNESCO/WMO/
QH WHO/IAEA/UN Joint Group of
545 Experts on the Scientific
.05 Aspec Impact of oil on the
I18 on. marine environment
1977

Cop.1

DATE			

FORM 125 M

BUSINESS/SCIENCE/TECHNOLOGY

The Chicago Public Library

JAN 9 1980

Received

© THE BAKER & TAYLOR CO.

M-45
ISBN 92-5-100219-3

The copyright in this book is vested in the Food and Agriculture Organization of the United Nations. The book may not be reproduced, in whole or in part, by any method or process, without written permission from the copyright holder. Applications for such permission, with a statement of the purpose and extent of the reproduction desired, should be addressed to the Director, Publications Division, Food and Agriculture Organization of the United Nations, Via delle Terme di Caracalla, 00100 Rome, Italy.

© FAO 1977

The designations employed and the presentation of material in this publication do not imply the expression of any opinion whatsoever on the part of the Food and Agriculture Organization of the United Nations concerning the legal status of any country, territory, city or area or of its authorities, or concerning the delimitation of its frontiers or boundaries.

REPORTS AND STUDIES No. 6

IMCO/FAO/UNESCO/WMO/WHO/IAEA/UN Joint Group of Experts
on the Scientific Aspects of Marine Pollution (GESAMP)

IMPACT OF OIL ON THE MARINE ENVIRONMENT

FOOD AND AGRICULTURE ORGANIZATION OF THE UNITED NATIONS
ROME, January 1977

PREPARATION OF THIS STUDY

This document is the edited and approved original English version of the Report of the GESAMP Working Group on the Impact of Oil on the Marine Environment, which met from 20 to 21 March 1974 in Geneva (preliminary meeting in conjunction with the Sixth Session of GESAMP), from 28 October to 1 November 1974 in Rome, FAO Headquarters, and from 17 to 24 September 1975, again in Rome. The following members participated in the preparation of the Report: R.A.A. Blackman, S. Genovese, P.G. Jeffery, A.B. Jernelöv, E.M. Levy, O.G. Mironov, K.H. Palmork, C.H. Thompson (Chairman), G. Tomczak, FAO (Technical Secretary) and S.L.D. Young, IMCO. Contributions to the work of the Working Group were also made by: H.A. Cole, R.R. Colwell, M.G. Ehrhardt, J. Farrington, C.R. Gentry, J. Schwartz, J.D. Walker and O.C. Zafiriou.

The Working Group was charged with identifying effects of specific concern to environmental scientists arising from any physical, chemical or biological impact of oil on the marine environment and its living resources, and with identifying specific references and the significant assumptions and short-comings present therein. The activities of the Working Group were financially supported by the United Nations Environment Programme.

For bibliographic purposes, this document may be cited as:

IMCO/FAO/UNESCO/WMO/WHO/IAEA/UN Joint Group of Experts on the Scientific Aspects of Marine
    1977   Pollution (GESAMP), Impact of Oil on the Marine Environment. Rep.Stud.GESAMP,
              (6):250 p.

CONTENTS

Part I: SUMMARY

		Page
	ABSTRACT	
1.	IMPLICATIONS OF OIL AND DISCHARGE SOURCES	1
2.	FATES OF OIL – ROUTES OF OIL EXPOSURE	2
3.	ANALYTICAL TECHNOLOGY	4
4.	CHEMICAL AND PHYSICAL EFFECTS OF OIL DISCHARGES	5
5.	EFFECTS ON MARINE LIFE FORMS	6
6.	HUMAN EFFECTS FROM OIL DISCHARGES	8

Part II: RESOURCE DOCUMENT ON IMPACT OF OIL ON THE MARINE ENVIRONMENT

CHAPTER 1   IMPLICATIONS OF OIL AND DISCHARGE SOURCES   13

    1.1 Descriptions of oil   13
    1.2 Oil types and significance to marine living resources   13
    1.3 Quantities of oil discharged into the marine system   18
    1.4 Rationale for classification of sources of oil pollution
        with respect to ecological damage   18

CHAPTER 2   FATES OF OIL – ROUTES OF OIL EXPOSURE   22

    2.1 Physical, chemical and microbiological factors influencing
        the fate of oil   22
        2.1.1 Spreading   22
        2.1.2 Evaporation   24
        2.1.3 Solution   25
        2.1.4 Emulsification   26
        2.1.5 Effects of dispersants   26
        2.1.6 Sedimentation   27
        2.1.7 Oxidation   28
        2.1.8 Chemical degradation   28
        2.1.9 Biological fates (general)   29
        2.1.10 Microbial degradation   29
        2.1.11 Tar balls   37
    2.2 States of oil available for uptake   38
        2.2.1 Uptake mechanisms   38
        2.2.2 Storage, metabolism, discharge   39
    2.3 Biomagnification   41

CHAPTER 3   ANALYTICAL TECHNOLOGY   42

    3.1 Collection of samples   42
    3.2 Isolation and separation   42
    3.3 Chemical analysis   45
    3.4 Reporting analyses   45
    3.5 Comments   46

CHAPTER 4   CHEMICAL AND PHYSICAL EFFECTS OF OIL DISCHARGES   48

    4.1 Gas transfer and deoxygenation   48
    4.2 Heating effects   50
    4.3 Pollutant sorption   51
    4.4 Hydrocarbon interference with carbon dioxide transfer   52

	Page

CHAPTER 5 EFFECTS ON MARINE LIFE FORMS ... 53

    5.1 Effects of oil on marine organisms ... 55
        5.1.1 Birds ... 55
        5.1.2 Mammals ... 58
        5.1.3 Fish ... 58
        5.1.4 Effects on fisheries ... 62
        5.1.5 Turtle fisheries ... 63
        5.1.6 Benthic and intertidal organisms ... 63
        5.1.7 Zooplankton ... 69
        5.1.8 Phytoplankton ... 70
        5.1.9 Macroscopic Algae ... 73
        5.1.10 Marine grasses ... 74
        5.1.11 Other marine plants ... 75
    5.2 Effects of oil on marine populations ... 75
        5.2.1 Birds ... 75
        5.2.2 Mammals and other higher marine life forms ... 76
        5.2.3 Plankton ... 76
        5.2.4 Microbial populations ... 76
    5.3 Effects of oil on the behaviour of marine organisms ... 77
    5.4 Miscellaneous topics ... 78
        5.4.1 Bioassay methods of determining the toxicity of oils ... 78
        5.4.2 Effects of oil on early stages of growth of marine organisms ... 86
        5.4.3 Habitat alteration and destruction ... 89

CHAPTER 6 HUMAN EFFECTS FROM OIL DISCHARGES ... 91

    6.1 Carcinogenic ... 91
        6.1.1 Problem – General ... 91
            6.1.1.1 Occurrence in oils ... 91
            6.1.1.2 Levels of PNAH's in oils ... 91
            6.1.1.3 Levels in other products and wastes ... 92
            6.1.1.4 Relative contribution of oil-derived PNAH's ... 92
            6.1.1.5 Levels in marine produce and other foods ... 93
            6.1.1.6 Summary ... 96
        6.1.2 Problem 1: What is the increased health hazard to man from oil-derived PNAH's ... 96
            6.1.2.1 Bioaccumulation and biomagnification ... 97
            6.1.2.2 Correlation of oil derived hydrocarbons with position in the food chain ... 100
            6.1.2.3 Summary of evidence for accumulation or discharge of PNAH's ... 108
        6.1.3 Problem 2: Threshold dose and hazard to man of elevated PNAH contents ... 110
        6.1.4 Problem 3: Induction of carcinomas in marine produce ... 112
            6.1.4.1 Summary of cancer induction in marine produce ... 115
        6.1.5 Mutagenicity and teratogenicity ... 116
    6.2 Loss of marine foods ... 116
        6.2.1 Definitions ... 116
        6.2.2 Problems ... 116
        6.2.3 Background ... 117
        6.2.4 Tainting ... 118
        6.2.5 Data ... 118
        6.2.6 Comments and conclusions ... 128

Part III: BIBLIOGRAPHY

		Page
1.	REFERENCES TO PART II	129
2.	ADDITIONAL RELATED LITERATURE	165
	2.1 Analytical	165
	2.1.1 Methods of taking samples	165
	2.1.2 Detection	165
	2.1.3 Chromatography	170
	2.1.4 Spectrophotometry	172
	2.1.5 General	173
	2.2 Effects	176
	2.2.1 Plants and plankton	176
	2.2.2 Shellfish	178
	2.2.3 Fish	180
	2.2.4 Other organisms	182
	2.2.5 Carcinogens	185
	2.2.6 Case histories	189
	2.2.7 Other effects and properties	196
	2.3 Dispersant/Detergents	200
	2.4 Biodegradation	207
	2.5 Levels of hydrocarbons	215
	2.6 Natural seeps/Offshore operations	217
	2.7 Oil lumps in the sea	219
	2.8 Properties of oil	220
	2.9 General/Discussions	223
	2.10 Symposia/Conferences	230
	2.11 Selected bibliographies	236
3.	AUTHOR INDEX	238

ABSTRACT

This Report has been prepared to enable the effects of oil in its several forms and kinds on the marine environment to be evaluated.

The original areas of interest as expressed in GESAMP VI/3 paragraph 8 were to assemble evidence upon the impact of oil on the marine environment and its living resources, covering tainting, deoxygenation, effects of microbial populations, carcinogenesis, lethal effects, sublethal effects, heating effects, routes into and within biosystems. GESAMP VI provided the additional guidance that work should include the effect of oil on sorption of chlorinated hydrocarbons and heavy metals and the interference with gas exchange, involving the transfer of carbon dioxide across the air/sea interface. GESAMP VII requested that the Working Group ensure their report be written bearing in mind the needs of governments of developing countries.

In response to this requirement, the impact of the variety of oils, as defined in Appendix I of Annex I of the International Convention for the Prevention of Pollution from Ships (IMCO, 1973), on the marine environment has been evaluated. An estimated total of 2-20 million tons of oil are discharged to the marine environment annually, with contributions from a number of sources. Of these, losses from marine transportation and from river run-off have the greatest and most direct impact. The fate of the oil can be estimated for times of a few hours to months and depends upon environmental factors such as sea state and temperature, upon the properties of the oil such as density, solubility of certain components, volatility, and upon the relative extent of processes such as photochemical oxidation and microbial activity. Analytical technology has not yet reached the state where standardized methods can be recommended, and difficulties are encountered in comparing one study to another. Evidence is not yet available to indicate that oil layers in the marine environment hinder gas exchange across the sea/air interface, that heating effects are more than a local problem or that a concentration of pesticides and heavy metals occurs in the oil. The effect of oil on marine life has been noted in laboratory and field studies. The most obvious global effect is the mortality of marine birds which, for certain species, can be serious. Likewise, bottom fauna and littoral ecosystems may locally be seriously affected. The loss of seafood due to tainting by oil is a matter of major concern to man. Evidence has not been found in this study that an unequivocal link exists between the presence in the sea of carcinogens found in the oil and carcinomas in marine organisms or of the presence in seafood of carcinogens from oil and cancer in man.

Part I: SUMMARY

Chapter 1: IMPLICATIONS OF OIL AND DISCHARGE SOURCES

Statement of the Problem

The Group considered the problems associated with definition and description of oil and the significance of the various sources of discharge to the marine environment. Consideration was given to the reliability of data indicating the total amount of oil discharged, the amounts of each type of oil, the amounts derived from particular sources and the significance of the toxic components from each source.

Summary of Observations and Facts

Pollution by crude oil arises from tanker accidents, deballasting operations and tank washing as well as from natural seepages and losses from offshore production. Tank washing and accidents also release fuel oil and other refined products to the marine environment.

Coastal refineries that utilize salt water cooling, discharge oil in cooling waters. The direct discharge of untreated municipal and industrial wastes contributes many types of refined and partly weathered oils, but not crude oil, to the marine environment. Surface waters discharged through storm sewers and rivers also contain small concentrations of similar refined and partly weathered oils but may contribute large amounts.

A number of authors have given estimates of the contribution of oil or hydrocarbons to the marine environment from different sources, ranging from 2 to 20 million tons per annum.

The contribution of oil from the atmosphere is less well understood, but is probably largely short-chain low-boiling hydrocarbons and compounds produced from them by oxidation. The atmospheric contribution of hydrocarbons is thought to have increased water solubility compared to other sources of hydrocarbons. Figures ranging from 0.1 to 10 million tons annually have been mentioned. The higher estimates are generally derived from an arbitrarily chosen figure on the percentage of evaporated oil hydrocarbons that "ought to return". The alternative method of measuring oil in precipitation and multiplying by overall precipitation over the oceans generally gives a lower figure. The Group accepted the latter approach and regarded the figure, presented by the U.S. National Academy of Sciences (1973-1975) oil pollution study group, of 0.6 million tons per year as being the best available estimate. The toxicity of oil from this source is a matter of concern. The short-chained low-boiling fractions of high volatility dominate. In any series, these are the ones of highest acute toxicity. However, after evaporation and passage through the atmosphere, they can be considerably altered prior to washout. The distribution patterns resulting from evaporation, atmospheric transport and washout is such that high local concentrations are unlikely to occur. The source "river run-off" poses a similar problem in regard to the toxicity. When reaching the coast, the oil has, in most cases, been subjected to weathering and thus lost a part of the more acutely toxic compounds. It is also "discharged" in a comparatively dilute form. On the other hand, the areas where it enters the ocean - the mouths of rivers - are frequently important breeding and feeding grounds for a variety of marine organisms. Different opinions exist as to the sensitivity of these ecosystems. However, compared to, for example, coral reef ecosystems, they are likely to be relatively tolerant of oil pollution.

Considering the list of oils[1], it is apparent that lack of internationally common terms describing various oils and lack of specific information on those oils make it very

---

[1] International Convention for the Prevention of Pollution from Ships, Annex I, Appendix I (IMCO, 1973)

difficult to even begin to estimate quantities of them. However, because there is more crude oil shipped and handled than refined products, it is highly probable that there is proportionally more crude oil directly discharged.

Toxicity has been related primarily to the aromatic fractions, and most of the limited data existing were developed under laboratory conditions using bioassay techniques on specific species which may not relate to the actual toxic or indirect toxic effects of oil discharges into complex ecosystems.

Findings of the Group

A categorization is essential for evaluating the fate and effect of crude oils and refined products to allow more efficient use of available and future data. Specifications must be provided for the evaluation of the list of oils[1]/.

A group should be established to examine the physical and chemical properties of oils and to develop a rationale for categorizing the materials. Information needed would include: origin of crude oils; quantity produced; quantity shipped and brief description of refining process used; quantities and origins of products; an exaplanation suitable for international understanding of the physical and chemical properties and uses of each oil, including total aromatic content, solubility, percentage composition by distillation range, viscosity, density, colour, sulphur content of tar fractions and phenolic content. It is recommended that workers in this field should include descriptions of the oils they are using. It is also recommended (Chapter 4) that an international oil reference material library should be established.

It is difficult to evaluate the relative significance of sources in terms of toxic effect. The low-boiling fractions of crude oils and refined products evaporate and become part of the hydrocarbon burden of the atmosphere. This source, although significant in size, is a diffused source and would not pose the same concentration and direct toxic threats to the biology of a given marine system as point sources.

Chapter 2: FATES OF OIL - ROUTES OF OIL EXPOSURE

Statement of the Problem

The Group considered the problem of the fate of oil, the pathways of exposure and degradation and focussed attention on the following problems:

- Routes into and within biosystems. - What are the physical and chemical factors that affect the availability of oil to biological systems?

- What are the factors which influence the rate and completeness of microbial degradation of oil in the marine environment? - To what extent is this due to a specific bacterial flora or to the adaptive capacities of numerous heterotrophic micro-organisms?

- Do petroleum and the products of petroleum biodegradation interfere with the activities of autotrophic and heterotrophic bacteria governing the fundamental cycles of carbon, nitrogen and phosphorus?

---

[1]/ see page 1

Summary of Observations and Facts

A variety of simultaneous physico-chemical processes govern the distribution and fate of oil immediately after a discharge or spill.

Spreading out on the surface is initially the dominating process. Evaporation of short-chained volatile compounds proceeds rapidly and this is highly temperature dependent. For example, at $20^\circ C$, compounds with up to 12 carbon atoms largely evaporate within a few hours. Partial loss by evaporation occurs within a few days for compounds with up to 22 carbon atoms.

The formation of emulsions of water-in-oil and oil-in-water depend on turbulence, but are of importance during the period days to weeks after the introduction of oil into the marine environment. The dissolution of oil components in water progresses continuously, but being dependent on the surface area of the oil it is most pronounced when a dispersion of oil-in-water is formed. Evaporation of lighter fractions and incorporation of denser suspended solids will eventually cause most oils to sink. This process is often most significant within weeks after the discharge.

Photochemical and other oxidative processes are at their greatest relative importance during the first weeks after the incident.

Over 90 species of micro-organisms including open ocean, coastal and estuarine bacteria, fungi and yeasts, capable of degrading petroleum by biological oxidation, have been identified. Their action does not become important until a week or so after an oil discharge. All kinds of oils are susceptible to microbiological degradation. Assuming the presence of appropriate micro-organisms and recognizing the other physical and chemical factors, the most important factor which influences the biodegradability of hydrocarbons seems to be the molecular configuration. Alkanes are attacked by more microbial species more rapidly than either aromatic or naphthenic compounds.

Tar balls, because of varying size and density, are distributed throughout the water column - from the surface to the sea bed. Some are formed soon after oil is discharged from tanker washings, others form from crude oil and heavy oil products over a longer period of time in the marine environment.

Bioconcentration has been demonstrated while bioaccumulation and biomagnification of oil are possible, but there is little convincing evidence of biomagnification.

Findings of the Group

It is concluded that the physical factors have the most significant initial effect upon oil discharged into the marine environment by spreading (both on the surface and in sediments), evaporation, dissolution and aerosol formation, emulsification, sorption onto particulate matter and settling of the oil, thereby altering the potential impact on the living marine resources. Chemical and photochemical factors such as oxidation, photolysis and polymerization are of next importance. Although biological activity occurs within a few days of a discharge, it is only significant after a period of weeks. The rate of degradation varies widely with the type of oil and the local conditions. The variations are large and the knowledge at present does not exist to combine the results into a reasonable estimate of total microbial degradation of oil per year in the global marine environment. Many different measurements are used to express the rate of degradation, e.g., 0.02 to 2.0 $g/m^2/day$; 350 $g/m^3/year$; 45% loss in 30 days.

Comparative studies are needed on the fate and rate of degradation of heavy oil fractions in sea water and bottom sediments. Studies are required of adsorption on particulate matter and sedimentation of slicks and dispersed oil under field or simulated conditions. Field studies should be conducted to enable the extrapolation of existing laboratory data to be verified.

Studies on the following topics should be conducted:

(i) the rate of formation and the persistence of the oil degradation products;

(ii) adaptation of bacterial mechanisms to hydrocarbon oxidation;

(iii) mechanisms and rate of degradation by micro-organisms under anaerobic conditions.

Chapter 3: ANALYTICAL TECHNOLOGY

Statement of the Problem

The Group considered the problem of analytical technology, including questions such as method of surface sampling and intercalibration of methods and techniques of measurement.

Summary of Observations and Facts

Analytical technology includes sampling, isolation, separation and methods of analysis of individual or groups of hydrocarbons (e.g., aromatics) in sediments, sea water, air and organisms.

Insufficient attention has been given to the collection of representative and uncontaminated samples. Analyses were carried out on these samples using insufficiently selective and sensitive analytical techniques. Current awareness of the problems involved has led to refined methods of sampling and the application of modern sophisticated analytical methods to environmental studies. For a more detailed review and assessment of analytical methods see Petroleum in the Marine Environment (National Academy of Sciences, 1975), Chapter 2.

One difficulty that the Working Group frequently encountered in the literature was that a number of authors used methods that are based on the determination of specific compounds or fractions of a reference oil to determine what they described as "total oil" or "oil equivalents". This approach, although being an acceptable working concept, should be applied with extreme caution in view of the variety of processes of degradation and differentiation of oil in the marine environment.

Further difficulties arise, for instance, in studies of microbial degradation, photochemical oxidation and other processes, where it is necessary to determine the nature as well as concentration of the metabolites.

Findings of the Group

It is widely recognized that there is a need in biological studies dealing with oil for both more analyses and more complete analyses of sample materials of all kinds.

For further understanding of the degradative and other processes involving oil in the marine environment, the development of analytical techniques must include not only more refined methods for oil compounds, but also better methods for the isolation, identification and measurement of those compounds that are formed in these processes.

Because the analytical procedures used are not specified or described in detail, it is often impossible to establish the validity of results quoted in papers. Unless this type of information is given, it is impossible to make an independent assessment or interpretation of the reported data.

A certain number of reference oils are now available in the United States and the United Kingdom.[1] Where appropriate, these should be used to establish the validity of the methods especially the effectiveness of the isolation and separation stages and the suitability of the measurement technique.

In view of the size and complexity of the marine environment and the usually limited investigative resources available, it is important that the sampling strategy is designed to be appropriate for the task in hand and subsequent reporting.

A commonly accepted method for sampling the oil from oil films on the surface of the sea has not yet been established. Methods of surface sampling should be studied and standardized.

The Working Group recognized the potential of instrumental methods such as the fully computerized combination of gas chromatography with mass spectrometry. Such methods provide more extensive, more accurate and more precise data than many of the older, simpler methods, and their use is to be encouraged.

Chapter 4: CHEMICAL AND PHYSICAL EFFECTS OF OIL DISCHARGES

Statement of the Problem

The Group considered the chemical and physical effects of oil discharges to include such topics as:

- Interference with gas exchange involving the transfer of oxygen and carbon dioxide across the air/sea interface;

- Effects on marine organisms and substrates due to elevated temperature produced by sorption of solar radiation by oil films;

- Absorption of pesticides and metals by oil films and layers on the sea surface and the effect this might have on marine resources.

Summary of Observations and Facts

In the laboratory, thick layers of oil have been found to interfere with gaseous exchange, but no indication was observed to support any suggestion that under open sea conditions, the presence of oil layers such as those that arise from pollution of the sea by oil, will give rise to a depletion of oxygen in sea water. No areas have yet been identified where, under local conditions, an oil slick has been associated with oxygen depletion, although it was recognized that such areas may exist.

Sorption of pesticides by surface oil slicks has not been observed, and the fear that it could have an effect on marine resources appears not to be substantiated. Similarly, the concentration of mercury in an oil slick remains to be substantiated. On theoretical grounds it could be expected that chlorinated hydrocarbons and mercury would accumulate in oil. No data have been found to show that this accumulation takes place in surface oil slicks in the marine environment. It has been reported that hydrocarbons in some marine sediments concentrate mercury and chlorinated hydrocarbons.

Findings of the Group

Few data were found on CO_2 transfer through oil specifically and the subject is still to be evaluated.

---

[1] Inquiries for reference oils should be directed to: The Director, Warren Spring Laboratory, Department of Industry, Gunnels Wood Road, Stevenage, Herts. SG1 2BX, England, or to: Dr. Jerry M. Neff, Biology Department, Texas A&M University, College Station, Texas 77843, U.S.A.

No experimental data have so far been found to support a significant heating effect due to absorption of solar radiation by oil films. Heating effects of stranded oil on corals and in intertidal zones have been noted. No measurements have been published to enable the albedo effect of oil in polar regions to be evaluated. It is likely to be confined to marginal areas. It was considered by the Group that, even if accumulation of materials in oil slicks should occur, it is not obvious that it would lead to elevated levels of chlorinated hydrocarbons in marine organisms.

No indication has yet been found to support any suggestion that, under open sea conditions, the presence of oil layers such as those that arise from pollution of the sea by oil spills, will give rise to a depletion of oxygen in sea water. It is likely that in local areas other effects of oil, such as acute toxicity or entrapment and smothering, would be important. Mangrove swamps and intertidal zones were noted in particular as areas where verification study is required to measure these effects.

Several unanswered questions and additional areas of study remain:

- Will albedo effects in oil films or slicks speed up microbial degradation?

- Would melting effects in polar regions result in a thicker layer of fresh water and thus produce regional climate modifications and affect the buoyancy of marine planktonic organisms?

- With greater attention paid to measuring effects in the top few millimetres of sea water below floating crude oil or refined products, would the heating levels and oxygen transfer rates become significant when oil is discharged into enclosed seas or estuaries?

- With further work can the effect of oil on the oxygen levels in marine sediments be determined and its environmental significance reported?

Chapter 5: EFFECTS ON MARINE LIFE FORMS

Statement of the Problem

The Group considered the problem of lethal and sublethal effects of oil on marine organisms.

Summary of Observations and Facts

- <u>Lethal Effects</u> (usually acute exposure)

Included in this study were the acute effects of toxicity, smothering and clogging resulting from massive oil dosages as occurring in spill incidents or deliberate discharges. The complete range of marine life - phytoplankton, zooplankton, microbes, algae, benthic animals (molluscs, crustaceans, echinoderms, etc.), fish, birds and mammals - in a variety of habitats ranging from intertidal to open ocean was covered. It was recognized that habitats could be subjected to chemical and physical alteration by acute incidents of lethal concentrations of oil to the extent that recolonization by plants and animals and restoration of the ecological balance is a very prolonged process. Acute lethal effects, when taken in context of the ecosystem in which they happen, have a long-term effect. Quantitatively, the data consist of field estimates of number and species of organisms killed in various oil discharge incidents, and laboratory attempts to measure toxicity through bioassays. The latter, in particular, are difficult to interpret because of the ways in which the tests have been carried out, and the non-uniform expression of concentrations in terms of the amount of oil added to the system rather than the concentrations and components to which the organisms were actually exposed. It was observed that toxicity is largely associated with the aromatic content of the oil.

- **Sublethal Effects** (usually chronic exposure)

In addition to their lethal effects on marine life, crude and refined oils affect marine organisms at concentrations that do not result immediately in death. Included in these effects are chronic toxicity, interference with feeding and reproduction, abnormal growth and behaviour, susceptibility to predation, interference with chemical communication, etc. These effects lead to changes in abundances and distribution of individual species and also to shifts in species composition within the oil affected area. The majority of available quantitative data have been obtained by laboratory biological testing methods for chronic toxicity, growth and reproduction rates and for effects on behavioural patterns of various organisms.

Findings of the Group

A great number of factors, acting both individually and in combination, govern the effects that an oil discharge may have on marine life. In general, the biological damage is more severe if the discharge occurs in a coastal or estuarine environment, especially if the intertidal zone is affected, than if it occurs in the open ocean.

Effects may be lethal or sublethal, acute or chronic with different organisms reacting in different ways.

Using laboratory tests of acute toxicity it is difficult to accurately predict field effects and sublethal effects because of the simplicity of laboratory systems compared to the complexity of marine systems. The laboratory measurements can provide exaggerated effects.

Sea birds are the only known group of marine organisms that have so far been affected by oil pollution to an extent sufficient to jeopardize local populations, which in some cases constitute the total world population.

There have been few instances in which extensive mortality of marine mammals from oil pollution has been reported.

Crude and heavy fuel oils appear seldom to cause extensive mortalities of adult fish. Light and refined oils have led to extensive mortalities. Planktonic eggs and larvae may be particularly exposed and sensitive to oil pollution especially to light refined oils. Kerosene-based dispersants with high aromatic content are especially dangerous in this respect and this type of dispersant has therefore been banned in many countries.

The use of existing fishery statistics to demonstrate the deleterious, beneficial or null effects of oil is fraught with dangers, due to the quality of the data and the potentially great number of unknown factors.

Subtidal and intertidal benthic organisms have suffered particularly heavy losses from discharges of light and heavy oils and their subsequent treatment. The long-term and ecological effects of oils and dispersants on these communities are the most extensively studied and documented. Complete recovery may take several years and therefore the effects of repeated oilings are particularly important. Destruction of marine grazing animals upsets ecological balance. In the intertidal zone, mortality of molluscs frequently results in an explosion of the population of attached green algae, which, in turn, affects other parts of the ecosystem. This type of secondary effect, is an aspect that has to be considered when damage due to oil is evaluated. A special case is offered by coral island ecosystems. If the corals are killed, the natural protection of the island from erosion will be lost. Further, because of isolation, most of the organisms in the coral island ecosystem will not be replaced. This would considerably prolong the duration of the effects even if the oil itself disappears or is removed.

The introduction of oil hydrocarbons to a population of marine bacteria selects those capable of utilizing this food source at the expense, at least initially, of the rest of the population. Evidence that low concentrations of crude oil and oil components inhibit bacterial chemotaxis has been questioned. It is noted that the number of micro-organisms able to use oil hydrocarbons as a source of carbon and energy increases from clean to oil polluted marine areas. This can be observed in inshore waters as well as in the open oceans.

Susceptibility of micro-algae to oils varies enormously. Effects may be long delayed in their appearance and result from very short exposures. Laboratory studies have indicated that very low concentrations of oil may stimulate primary production by phytoplankton, but that higher concentrations lead to reduction of carbon fixation, arrest and finally mortality. Macro-algae, if coated with oil, may be mechanically stripped from their substrate. Diesel oil inhibited the photosynthetic activity of young kelp blades. Some lower forms are resistant to oil pollution and thrive in polluted environments. Low concentrations of oil have been shown to depress the growth of red-algae sporelings. Experiments appear to show that higher levels of oil inhibit biosynthesis of nucleic acids and their polymerization in macro-algae. Oil penetrates the higher forms of plant life, blocking intercellular spaces, increasing respiration and decreasing transpiration rates and affecting flowering and reproduction. Some littoral or salt marsh plants may tolerate repeated light oilings but heavy fouling often leads to mortality. These effects may take several years to appear.

Mortality of some zooplankton species, including pelagic fish larvae, occurs from oil slicks at sea, with unknown ecological significance.

The combination of the variety of sublethal responses of marine organisms to crude and refined oils and their ecological implications are not fully understood. Effects of oil should be studied at the ecosystem level, rather than by single species bioassay. Attention should be paid to chronic and sublethal effects using, for example, histopathological techniques. Genetic alterations and other effects on single species should not be ignored.

There is a need for intercalibration of bioassay methods and for agreed methods of quantifying and describing the oils or oil components presented.

Chapter 6: HUMAN EFFECTS FROM OIL DISCHARGES

Statement of the Problem

The Group considered two topics which have a bearing on the susceptibility of marine food products, namely: the possible risk to man from the consumption of sea foods contaminated by oil-derived carcinogens such as the polynuclear aromatic hydrocarbons (PNAHs) and the loss of sea foods due to tainting as a result of oil pollution.

The Group considered that their coverage of these two topics should include:

(i) an examination of the following hypotheses in connexion with the danger to man posed by PNAHs:

- that oil-derived aromatic hydrocarbons including PNAHs are bioconcentrated in the tissues of marine produce exposed to even low ambient oil levels until significantly elevated concentrations are reached, and that such bioaccumulated contents are transferred to higher members in the food chain, leading, by a process of biomagnification, to significantly elevated concentrations in the tissues of produce not directly exposed to oil;

- either that there is a threshold concentration of PNAH carcinogens to man, above which the carcinogen takes effect, and that this threshold can be exceeded in marine produce consumed by man, due to these processes of bioaccumulation and biomagnification; or that there is no threshold concentration to man for carcinogens, i.e. that there is no safe level;

- that oil-derived PNAHs induce carcinomas in marine produce and that ingestion of such cancerous tissues would present a significant health hazard to man.

(ii) an examination of the following factors in connexion with the loss of marine foods through tainting:

- the components which can cause tainting

- the threshold levels at which taint can be detected

- a quantification of the scale of the problem.

Summary of Observations and Facts

- Carcinogenesis

Polynuclear Aromatic Hydrocarbons (PNAHs)[1]/ of known mammalian carcinogenecity occur in crude and, particularly, in refined oils. Reported levels of named compounds, such as 3,4 Benz-pyrene (BaP) and 1,2 Benz-anthracene (BaAnth), vary widely from 0.005 ppm_3 to over 3 ppm, but the residues from catalytic cracking and pyrolysis may contain over 10^3 ppm.

Compared with biosynthesis and terrestrial run-off, oil does not provide a significant proportion of the PNAH input to the marine environment on a global scale, but can be a major contributor on a local scale, particularly in sewage discharges and refinery effluents. The reported incidence and levels of PNAHs in marine produce shows that relatively high levels can occur, particularly in molluscan shellfish, and that these high levels are frequently, but not necessarily, associated with known sources of terrestrial pollution, including oil. Fish and crustacea appear to possess the requisite systems for the metabolism of PNAHs and their excretion as the more water-soluble hydroxylation products. Molluscs appear to lack these systems.

There is a greater storage and persistence of aromatics and PNAHs in lipid-rich than in lipid-poor fish types; in lipid-rich than in lipid-poor oyster populations; in lipid-rich gonad of oysters and mussels than in their lipid-poor muscles.

The uptake of aromatic hydrocarbons (including PNAHs) is faster than paraffins. Concentrations in the range of a hundred to a thousand times background levels may be observed in the laboratory with the greater part of the aromatics being quickly discharged on return to clean water conditions. However, some 1 to 10 percent of the maximum uptake may persist for longer periods.

PNAHs in marine produce will be ingested, usually after cooking, and may or may not form a regular proportion of the daily food intake. PNAHs are poorly absorbed by the mammalian gastro-intestinal tract when added to the diet. It is apparently still a matter for debate whether there is any dose-response relationship for chemical cancer-induction in man, or a threshold dose below which carcinogens do not induce cancer. No reports have been found of epidemiological studies which link gastro-intestinal cancers in man with the

---

[1]/ Where the term PNAHs is used in this report in reference to biosynthesized versus petroleum-derived compounds it refers to total PNAHs (of which not all are carcinogenic), but, in the majority of the text, the term refers to the few carcinogenic PHAHs, specifically analysed for, such as 3,4 Benz-pyrene (BaP) and 1,2 Benz-anthracene (BaAnth)

ingestion of oil-contaminated marine fish or shellfish. There are, however, indications of higher frequencies of stomach cancer associated with consumption of smoked fish (which does contain high levels of carcinogenic PNAHs) in Iceland.

- Loss of Marine Foods

The extent of losses proven or alleged to be due to oil pollution has been impossible to evaluate fully, due to the inadequate documentation of incidents, claims, closures, or condemnations of produce. However, a few examples have been compiled which serve to indicate the probable extent of the problem. Examples have been found of produce being rendered unacceptable on the grounds of altered appearance, and of the closure of fisheries on the grounds of governmentally determined health risk, but most of the existing data and documentation refers to the problem of tainting.

It was established that (1) crustaceans, fish and molluscs exposed to oily conditions can acquire an objectionable oily taste; (2) the ability to taste oily produce is intimately associated with the presence of volatile compounds derived from oils, refined products or dispersants; (3) the range and quantity of odorous compounds varies with the type of oil and refining process, with the middle distillate fractions, like diesel oil, containing the greatest number.

The use of dispersants facilitates the uptake of oil and its components and their introduction to the lipid pool of the organism. The use of oil dispersants increases the likelihood of tainting. The solvent fractions of older dispersants contained tainting compounds of the same nature as those found in diesel and crude oils. There are indications that the tissue lipid content and the amounts of free lipids increase the susceptibility for tainting. Fatty fishes and lipid-rich organs, such as gonads, will become more strongly tainted and perhaps for a longer period of time. Therefore, the seasonal condition of the produce will also affect the susceptibility and the strength of taint, as lipid contents and metabolic rates vary.

External contamination with oil does not necessarily mean that tainting of flesh has occurred, although visible external contamination may in itself be a reason for rejection of produce. Even ingestion of oil by marine organisms does not necessarily cause tainting of flesh, but some species of crustacea and molluscs are consumed together with their gut contents, which may lead to rejection of produce. Cooking of whole animals which have been internally or externally fouled with oil may lead to tainting of the meats. There have been too few studies on the tissue levels of oil components in tainted produce for any tainting threshold levels to be established from them. A threshold of 10-30 ppm in tissue spiked with a North Sea crude oil has been reported with an upper limit of 200-300 ppm, beyond which no further increases can be sensed by a trained taste panel. Threshold levels of 5 ppm gas oil in spiked mussel tissues, and 4-12 ppm extractables from diesel oil in lobsters have also been reported. Exposure to ambient water concentrations as low as 0.01-0.02 ppm oil can lead to tainting of meats.

Findings of the Group

- Carcinogenesis

In some polluted and some unpolluted areas, levels in some filter-feeding bivalve molluscs, and perhaps in some pelagic fish, exceed those in unsmoked foods, and only the notably polluted molluscan levels greatly exceed the levels in smoked foods. Marine fish and shellfish do concentrate PNAHs to high levels within their tissues when exposed to oil but not indefinitely. It is argued that a threshold is reached. Initial rates of PNAH elimination are high, but some 1 to 10 percent remain after a period of days, which may still be 10-30 times the background level and which is slowly discharged, if at all. The slow rates of discharge of the remaining 1 to 10 percent hydrocarbons, particularly in shellfish, provide a potential for biomagnification by predation.

Systematically obtained data are needed on the contents of specific carcinogenic PNAHs in marine produce taken from clean and polluted waters using standardized methods of tissue sampling and analysis. Ambient water or sediment contents, and annual variations in tissue contents are also required since reported contents for produce vary so greatly.

More careful studies are required on bivalve rates of uptake and elimination of PNAHs found in oil, in oil dispersion, and sorbed on particulate matter. Sufficiently precise and accurate methods of analysis should be used to study amounts of PNAHs in the surrounding water or sediment, tissue and gut contents. Evidence should be sought for the metabolism and excretion of PNAHs by bivalves.

During and after exposure the effects of feeding on the translocation, metabolism, and excretion of oil-derived PNAHs in fish and crustacea should be investigated. The relationships between lipid content and gametogenesis on levels of uptake and storage of PNAHs in marine produce should be investigated.

The effects of the use of oil dispersants on PNAH uptake and elimination from oils should be investigated.

The collated evidence illustrates the impossibility of predicting the behaviour of the aromatic fraction of oils in water and tissues from that of the paraffinic fraction. In view of the relative wealth of information on the lower aromatics compared to the PNAHs, competent scientific opinion should be sought on the degree to which the behaviour of the PNAHs can be deduced from that of the lower aromatic fraction. The possibility of selective uptake and accumulation of PNAHs with their homologues needs examination.

The point at which increased polynuclear aromatic hydrocarbon (PNAH) levels in marine organisms become significant as a health hazard to man as a consumer still appears to be a matter for medical debate. The problem seems particularly associated with bivalve molluscs which usually form a very small proportion of the total diet.

No unequivocal association between oil and neoplasms has been found. Chemicals likely to occur in fresh or partly degraded oils have been shown to cause growths on edible algae under field conditions. There have been reports in areas of oil spills showing greater than background incidences of neoplasms in molluscs.

Medical opinion competent in the field of carcinogenesis is required (a) to evaluate the health risk from the few reported field and laboratory tissue levels of PNAHs in produce, (b) to assess the additional risk of increased PNAH levels above any threshold and (c) to set any PNAH levels above which a fishery should be closed and which must be regained by depuration in the field or under transferred stock-holding conditions before such produce could again be exposed for consumption. Medical consultations covering the field of experimental carcinogenesis should review the evidence and prepare future evaluations and conclusions, recognizing that the necessary experimental data, such as histopathological and detailed chemical analysis and their correlations, are probably not available at present. Such a task must include a decision as to which "indicator" PNAH compound, or spectrum of compounds, should be analysed for, and by which methods, in a sample of produce suspected of contamination. This leads to the important consideration of the relationship of oil "taint" in produce to PNAH levels in tissues. Taint in itself may be a reason for rejection of produce but its presence or absence is not necessarily an indication of the PNAH levels. The concentration of selected carcinogenic PNAH compounds or assemblages must be determined by direct analysis. These levels should form the basis for closure or re-opening of a fishery or the release for consumption of marine produce which has been affected by oil pollution, but which has subsequently been transferred to cleaner waters.

In the absence of a consensus of qualified medical opinion (yet to be represented in this study) it is irresponsible to conclude whether ingestion of truly cancerous tissue would present any cancer risk to man. In itself, the possibility of transfer is very remote.

However, if viruses are proven to be causal agents (or co-agents under PNAH contaminated conditions) in the induction of cancers in man and neoplasms in marine organisms, then transmission by ingestion may be possible; in any case, ingestion of cancerous tissue may involve ingestion of unmetabolized carcinogens which are still present.

- Loss of Marine Foods

Accumulation of oil and/or oil compounds in marine organisms takes place. Edible crustaceans, fish, molluscs and seaweeds exposed to oily conditions can acquire an objectionable oily taste.

When contaminated organisms are placed in clean water, the accumulated oil and oil components can be eliminated to a significant degree. The ability to taste or smell oil is intimately associated with the presence of the volatile compounds in the oil, but establishment of threshold levels in marine organisms is yet to be carried out. Major tainting components are thought to be phenols, dibenzothiophenes, naphthenic acids, mercaptans, tetradecane, naphthalene and the methylated naphthalenes. Chemical verification of oily taints should therefore involve analysis of these components.

Taste panels have traditionally been used to establish whether marine produce is tainted. This method will also have to be used in the immediate future but the observations should be confirmed with sophisticated chemical analyses. Although there is no evidence to suggest that taste panel members are exposed to any significant health threat, routine monitoring of the health of panel members should be carried out to establish levels of risk, if any.

The global estimates of marine produce which is rendered unavailable to consumers due to official closures of fisheries or banning of sales of marine plants and animals were determined to be beyond the scope of this study. The scientific literature was determined as not being the most useful source for these data which can only be gathered through international cooperation of governments reporting these events to a central file.

Part II: RESOURCE DOCUMENT ON IMPACT OF OIL
ON THE MARINE ENVIRONMENT

Chapter 1: IMPLICATIONS OF OIL AND DISCHARGE SOURCES

1.1 Descriptions of oil

The Working Group was given as terms of reference the very broad list of oils contained in Appendix I of Annex I of the 1973 IMCO Convention on Prevention of Pollution of the Sea by Ships. This list of oils and relevant physical chemical properties is given in Table 1-1.

It was recognized that to evaluate the fate and effects of all "oils" in the marine environment on biological systems would severely stress the limited amount of information. Further it was thought, as far as assessing the global impact of oil on the biology of the oceans, that some careful generalization may be appropriate.

1.2 Oil types and significance to marine living resources[1]

Mineral oil enters the marine environment mainly from four different sources: accidental and intentional release of fossil fuel during production, transportation, and use, advection through land run-off and domestic and industrial sewage, precipitation from the atmosphere, and natural submarine seeps. Submarine oil seeps were thought to be minor contributors (Blumer, 1972), however, Wilson et al. (1974) estimate 0.6 million tons annually. Thus, the greater part of the oil is introduced at or near the sea surface or, in the case of leaks or accidents involving offshore oil wells, at the bottom of shallow sea areas. After release to the sea surface, low boiling components of the oil will soon evaporate (Mackay and Matsugu, 1973). How much of this material will return into the sea and in which form is unknown. The remainder of the oil appears in five different forms:

as large particles (tar balls),

as microparticles (fine droplets),

adsorbed onto particulate matter such as silt, detritus and phytoplankton,

dissolved in sea water,

water-in-oil emulsion.

Large floating aggregates, although they are a nuisance in many respects, are themselves not taken up by marine organisms and are thus not regarded as an introductory pathway of oil into the marine ecosystem. However, semi-solid oil residues release components into solution over extended periods of time (Blumer, Ehrhardt and Jones, 1973). Microparticles, by virtue of their higher surface/volume ratio are probably depleted of most soluble components. Spooner and Schramm reported that these particles are directly ingested by zooplankton and filter feeders, respectively (Ehrhardt, pers. comm.). They are excreted seemingly unchanged as pseudo-faeces. Whether or not components of these particles are taken up by the organisms whose digestive tracts they traverse is unknown.

Adsorption of oil components onto particulate matter may play an important role in their vertical transport, sedimentation, and uptake by organisms. However, little if anything is known about the qualitative and quantative aspects of these mechanisms. Thus, oil in the sea prior to incorporation into organisms essentially exists in two forms:

in solution

as microparticles.

---

[1] Chapters 5 and 6 will explain the effects of oil discharge on marine biota and man in more detail

Table 1-1

Physical chemical properties

Oil type[1]	Flash pt. deg. F	Pour pt. deg. F	Water and sediment % vol.	Distillation temp. 90% max. deg. F	S^e Saybolt visc. univeral at $100°F$, min.	Gravity deg. API min.
Fuel oil no. 1	100	0	trace	550		35
no. 2	100	5	0.05	675	32.6	26
no. 4	130	20	0.50	420 – 683	45	36
(light) no. 5	130		1.00		150	24
(heavy) no. 5	130		1.00		350	24
no. 6	150	60+	2.00	492 – 1262	(900)	23
Diesel no. 1-D	100	-40	0.05	550		
no. 2-D	125	-10	0.05	640		
Aviation gasoline						
JP-5	140		1.50	550		36 – 48
JP-6			1.50	470		45 – 57
Motor gasoline				356		57
Gas turbine fuel						
oils no. 1-GT	100	0	0.05	550		35
no. 2-GT	100	20	0.10	540 – 675		30
no. 3-GT	130		1.00		45	
no. 4-GT	150		1.00		45	
Solvent Naphtha						
Refined				145		
Crude, light				160		(30 – 53)
Crude, heavy	110	-50		200		(45 – 75)
Petroleum spirits	100					
Asphalt, grade 60-70	450+			≫500 – ≫1300		-8 – 18
grade 40-50	450+					
Electrical insulating oils						
mineral oil uninhibited	295	-40			65 max.	
low pressure cables	300	-40			98 – 108	
high pressure cables	380	-5			750 – 800	
mineral oil for capacitors	455	23			2000 – 2600	
Crude oil, Louisiana				>850	46	34.4
JP-3			1.5	470		50 – 60
JP-1	110		1.5	490		35
JP-6			1.5	500		37 – 50
Distillate Heating oils						
grade 1				533-E.P.		42.6
grade 2				629-E.P.		34.9
grade 4				754-E.P.		21.2

[1] Many workers provide much more chemical detail relative to components making up each of the oil types.

(Table 1-1 continued)

Residual heating oils					
grade 5			0.16		
grade 6			0.15		
Kerosene					
Kerosene	115			572-E.P.	42.0
300 mineral seal	250				
long-time burning	115			599-E.P.	
Petroleum spirits	100			410-E.P.	
Heavy pet. spirits	125			487-E.P.	
Diesel oil, marine	150	0		675	33 – 45
Cleaning cmpd.,solv.	180	10			
Burner oil, special	150	15	0.5		11.5
Burner oil, heavy	150	50			10.0
Corrosion preventive					
aircraft, engine	400	10			
Cleaning oil, turbine	250	–15			
Internal combustion					
engine, diesel					
Heavy-duty 9005	350				44
9020	360	0			50 – 58
9030	390	10			58 – 70
9040	400	15			70 – 85
9050	400	15			85 – 110
Lubricating oil,					
aircraft instrument					
low volatility	270		70		
Lube oil, gear pet.					
base	280	–40			
Rocket fuel, RP-1	110			525-E.P.	42
Insulating oil	275	–40			65
Kerosene	115			572-E.P.	42.0
Motor oil				640 – 879	24 – 30
White oil					29 – 32
Gas oil				400 – 800	30 – 33
Casinghead (nat.)					76.5
SAE lube oils				58 – 2115	19.0
					31.0
Bunker C (max.)				300 at 122F	8.0

Oil in solution is added to the pool of dissolved organic material. Interactions of dissolved organic matter with marine organisms occur at all trophic levels in an extremely complicated manner which we are just beginning to understand. Fig. 1-1 shows a simplified block diagram of the fluxes of dissolved and particulate organic material in the sea. Many organisms have been shown to take up organic solutes including petroleum. The molecular composition of the dissolved organic material is largely unknown. To date the chemical structures of only some 10 per cent by weight of this material have been elucidated. Among them are carbohydrates, amino acids, carboxylic acids, hydrocarboxylic acids, ketones, alcohols, amines, phenols, hydrocarbons, vitamins, antibiotics, growth promoting and growth inhibiting compounds to name just a few. Very little is known about the physiological roles of these compounds. The nutritional value of amino acids and carbohydrates is certainly important for some heterotrophs. Organic chelators condition sea water for phytoplankton growth (Barber et al., 1971), and it is quite conceivable that species succession in plankton blooms is controlled by organic compounds. Moreover, recent experiments reveal that a significant part of information transfer in aquatic life is based on chemical clues (Aubert, 1971; Aubert et al., 1972; Blumer et al., 1973; Mincemin, 1971; Mitchell, Fogel and Chet, 1972; Müller and Jänicke, 1973; Todd, Atema and Boylan, 1972; Wayne and Mitchell, 1972; Whittle and Blumer, 1970; Zafiriou, Whittle and Blumer, 1972).

- 16 -

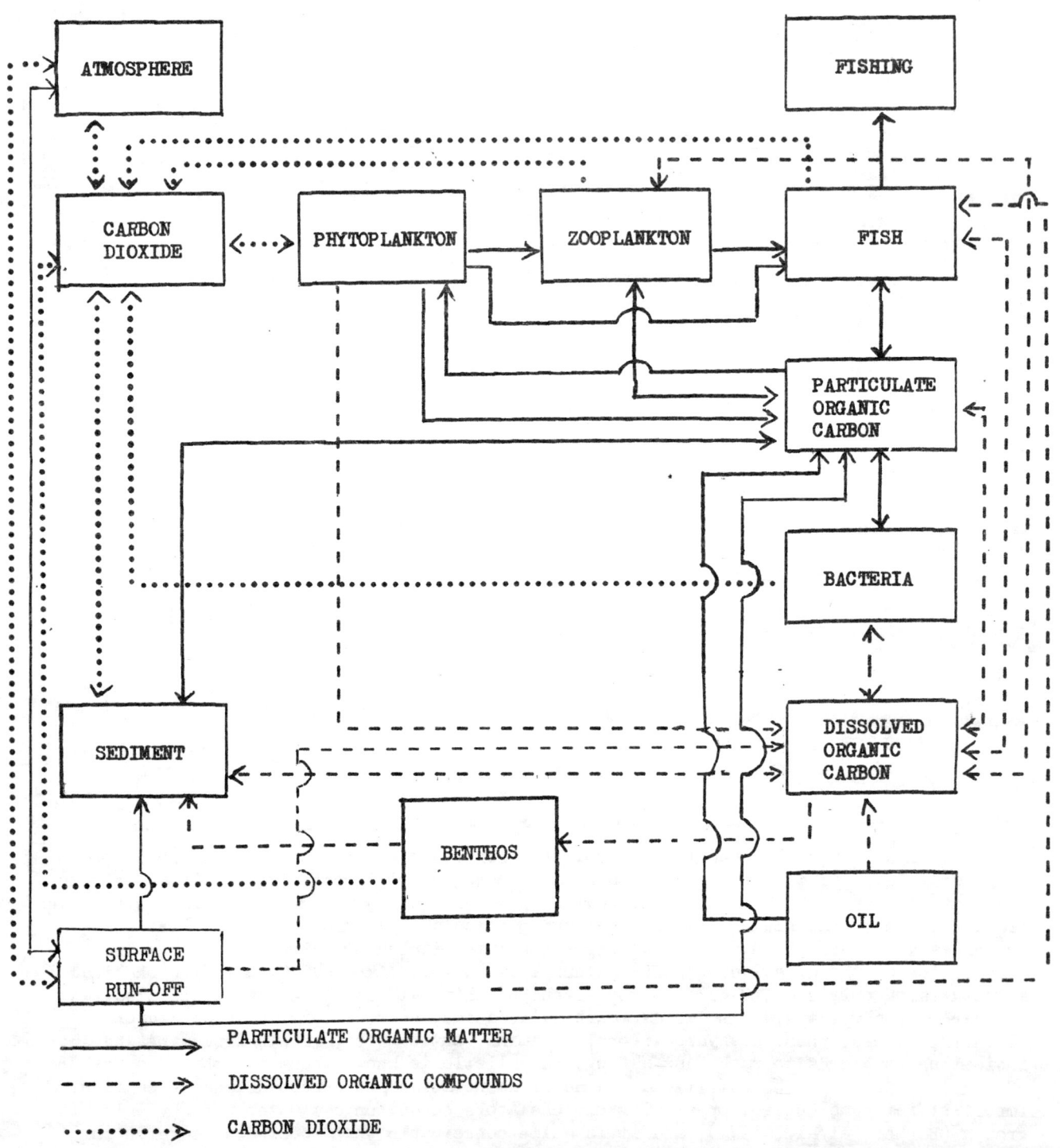

Fig. 1-1

Fluxes of dissolved and particulate organic carbon in the sea

Fucoserratene, the female sex attractant of Fucus serratus, L., has been identified tentatively by Müller and Jänicke (1973, loc. cit.) as the hydrocarbon all-trans-1,3,5-octatriene. Mitchell, Fogel and Chet (1972, loc. cit.) showed that chemoreception of marine bacteria is totally inhibited by high levels of certain hydrocarbons; however, this has been studied further and is reported in chapter 5 (section 5.2.5). Youngblood et al., (1972) found a peculiar composition of hydrocarbons in the reproductive tissue of brown algae. This observation suggests a possible participation of hydrocarbons in the reproduction of these plants.

These examples may suffice to show that the term "uptake" of oil components by marine organisms must include the uptake by sensory organs. Interference with, or disruption of, chemical communication may be caused by minute concentrations which may thus unbalance an ecosystem.

Of course, the concentration and the composition of oil in the environment changes due to the rate of inputs, physical mixing (Abraham and van Dam, 1972; Talbot, 1972; Wolff, Hansen and Joseph, 1972; Weidemann and Sendner, 1972; Zats, 1972), and varying rates of biological degradation of its components (Ehrhardt and Blumer, 1972). This applies as well to biosynthesized organic compounds in sea water, but a number of observations (Blumer, Souza and Sass, 1970; Blumer et al., 1970, 1971) confirm that the half-life of many oil components is measured in terms of years in the environment under certain conditions. In contrast, biosynthesized organic components of sea water are turned over comparatively rapidly (Duursma, 1963). Care must be taken to recognize that differences in half-lives may also be due to oil being buried in poorly aerated sediments or oil existing in the water column.

The mechanisms of uptake of petroleum from solution or from fine dispersion by marine organisms is subject to speculation. Early studies of qualitative analyses of petroleum components extracted from organisms give no indication of selectivity, but as discussed in chapters 5 and 6 evidence does exist. Hydrocarbons isolated from oysters (Blumer, Souza and Sass, 1970; Cahnmann and Kuratsune, 1957; Ehrhardt, 1972), from plants and animals of a Sargassum community (Burns and Teal, 1973), from clams (Mercenaria mercenaria)(Farrington and Quinn, 1973), from the adductor muscles of scallops (Aequipecten irradians) (Blumer et al., 1970), from mullet (Mugil cephalus) (Shipton et al., 1970), and from mussels (Mytilus galloprovincialis) (Fossato and Riviero, 1974) covered a wide range of molecular weights and structures characteristic of mineral oils. Components boiling below 250°C at atmospheric pressure were usually missing. However, toward increasing molecular weights the entire spectrum of oil components liable to chromatographic separation techniques was found. On the other hand, Blumer, Mullin and Guillard (1970) reported the ability of the copepod Rhincalanus nasutus to accumulate selectively the biosynthesized polyolefinic hydrocarbon 3,6,9,12,15,18-heneicosahexene.

Once oil is taken up by organisms little is known of its fate, particularly as it relates to this oil being a significant secondary source to the marine environment. The work of Blumer established the refractory nature of biosynthesized hydrocarbons (Blumer, 1967; 1969; Blumer, Mullin and Guillard, 1970, loc. cit.; Blumer and Thomas, 1964, 1965; Blumer et al., 1969). Petroleum hydrocarbons have been shown to be highly persistent in organisms at low concentrations (Blumer, Souza and Sass, 1970, loc. cit.; Blumer et al., 1971; Blumer and Sass, 1972c). Blue mussel (Mytilus edulis) experimentally exposed to high concentrations of petroleum components discharged most of the hydrocarbons in clean water (Lee, Sauerheber and Benson, 1972). A similar experiment (Stegemann and Teal, 1973) demonstrated that oysters (Crassostrea virginica) rapidly accumulate No. 2 fuel oil from solution (106 µg/l). Exposure to clean water caused the concentration of oil within the oyster to drop to approximately 30 µg/g of wet weight. This concentration was subject to little further change over a further three-and-a-half months. Similar concentrations of a No. 2 fuel oil were found in a variety of shellfish eight months after a spill (Blumer et al., 1970). However, this phenomenon was not found by others as is explained in chapter 6.

An interesting mode of introduction of oil into the marine environment is suggested by observations of Hansen (1975). Photochemical degradation of crude oil fractions under sterile conditions lead to the formation of long-chain carboxylic acids. These compounds are effective emulsifiers (Barger and Garrett, 1970; Garrett, 1970) which should tend to increase the water solubility of oil components because of their dispersant activity.

1.3 Quantities of oil discharged into the marine system

Estimates of oil discharged globally are difficult and, on a worldwide basis the annual contribution of petroleum to the pool of dissolved organic matter could appear to be insignificant. This might be borne out by oversimplified calculations based on the following data: the volume of the world's oceans is roughly 1.4×10^9 km^3 (Dietrich and Kalle, 1965). The average concentration of dissolved organic carbon is assumed to be 1.0 mg/l. Thus, the total amount of dissolved organic carbon is 1.4×10^{12} metric tons. The annual input of petroleum hydrocarbons was conservatively assumed for these calculations to be near 5×10^6 metric tons (this working group used the range of 2 to 10×10^6 tons annually elsewhere) or 4/1 000 000 of the total amount of dissolved organic matter in the ocean.

However, average carbon-14 age of dissolved organic matter has been found to be approximately 3 500 years. This means that the present concentration of biochemical-dissolved organics has accumulated over a considerable length of time. It remains to be seen whether or not petroleum constituents accumulate more rapidly.

Moreover, petroleum is introduced mainly into the biologically productive surface layer and near-shore areas. Here, concentrations of 10 μg/l are frequently encountered; Iliffe and Calder (1974) showed 12 - 47 μg/l at three stations in the Gulf of Mexico. Assuming a higher concentration of natural dissolved organic carbon in the surface layer and in near-shore waters of approximately 3 mg/l, petroleum hydrocarbons could annually contribute roughly 3/1 000 of the total concentration by this hypothetical analysis.

Other approaches have been taken to establish the estimates of the total amounts of hydrocarbons discharged into the marine biosystem which vary from 2 to 20 million tons per annum. Typical examples of these estimates are given in Table 1-2.

Another approach to the problem of illustrating the sources and significance of oil or hydrocarbon discharges into the marine environment may be seen in Table 1-3. Crude oils have been divided into 5 categories (based on boiling ranges) from light to heavy, and refined oils into 6 categories from light distillates to residuals. In the case of atmospheric rainout the possibility of the input having undergone substantial photo-oxidation is indicated by the term (ox.). If it were possible to clearly define categories of crude oil and refined products which were internationally understood and accepted, then the priorities for prevention and control could be directed at the most environmentally impactive categories of oils and their sources. Until this is done, however, Table 1-3 can only be used as a qualitative illustration to such concepts suggesting that urban runoff will be composed mainly of the heavier fractions of refined oils due to the marketing and use patterns and the evaporation and degradation action which should take place. During this study, classification of crude and refined products were proposed, but were found unsatisfactory.

1.4 Rationale for classification of sources of oil pollution with respect to ecological damage

From the previous discussion, the general sources may be seen. However, to evaluate the impact of oil on the marine environment a technique should be used to allow the relative comparison of one oil and a source to another oil and a source. Evaluating biological impacts of different sources of oil pollution or the prediction of effects of corrective measures is a multidimensional quantitative problem that requires the use of systems analytical techniques. A full solution requires quantification of the important factors (e.g., persistence, toxicity, global load), the derivation of covariance matrices with the aid of causal mathematical models (or probabilistic models), and the projection into a one-dimensional ranking list.

TABLE 1-2

Comparison of estimates for petroleum hydrocarbons
annually entering the ocean, Circa, 1969-1971

Source	Authority (millions of tons per annum)		
	MIT SCEP Report (1970)	USCG Impact Statement (1973)	NAS Workshop (1973)
Marine transportation	1.13	1.72	2.133
Offshore oil production	0.20	0.12	0.08
Coastal oil refineries	0.30	-	0.2
Industrial waste	-	1.98	0.3
Municipal waste	0.45	-	0.3
Urban runoff	-	-	0.3
River runoff [a]	-	-	1.6
SUBTOTAL	2.08	3.82	4.913
Natural seeps	?	?	0.6
Atmospheric rainout	9.0 [b]	?	0.6
TOTAL	11.08	?	6.113

[a] PHC input from recreational boating assumed to be incorporated in the river runoff value

[b] Based upon assumed 10 percent return from the atmosphere

From: National Academy of Sciences (1975)

TABLE 1-3

Sources of oil fractions

	CRUDE OIL light ⟷ heavy	REFINED OIL light ⟷ residuals dist.	Total amount million tons/year*
Marine transport LOT	X X X X X		0.6
Marine transport non LOT	X X X X X	X X X X X X	1.5
Offshore oil production	X X X X X		0.08
Coastal refineries		X X X X X X	0.2
Industrial waste		X	0.3
Municipal waste		X X X	0.3
Urban runoff		X X X	0.3
River runoff	X	X X X	1.6
Natural seeps	X X X X X		0.6
Atmospheric rainout		X X (ox)(ox)	0.6

* National Academy of Sciences (1973); see also Table 1-2

An attempt has been made to rank the relative impact of various oils with the aid of some general assumptions and a table where seven ecologically significant factors are evaluated for each of ten sources of oil pollution. The general assumptions which follow are necessary to make compatible different geographical areas, types of oil, etc. These are:

1. Coastal marine areas receive the main part of the released oil.
2. Coastal marine areas are more productive than are offshore areas on the same latitude, and hence of greater economic importance.
3. Coastal marine areas are more sensitive to pollution than are offshore areas on the same latitude.
4. Volatile components of oil are usually more acutely toxic than are non-volatile components.
5. Volatile components of oil are generally less persistent when discharged into the sea than are non-volatile components.
6. The present trend toward super-tankers will result in a different proportion of incidents due to improved safety of operation, but much larger amounts of oil will potentially be discharged.
7. Different climatic conditions greatly influence the rates of oil degradation and its biological effects.
8. There is at present no satisfactory oil spill cleanup technology for combatting large oil spills at sea.

Two matrices are given in Table 1-4: one for highly toxic compounds and one for compounds of low toxicity. This arrangement has been necessary to enable the weighting since toxicity is a factor of ecological importance.

The matrix rows are sources of oil pollution and the matrix columns are ecological criteria. The figures in columns I-V are estimated importance (taking values of 1,2 or 3 from least to most important) of each criterion to that source. Summing each row over all five columns thus yields a figure indicative of the relative estimated ecological and economic impact of each source on the marine environment.

TABLE 1-4

Relative impact of oils and sources

	I Total amount released	II size of geogr. area affected	III duration of damage caused	IV economic importance of area affected	V possibility to influence present situation by means of new technology or admin. decision	VI sum of criteria I to V	VII degree of certainty of knowledge
a) Compounds of high toxicity							
A transportation, crude oil	3	2	1	3	2	11	2
B transportation, light oil	1	2	1	3	2	9	2
C transportation, heavy oil	1	2	2	3	2	10	2
D refineries	1	1	3	1	3	9	2
E offshore production	1	1	2	3	2	9	1
F submarine seeps	1	2	2	2	1	8	1
G harbour and terminal	1	1	3	1	3	9	3

(Table 1-4 cont'd)

	I	II	III	IV	V	VI	VII
b) Compounds of low toxicity							
H atmospheric	1	3	3	1	1	9	1
I municipal and urban runoff	2	1	3	2	2	10	2
J river runoff	3	2	3	3	2	13	1

(Figures in columns I to V and VII are arbitrary values. I to V: from least (1) to most important (3); VII: estimated values from least (1) to most certainty (3).)

Chapter 2: FATES OF OIL - ROUTES OF OIL EXPOSURE

Definitions: For the purpose of this discussion, the following definitions have been used:

Routes are the means by which inputs of oil become available to marine organisms.

States are the form (whether dissolved, dispersed, absorbed on particulate matter, etc.) in which the oil is available for uptake by marine organisms.

Oil is understood to include crude oils and refined products but not petrochemicals. It is defined in the broadest sense, but specific, predominant hydrocarbons and chemical groups are noted.

Microbial populations are defined to include open ocean, coastal and estuarine bacteria, viruses, fungi and yeasts.

The topic of degradation is understood to include the rates and mechanisms, by which the original pollutants are altered, and the chemical products of any particular mechanism.

Problem and Controversy: The routes by which oils move through the marine environment, the forms in which they are presented to marine organisms, and the mechanisms by which they are taken up, metabolized, stored, and discharged by marine organisms are not adequately understood. In particular, considerable controversy exists over whether or not oils are accumulated within marine organisms or biomagnified in the marine food chain, and, if so, whether it is the whole oil or certain fractions of it. In addition, the problem must be focused upon the sequence of events occurring at the microbiological level upon introduction of petroleum to the marine environment. The questions examined in the following pages include:

What fractions of petroleum-derived oils are degraded by micro-organisms, to what extent, at what rate, at what locations, and at what seasons of the year?

What are the controlling physical and chemical factors governing the rate of degradation?

Can specialized and preserved cultures of micro-organisms be used to combat oil spills?

Are long chain fatty acids always products of degradation and, if so, are these the result of photochemical or microbial action; will they interfere with a complete biodegradation of petroleum?

2.1 Physical, chemical and microbiological factors influencing the fate of oil

Petroleum, upon leaking or spilling into the sea, is subject to a series of diverse processes that distribute the product in the environment (see Fig. 2-1) and at the same time cause it to age or gradually convert, thus changing its physical and chemical characteristics.

2.1.1 Spreading

The first process after the spill is that of the extension, or spreading, of the product over the surface in the form of a thinning film. The spread of oil into a calm sea was considered (Fay, 1969; Fay and Hoult, 1971) and concluded that gravitational effects controlled spreading in the initial stages of a spill. Other factors dictate spreading characteristics as the oil layer thins. For most spills, after the first hour or two, slick size is generally controlled by a surface tension-viscosity relationship which is independent of spill volume. The area occupied by the slick increases rapidly under the influence of both hydrostatic and surface forces. This area eventually becomes constant as further enlargement is limited and offset by natural forces.

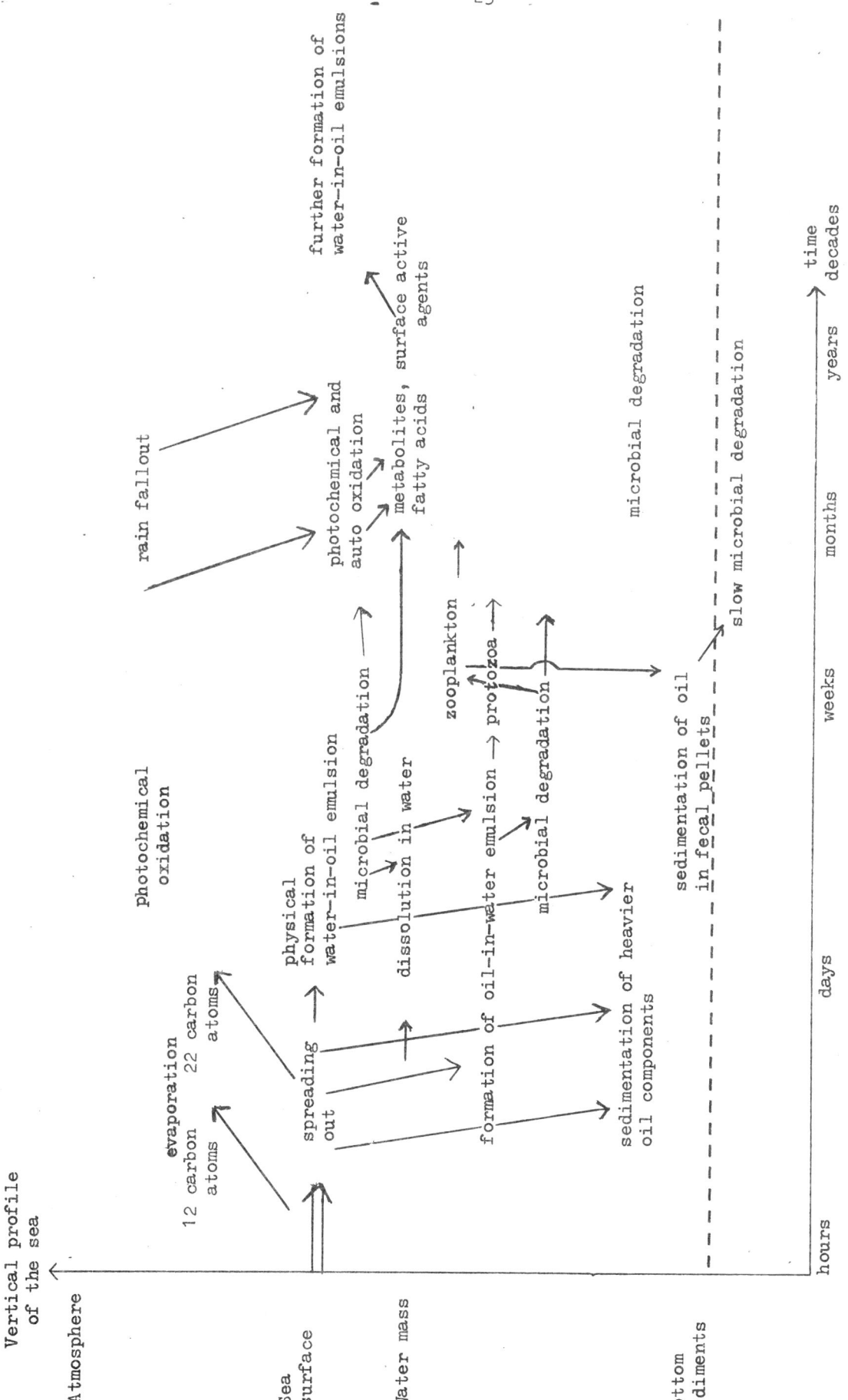

Fig. 2-1: Diagrammatic summary of fate of oil

As more data on accidental spills become available, it is apparent that even the viscous crudes spread rapidly into thin layers which come under the influence of surface effects. Even the great spills of the last few years eventually decreased in thickness to a point where surface forces would determine their characteristics. Thus, even the massive spills eventually spread into thin layers, not only by the spreading mechanisms but also through the action of the manifold natural dispersive forces which act to remove oil from the water surface. Once a spill has thinned and surface forces begin to play an important role, the oil film is no longer continuous and uniform but becomes fragmented by wind and waves into patches and windrows.

The influence of surface forces on the spreading of thin oil films becomes increasingly apparent as a spill spreads. The major components of petroleum do not spread spontaneously onto a water surface but hydrocarbons with low molecular weights do not remain as a liquid lens and spread rapidly.

The sulphur- and oxygen-containing compounds are surface-active. The hydrophilic polar portion of the molecule is associated with the water surface and the non-polar or hydrocarbon portion is oriented toward the petroleum oil. These surface-active constituents cause the otherwise nonspreading hydrocarbon oil to thin out into expanded slicks surrounded by thin films which approach monomolecular dimensions. The extent of spreading is a function of the quantity and nature of the surface-active components of the oil.

Spreading must compete with emulsification. The effect of water-in-oil emulsification increases with time and greatly increases the viscosity of the remaining slick, which reduces the tendency of the slick to spread. Thus, spreading is a self-retarding process; it accelerates evaporation and leads to increased viscosity and increased pour point of the oil (Berridge, et al. (1968).

2.1.2 Evaporation

Evaporation is the process by which the low to medium molecular weight components of relatively low boiling point are volatilized into the atmosphere. The rate of evaporation is a function of the vapour pressure of each component in the petroleum, the concentration of that component, the surface area and thickness of the spill, wind, and temperature. While still substantial for some oils after several months, evaporation is most intense during the first few hours to weeks after the spill.

The percentage of components lost to the atmosphere by evaporation correlates well with the carbon number. Figures vary, but it is generally agreed upon that all hydrocarbon fractions containing approximately 13 carbon atoms or less are subject to great losses after the first few days, with heavier fractions, up to about 20 carbon atoms, evaporating after a few weeks. Evaporation therefore selectively depletes the lower boiling components of the oil, increasing the specific gravity as the oil loses its volatile fraction. The process can contribute to the formation of thick residuals, oil sludges, and eventually, the possible formation of tar balls. In either case, the specific gravity of the remaining oil will increase, and the residual oil may become denser than sea water, increasing the possibility of sinking.

Evaporation can be an important means for preventing toxic components from entering the marine ecosystem. However, higher ring aromatics, some of which are thought to be involved in the long-term toxicity, are less volatile, so that these potentially hazardous compounds, e.g. PNAHs, are practically unaffected by evaporation for long time spans. It must also be understood that those lower molecular weight hydrocarbons which are susceptible to evaporation loss are also the most soluble. Although evaporation and dissolution leave behind the same type of residue, their effects upon the marine ecosystem are entirely different, as those products which dissolve into the water are available for uptake by marine organisms. Thus, although evaporation and dissolution are responsible for the removal of the same types of products, they are competitive processes and one must be cautious not to consider one to the exclusion of the other.

2.1.3 Solution

Solution is the physical process by which the low molecular weight hydrocarbons, as well as some of the more polar nonhydrocarbon compounds, are lost by the oil to the water. The rate of this process is governed by wind and sea state and certainly by the properties of the petroleum material (chemical composition, specific gravity, viscosity, pour point, surface tension, solubility, etc.). Although this solution process does start immediately, it has long-term effects as well. This is true because oxidation processes constantly produce polar compounds (fatty acids, fatty alcohols, etc.) from hydrocarbons in the oil, which are more soluble than the original product.

For the low molecular weight components, the aromatic class of compounds has the greatest solubility. Small amounts of polar, high molecular weight components of high boiling point may also partition from petroleum. The rates at which the soluble components are selectively partitioned from a petroleum are mainly unknown, resulting from difficulties in analysing extremely dilute concentrations of these components. However, limits can be set. Over short periods of time, losses of the highly soluble aromatics through solubility will be minor in comparison with evaporation. Over longer periods, while mixing and contact with larger amounts of water are minimal, the losses due to solubility are small owing to the low solubility of petroleum's high molecular weight components. For prolonged periods of time, there is considerable uncertainty about the magnitude of the losses. These comments must be evaluated in light of the fact that although the solubility of petroleum's components, as reported in the literature, is considered "small", sea water extracts of solubles from several oils have shown a significant toxicity to some marine life.

As far as the solubility of specific oils is concerned, Anderson et al. (1974) examined oil-water dispersions and water-soluble fractions of No. 2 fuel oil, Bunker C residual oil, Kuwait crude oil, and South Louisiana crude oil. The two crude oils gave more concentrated water-soluble fractions than did the water-soluble fraction prepared from the two refined oils; they had significantly greater concentrations of di- and tri-aromatic hydrocarbons than did the latter. The crude oil water-soluble fractions, on the other hand, were particularly rich in the light aliphatics propane through isopentane, and the light aromatics benzene, toluene and xylene. These observations strongly suggest that the acute toxicity of an oil is largely a function of its di- and tri-aromatic hydrocarbon content.

Since the soluble fraction of most petroleum products consists mainly of medium weight aromatics and this fraction contains those compounds known to be the most toxic, persistence of the soluble fraction is an important property to consider in determining the relative harmful effects of petroleum products. Frankenfeld (1973) investigated the weathering of No. 2 fuel oil, Bunker C residual oil, and Venezuelan crude oil under simulated natural conditions. After a week's weathering in laboratory simulators, the dissolved fraction of the No. 2 fuel oil was approximately 3.5 times that found for the heavier oils. This belies the idea that light oils disappear almost entirely due to evaporation shortly after being introduced in the marine environment. The above evidence indicates that lighter oils are in the water column more readily than are heavy oils and that light oils persist in the water column whereas heavy oils form surface slicks. Additional field studies conducted by Sivadier and Mikolaj (1973) suggest that the evaporation of light fractions is quite significant.

In addition to evaporation and dissolution, petroleum hydrocarbons are removed from the sea surface by wave-produced spray and bursting bubbles. The transfer into the atmosphere depends on wind speed, sea state, and the extent to which wave breaking and whitecap formation are suppressed by oil films, which in turn is dependent largely on film thickness and horizontal extent of the film. The process seems to be accelerated by solar radiation, which can cause the formation of polar, surface-active molecules in the film. Sea/air transfer processes of this type are most effective for the removal of relatively thin films. It should be emphasized that, except in coastal areas with onshore winds, removal of hydrocarbons from the ocean surface by this mechanism is temporary. Most particles will be redeposited in the ocean at distances ranging from a few metres to several hundred kilometres from their point of injection into the atmosphere, although their components may have undergone photochemical breakdown while airborne.

2.1.4 Emulsification

Rough seas tend to create emulsions of the oil-water phase. This process may take two forms: oil-in-water emulsion, where the sea is the continuous phase, or water-in-oil emulsion, where the stable floating emulsion contains about 30 to 80 percent water. Berridge, Thew and Loriston-Clarke (1969) studied the formation and stabilities of water-in-oil emulsions and found that they were stable for periods ranging to greater than 100 days. Since refined products did not form water-in-oil emulsions, they concluded that the ability to form water-in-oil emulsions was related to the amount of nonvolatile residues, particularly the asphaltenes in the crude oils. According to Berridge, water-in-oil emulsions do not require the addition of external dispersing agents but form naturally on a dynamic sea surface. It is generally the case that the generation of this type of emulsion requires violent agitation at the air/sea interface and relatively thick oil films. Under these conditions even gasoline may form temporary emulsions (Blackman et al., 1973).

Oil-in-water emulsions occur with lighter petroleum distillate products when stresses at the sea surface force oil into water as small drops. Dispersants used in oil spill clean-up have a large hydrophilic nature and lead to the formation of oil-in-water emulsions. Emulsification of crude oil into sea water may also be promoted by surface-active materials produced during degradation of the oil by certain micro-organisms.

The emulsification promotes solution of the more soluble components in the petroleum or petroleum product.

2.1.5 Effects of dispersants

The massive applications of toxic "detergents" during the TORREY CANYON incident initiated and provoked the cloud of controversy which exists today concerning the use of these products. Chemical dispersants, or emulsifiers, are liquids, the base of which is organic solvents and surface-active products, the purpose of which is the immediate elimination of the oil on the surface by converting it into a fine oil-in-water emulsion.

The organic solvent reduces the viscosity of the oil and aids in distributing the surfactant more uniformly to the oil layer. The surfactant lowers the interfacial tension because of its amphiphatic nature, i.e., partly oil-soluble and partly water-soluble. When mixed, fine oil droplets readily form, the same surface-active agent preventing the coalescence of the droplets. In essence, the surfactant acts to fend and physically parry droplet collisions. This same property reduces the tendency of droplets to stick to and thereby wet immersed solid surfaces.

In discussing the fates and effects of dispersants, there are two facets which are of interest: (a) the effects of the dispersant itself and (b) the effects of the dispersant/ oil mixture. The concern and conclusion that all chemical dispersants are in themselves inherently toxic is incorrect. Some of the most effective emulsifiers/ dispersants available are those derived from and found in the natural environment. The toxicity of the dispersed oil itself is a more valid concern.

Battelle Memorial Institute (1974, 1974a) conducted extensive studies concerning the effects of chemically dispersed oils on selected marine biota. Oils used were No. 2 fuel oil and Kuwait and South Louisiana crudes. Toxicity was found to be greater for dispersed than undispersed oils. As stated in the report, this does not indicate a greater inherent toxicity for dispersed oils; rather, the hydrocarbon concentrations within the water column were proportionately greater for the dispersed oils. The exception to this was the dispersion of Kuwait crude, and this was attributed to the lower effectiveness of the dispersant with this oil. The crude oils were more toxic in the dispersed states, and carbon-14 bioassay of the dispersant indicated that the dispersant contribution to the observed toxicity of dispersed crude oil was minimal. The increased toxicity is more likely a direct result of increased exposure to the toxic components of the oils in question or increased exposure to mechanical damage of the algal cells tested, as a result of contact with insoluble oil globules. For example, a covering of oil may block gaseous exchange thereby inhibiting the uptake of carbon dioxide. Also damage to cell membranes may cause hydrocarbon molecules to penetrate the cell and cause leakage of cell contents. It was also found that oil-dispersant mixtures restricted feeding activity in oysters. Possibly coated food organisms of microscopic size were rejected as pseudofaeces and not ingested. In tests performed with the dispersant alone, proliferation of microbiological growth was stimulated, due to the dispersant serving as a readily oxidizable nutrient source. However, when dispersed No. 2 fuel oil was tested, growth of the microbiological community was inhibited. This may be attributed to the presence of a greater concentration of aromatic hydrocarbons in the No. 2 fuel oil, which are generally considered to be the most toxic constituents of oil. When South Louisiana crude was used, microbial growth was stimulated to a greater extent by the dispersed than the undispersed oil. In a separate experiment, it was found that chemical dispersion increased by two times the maximum level of n-alkanes and methyl substituted naphthalenes in oyster tissues. Emulsifying the oil therefore increased the leakage of the soluble components of the oil to the sea water.

Two recent reports (Atlas and Bartha, 1973; Mulkins-Phillips and Stewart, 1974) describe effects of dispersants on the biodegradation of oil under laboratory conditions. Atlas and Bartha found that dispersants increased the rate, but not the extent, of mineralization, i.e., components of petroleum persisting without the addition of dispersants also persisted when dispersants had been added. Dispersants did not alter the microbial capability, but, most likely because of the enhanced exposure of the oil to microbial attachment and subsequent utilization, the rate of degradation was enhanced. Mulkins-Phillips and Stewart examined four dispersants and found only one that increased degradation of n-alkanes in the crude oil they studied. An oil dispersant was observed to have no obvious deleterious effects, either alone or in combination with Kuwait crude oil, on the microflora of beach sand (Bloom, 1970).

Although dispersing oil in water generally promotes microbial degradation, certain chemical dispersants will inhibit microbial activity in various ways. Some dispersants are toxic due to damage to cell membranes and cell permeability or to essential enzymes, causing irreversible denaturation. Some dispersants inhibit microbial activity by reducing the surface tension to levels inhibitory for growth, i.e., surface levels less than 45 dynes/cm are often inhibitory, with oil-degrading bacteria having an optimum surface tension of 45-60 dynes/cm, which is considerably less than the surface tension of normal average sea water (73-74 dynes/cm at $20^{\circ}C$). Thus, some surface tension depression may be advantageous for microbial degradation, but too much may be lethal or, at the least, may act to retard bacterial growth (ZoBell, 1969).

Therefore, the overall effect of the chemical dispersion of oil is the introduction of dispersed oil droplets several feet or more into the water column. In this form, the oil is more susceptible to microbial degradation than it was in the original form. However, in this form, the oil is also made available to other types of marine life in addition to the hydrocarbon-oxidizing bacteria. Nekton and other filter feeders may now come into contact with dispersed oil droplets that they might have otherwise escaped as an oil film on the surface of the water. The fact that droplets of this small size can be maintained in suspension in the sea is most relevant. Since the basis for the efficiency rating of the chemical dispersant is the stability of the oil-in-water emulsion and the finer the oil droplet, the more stable the dispersion, the most efficient dispersant is that producing the finest droplets. However, finer droplets have greater immediate acute toxicity to marine life, particularly in the ten-micron range. Probably of greater concern than the acute effects is the fact that the finely dispersed oil droplets become a more widespread contaminant and may cause long-term effects.

2.1.6 Sedimentation

Rough seas may also increase the chances of the petroleum to be absorbed on or mixed with particulate matter (sand, silt, clays, shell fragments, etc.), and eventually settle to the bottom when the seas become calmer. Its absorption by solids renders the oil more susceptible to auto-oxidation and particularly to microbial oxidation, but little oxidation takes place after hydrocarbons are buried in reducing sediments.

Blumer and Sass (1972c) presented data showing the extremely slow rate of bacterial degradation of polluting oil trapped in a marine sediment. Thus, ten months after the accident at Cape Cod, the pollution of the bottom sediments covers an area that is much larger than that immediately after the spill.

Hydrocarbons from a No. 2 fuel oil spilled in Buzzards Bay, Mass., were detected at a depth of 7.5 cm below the sediment-water interface in the offshore environment (Blumer and Sass, 1972) and some of the low boiling C_{12} and C_{13} components were present more than two years later. Kerosene has also been recovered from river sediments containing materials below C_{10} (Connell, 1971).

The mechanisms by which presumably sedimented oil is resuspended and spread over wide areas, considerably after the spill, are not well understood. Various resedimentation and suspension processes likely occur: current movement, suspension by pressure waves, resuspension by organic reworking of surface sediments. To understand the spread of oil in bottom sediments, it is necessary to know the exact form of the oil on the bottom and in the sediments. Presumably, the oil occurs initially as a surface layer of dispersed particles, combined with varying amounts of sediment. Reworking on and in the bottom sediments should lead to an increase in the sediment content of the oil-rich phase. Biogenic reworking by epifaunal and infaunal bottom living animals should lead both to resuspension and to downward penetration of oil into the sediments. The extreme longevity, and presumed slow rates of degradation, of hydrocarbons introduced by spills into shallow water sediments (and presumably into the deep sea as well) may be due to the reworking of oil into subsurface, anaerobic layers of the sediment. In this regard, detailed studies are needed of the textures and fabrics of oil-polluted sediments, to determine the exact position and nature of hydrocarbons in the sediments. Of all the reservoirs considered in this study, the fate of oil in bottom sediments is understood the least. Studies such as reported in the paper by Shelton and Hunter (1975) are needed and assist greatly in understanding the role of anaerobic bacteria in oil degradation.

2.1.7 Oxidation

The chemical reactions occurring in the petroleum are mostly oxidative in nature. Since much of the petroleum material floats on the surface, a major portion of the oxidation occurs here. Additional oxidation reactions likely occur in the water column. Reduction reactions probably occur when the material is carried to or released from the ocean bottom and the oxygen content in the overlying water column is extremely low.

The oxidation reactions are either chemical or biological in nature. The term auto-oxidation is applied to oxidation processes brought about by the oxygen of the air at ordinary or slightly elevated temperature. Auto-oxidations are accelerated by light and the catalytic action of metals. Sulphur compounds are likely to decelerate the auto-oxidations.

2.1.8 Chemical degradation

The nature of the chemical degradation process is not known precisely, but would appear to be largely the result of the photo-oxidation of hydrocarbons, with the general result being that the oxidized compounds are more soluble in water than the original ones. Perhaps the most notable effect is the increase in the resin and asphaltene contents. It would appear that aromatic hydrocarbons of medium and high molecular weight have a basic part in this process. This increase is most noticeable in the case of petroleum products that have been subjected to heating, as is the case, for example, with fuel oils.

Photolysis (above 3 200 Å) is thought to afford an initiating mechanism for the oxidation of larger more complex molecules (auto-catalytic oxidation), as well as polymerization reactions, hydrocarbon formation and removal, and tar ball formation (Feldman, 1973). The effects of photochemical degradation only become significant after the first day or so.

Chemical weathering has been reported by Hellman (1971, 1971a) to influence viscosity. After an exposure of a few days, several crude oils increased their viscosities by one decimal power, presumably through evaporation and emulsification. After several weeks of exposure, their viscosities increased by two or three additional decimal powers. Solar radiation was suggested as causing polymerization of oil hydrocarbons.

2.1.9 Biological fates (general)

An evaluation of the biological fates of oil must stress two aspects. The first concerns those factors which change and remove hydrocarbons and hydrocarbon products from the water environment. The second concerns the biota as a physical reservoir that takes up and holds the oil. The indigenous microflora assume a dominant role in the first case, and the macro-communities are most important in the second case.

Some components of petroleum are readily evaporated or biodegraded but some components of petroleum persist in the marine environment (Hartung and Klingler, 1968; Holcomb, 1969; Horn, Teal and Backus, 1970; Blumer, 1971, 1971a; Blumer et al., 1971; Blumer and Sass, 1972, 1972a,b; Blumer, Ehrhardt and Jones, 1973; Blumer et al., 1972). Thus, it is necessary, as a first step, to categorize those components of petroleum that are biodegradable and those that are recalcitrant (Alexander, 1971).

It is relatively easy to demonstrate in the laboratory that many marine and estuarine bacteria, under optimal conditions, can remove selected fractions of oil, usually the n-alkanes, in a matter of days or weeks so that a certain percentage of oil by weight will disappear in a given period of time, but all of the oil will not disappear in a proportionate time, since the remaining fractions may be more refractory to microbial attack (viz., Walker Colwell and Petrakis, 1975). In nature, conditions rarely are favourable for maximum biodegradation, i.e., in the deep ocean, temperatures are less than or equal to $4^\circ C$ and nutrients are present in concentrations of mg-μg per litre, whereas in the laboratory, flask cultures often are incubated at $25-37^\circ C$ with g/l quantities of nitrogen, phosphate, and carbon sources provided. Hence, the rate of degradation can be markedly slow in the natural environment and oil may persist in the ocean for a very long time. Floodgate (1972, 1972a) surmises that some components of the 10 million tons of oil that found its way into the sea in the second world war are still in the oceans. Considering the very slow degradation rates observed for aromatics and resins found in certain crude oils (Walker, Colwell and Petrakis, 1976), this seems to be a valid statement. Points in question are several: will in vitro degradation occur in vivo; will oil reach depths of greater than or equal to 1 000 metres and, if so, will microbial activity at depths greater than 1 000 m be efficient and rapid enough to degrade the quantities of oil reaching these depths. Furthermore, will concentrations of oil above and/or below a threshold level be degraded. Jannash (1970) noted that threshold concentrations of carbon in sea water may act to limit bacterial growth. That is, if the level of carbon compound is below a minimum concentration, microbial growth is not sustained, or at least, cannot be considered to be effective.

The problems associated with microbial degradation of oils require attention because of the increasing amounts of petroleum-derived hydrocarbons discharged to the sea. There is a genuine and increasing concern that the marine environment does not have an infinite capacity for degradation or "removal", i.e., rendering these materials harmless. Adequate and relevant information is required. Several general assessments of the problem have been reported (ZoBell, 1973).

2.1.10 Microbial degradation

As the residence time of the oil on the water increases, biological processes begin to operate and rapidly gain in significance (Templeton, 1971; Miget, 1973; and Pilpel, 1968). Over 90 species of micro-organisms (bacteria and fungi) capable of subsisting in petroleum and therefore capable of degrading it by biological oxidation, have been identified. The extent of such degradation in the sea is dependent upon water temperature, and on the presence of certain volatile fractions (naphtha and kerosene) in the spill, some components of which are bacteriocidal or bacteriostatic. When these fractions are removed the residues are therefore more biodegradable than are the crudes from which they originate, although this will not actually become apparent until several months after the spill has occurred.

Virtually all kinds of oils are susceptible to microbial degradation, but the most important factor which influences the biodegradability of hydrocarbons seems to be the molecular configuration. Alkanes are attacked by more microbial species, more rapidly and with more support of growth, than either aromatic or naphthenic compounds. Within the alkane series, normal compounds are more susceptible to microbial oxidation than branched chain compounds. With increasing chain length, alkanes seem to be increasing refractory to microbial oxidation, which may be attributed more to lower solubility in water than lack of vulnerability to enzyme action.

Qualitative and quantitative differences in hydrocarbon content of petroleum influence its susceptibility to degradation. Two crude oils and two refined oils were examined for susceptibility to degradation by bacteria present in water and sediment samples from Baltimore Harbor, an oil-contaminated environment. One of the oils studied, South Louisiana crude, contained a high percent of saturates (56 percent). The other crude oil was Kuwait crude, containing 34 percent saturates and a significant amount (18 percent) of resins (pyridines, quinolines, carbazols, sulfoxides, and amides). One of the refined fuel oils studied, the fuel oil No. 2 was composed of saturates (61 percent) and aromatics (39 percent). The other one, the fuel oil No. 6 (Bunker C) contained 60 percent asphaltenes (phenols, fatty acids, ketones, esters and porphyrins) and formed globules (Walker, Colwell and Petrakis, 1975). The total amount of each of the four oils remaining after exposure to bacteria from the oil-contaminated Baltimore Harbor sediment for seven days was approximately the same, indicating similar quantitative degradation. However, comparison of the degradation of selected components of the four oils indicated that there were differences in susceptibility to degradation of each of the components. Quantitative estimates of the biodegradation of the individual components were obtained by column chromatography and computerized mass spectrometry (Walker, Colwell and Petrakis, 1975). Thus, biodegradation is an uneven process at best, some components of the petroleum being selectively degraded and others persisting.

Mulkins-Phillips and Stewart (1974; 1974a) compared degradation of Venezuelan and Arabian crude oils by a Nocardia species and found that 77 percent and 13 percent of the n-alkanes and unresolved fraction of the Venezuelan crude, compared with 94 percent and 35 percent of these components respectively, in an Arabian crude oil, were degraded. Atlas and Bartha (1973a) demonstrated that two paraffinic crudes, Sweden and Louisiana, were degraded similarly (70 percent) by mixed bacterial cultures grown in sea water. Jobson et al. (1974) compared growth of micro-organisms on what they termed an "inferior" and a "high grade" crude oil. Westlake et al. (1974) reported that crude oil composition affected biodegradation, specifically the n-saturate fraction, resins, and asphaltenes. Thus, there is evidence that each oil has a characteristic composition which affects microbial degradation, some oils being degraded more completely than others. Some fractions of petroleum are degraded readily and relatively quickly (n-alkanes), whereas others persist and are slowly degraded, if at all.

The microbial flora of an oil polluted environment differs from that of a pristine environment and will degrade petroleum differently and at a different rate. If oil is already present in an environment, there will have been an ongoing selection for oil-degraders, and the degradation of oil may occur more quickly. However, the degradation is not necessarily more complete. For example, two environments were studied, one polluted with oil and the other unpolluted. Samples were collected from two areas of the same geographic location in Chesapeake Bay, a major estuary on the east coast of the United States. Water and sediment from an oil-polluted environment, Baltimore Harbor, and the same from an oil-free environment, Eastern Bay, were used as inocula for determining the degradative capability of micro-organisms in the water and sediment at the two sites. Bacteria from Baltimore Harbor, previously exposed to oil, were more effective in degrading crude oil. Data for weathered samples showed that components could be lost by evaporation, but that microbial biodegradation accounted for the significant loss, i.e., metabolism, of certain fractions of the South Louisiana crude oil. Sediment inocula from two oil-polluted harbours in different geographical locations were compared for ability to degrade a South Louisiana crude oil. Cultures from Baltimore Harbor were found to be more effective in removing components of the oil than cultures from San Juan Harbor sediment and water samples. After growth for several weeks under optimal conditions in the laboratory, an increase in resins and asphaltenes in Baltimore Harbor cultures was noted, when compared with the San Juan cultures. Thus, the microbial populations from the two harbours were, indeed, different.

Components either slowly degraded or not degraded at all may accumulate as "refractory" components, or they may arise during, and as a result of, microbial synthesis (Walker and Colwell, 1976a). In fact, Atlas, Bartha and van Leeuwenhoek (1974) reported that accumulation of compounds such as fatty acids, in flask cultures, may prevent complete biodegradation of oil. In a study reported by Walker and Colwell (1976), four crude oils, South Louisiana, Brass River Nigerian, Anaco Venezuelan, and Altamont crude, were subjected to microbial biodegradation by five cultures, Acinetobacter lwoffi, Pseudomonas sp., Nocardia sp. Corynebacterium sp., and an unidentified strain. A mixed culture, i.e., a water sample from Colgate Creek in Baltimore Harbor, Chesapeake Bay, was also examined. Tar balls from Miami Beach, Florida, and the Sargasso Sea, collected during oceanographic cruises were extracted. (Brunnek, Duckworth and Stephen, 1968; Ramsdale and Wilkinson, 1968). An increase in n-alkanes C_{18} to C_{36} was noted during growth of the Nocardia sp. on Altamont crude oil. A waxy residue which remained after biodegradation of A. lwoffi, Pseudomonas sp., the unidentified Nocardia strain, and the mixed culture of Altamont crude oil contained no low-boiling n-alkanes, but did contain significant quantities of high boiling n-alkanes C_{27} to C_{36}, as well as C_{38} to C_{43} n-alkanes. Tar balls collected off Miami Beach, Florida, also contained long-chain n-alkanes, whereas the tar balls collected at the deep ocean station in the Sargasso Sea showed only trace amounts of long-chain n-alkanes. Appearance of long-chain n-alkanes in a waxy residue can result from microbial degradation. The composition of tar balls may, in fact, indicate the degree of microbial degradation to which they have been exposed.

Unpolluted areas that are geographically separated may also have unique microbial flora that respond differently to oil input. Two sediment inocula from nonpolluted environments of different geographic locations, an estuarine area and a deep ocean site, were compared for ability to degrade South Louisiana crude oil. The deep ocean sediment samples demonstrated greater petroleum-degrading potential than did the sediment from the Eastern Bay estuary, an unpolluted site in Chesapeake Bay. When these samples were compared with samples of sediment collected from a polluted harbour, the harbour sediments clearly showed the greatest degradative potential. As was noted for estuarine and deep ocean sediments, coastal zone and deep ocean water and sediment will differ in microbial degradative potential. Differences between neritic and deep ocean environments, especially with respect to such parameters as temperature and pressure, may have a marked effect on hydrocarbon degradation. Unfortunately, there are no data for petroleum which deal with this point.

Experiments using hydrocarbons have been performed by Schwarz, Walker and Colwell (1974, 1974a) using carbon-14 labelled hydrocarbons and a mixed hydrocarbon substrate. Maximum growth of hexadecane by the culture was reached within four weeks at 1 atm pressure and $4^{\circ}C$, stationary phase, but at 500 atm, maximum growth was attained only after incubation for 32 weeks. Utilization of 94 percent of the hexadecane was accomplished after eight weeks at 1 atm, but only after 40 weeks at 500 atm. Thus, at low temperature and high pressures, utilization of hexadecane was considerably restricted. (Schwarz, Walker and Colwell 1974, 1974a, 1975). These data strongly suggest that oil entering the deep ocean will be degraded only very slowly by the microbial populations found therein.

Most of the data in the literature deals, of necessity, with batch or flask cultures of petroleum, or its fractions, and either mixed or pure cultures of bacteria. Although such data offer useful information and provide model systems for such events in nature, there are complexities of the natural ecological system that must be ignored in order to establish individual parameters governing microbial degradation of petroleum in the marine environment. The process of oil degradation in batch culture, therefore, as carried out in most laboratories, involves a number of limitations. For example, in a flask or batch culture, exhaustion of inorganic nutrients, production of toxic substances, and creation of a refractory organic complex after oil degradation, may occur. There is evidence that biodegradation can be limited by fatty acid production during biodegradation in laboratory studies (Atlas, Bartha and van Leeuwenhoek, 1974). These components most likely interfere with degradation by accumulation and repression of metabolic activity. Increase in asphaltenes fractions, which are organic complexes and refractory to further degradation, may also result from microbial action on oil. However, this aspect of petroleum degradation has been neglected, and analytical methods will have to be developed in order to learn more about specific refractory materials resulting from degradation.

Nutrient limitation, specifically by nitrates and phosphates, is a major factor in the degradation of petroleum in the marine environment. Evidence is available showing a lack of significant petroleum degradation in laboratory studies simulating field conditions (Atlas and Bartha, 1972). An eight to nine-fold increase in biodegradation was observed in flow-through sea water tanks when nitrate and phosphate were added to oil-water mixtures (Atlas and Bartha, 1972; 1973b). An artificial sea water medium with one percent filter-sterilized Sweden crude oil was added as the sole carbon source. A combination of octyl-phosphate and a slow-release paraffinized urea fertilizer were found to be as good in laboratory studies as nitrogen or phosphorous salts. Increased utilization of mixed hydrocarbon substrate in batch culture by bacteria in Chesapeake Bay water, which is of estuarine salinity (14-18 parts per thousand) as compared to sea water (35 parts per thousand), containing increased amounts of inorganic nitrogen has been demonstrated (Walker and Colwell, 1975a). Similar effects have been observed with bacteria cultured in sea water (Walker et al., 1975). Based on the data now available, the N:P:C: ratios for degradation of petroleum can be approximated to enhance biodegradation of petroleum in field studies (Walker and Colwell, 1975).

Climatological conditions affect microbial degradation of petroleum. Two stations in Raritan Bay were monitored at five intervals between July 1971 and May 1972, and it was discovered that the highest counts of petroleum degraders were observed in July (Atlas and Bartha, 1973a). The temperature of the water column is highest in the summer months. In conjunction with other climatological conditions, this strongly influences rise and fall in selected microbial populations (Kaneko and Colwell, 1972).

However, increased numbers of petroleum degraders were found at two different time periods, December-February and June-July, in Chesapeake Bay (Colwell et al., 1974). To determine if enrichment for psychrophilic petroleum degraders occurred during the winter months, sediment bacteria from an oil-polluted environment in Chesapeake Bay were sampled during January and September. These samples were tested for the presence of petroleum-degrading bacteria capable of growth and oil degradation at $0^{\circ}C$. No significant differences were observed in the amount of oil degraded by micro-organisms in samples collected in January, compared with samples collected in September, suggesting that enrichment for psychrotrophic and/or psychrophilic petroleum degraders during winter months does not occur. It is more likely that eurythermic petroleum-degrading micro-organisms are present in Chesapeake Bay. There is selection for petroleum-degrading bacteria by introduction of oil into the environment and the above experiment indicates that temperature would not be a limiting factor with respect to developing populations of micro-organisms capable of degrading petroleum. Nevertheless, the overall _rate_ of degradation would be lower in the winter than in the summer.

The general factors influencing microbial degradation in temperate and arctic zones are similar to those for seasonal conditions, except that in temperate or arctic zones, specific conditions are more or less established for given geographical areas within the zones. For example, oil entering a permanently cold environment would be subjected to degradation by the pre-existing psychrophilic or psychrotrophic microbial populations present in and adapted to that environment. The principal physico-chemical effects that must be considered at low temperatures are the decreased volatilization and increased water solubility of volatile petroleum hydrocarbons (Atlas and Bartha, 1972a; Walker and Colwell, 1974). There is an indigenous microbial population in each given area of the world oceans, with more abundant populations found in coastal zones and in estuarine waters than in the deep ocean (Colwell, Walker and Nelson, 1973; Colwell et al., 1974, 1976). Furthermore, populations are adapted to the immediate environment. The introduction of petroleum alters the environment, causing a selection for those micro-organisms capable of degrading petroleum. Degradation will proceed quickly or slowly depending on nutrients (N,P), temperature, mixing of the oil-water substrate, and other parameters. Microbial degradation in arctic environments will involve psychrotrophic and psychrophilic micro-organisms indigenous to the area and degradation will proceed, albeit much more slowly than during the warm season in the temperate zone. These considerations must be accommodated in developing a strategy for combatting an oil spill that involves microbial degradation.

The separate and interactive roles of bacteria, yeasts and filamentous fungi in degradation of petroleum occurring in situ must be elucidated, because pure cultures rarely, if ever, occur in nature, and it is the mixed population that is responsible for degradation. It must be established whether each of these groups of micro-organisms is antagonistic or synergistic. The data available indicate that the total numbers of petroleum-degrading bacteria, yeasts and fungi in an oil-exposed area correlate with petroleum-degrading activity, but that biomass, i.e., total viable populations, does not. Hence, the population of all types of petroleum-degrading micro-organisms must be studied, since the pattern of degradation differs from one to another (Walker, Austin and Colwell, 1974). A petroleum-degrading achlorophyllous alga was recently discovered, and its pattern of oil degradation has been described (Walker and Colwell, in press; Walker, Colwell and Petrakis, 1975a). Thus, even algae must be considered in determining the capability of microbial populations to degrade oil.

Miget et al. (1969), Atlas and Bartha (1972b), Reisfeld, Rosenberg and Gutnick (1972), and Cerniglia and Perry (1973), report utilization of petroleum by pure cultures of bacteria and fungi. Bacterial species (Flavobacterium, Brevibacterium and Arthrobacter spp.) were reported to utilize 35-70 percent of paraffinic crude oils, whereas fungi (Penicillium and Cunninghamella spp.) utilized 85-92 percent of paraffinic crude oils. Comparative studies by Walker and Colwell (in press) demonstrated that yeasts (Candida) and fungi (Penicillium) degrade South Louisiana crude oil more extensively than bacteria (Pseudomonas and coryneforms) but in none of the reports cited was the oil completely degraded. This is a most important fact to be considered. Laboratory experiments have been initiated to determine the extent and sequential events occurring during the biodegradation of crude oil by combinations of bacteria, yeasts and fungi (Conrad, Walker and Colwell, 1976). Preliminary data indicate that there is a sequence of degradation, approximately as follows: Bacteria initiate the degradative cycle, utilizing mainly the alkanes and selected cycloalkanes of the crude oil. The yeasts and fungi achieve ascendancy in population numbers and in degradative function subsequent to the bacterial activity, having the capability to degrade the cyclo alkanes and aromatic fractions.

Experiments have been done using a mixed hydrocarbon substrate (Walker and Colwell, 1974, 1974a). Growth of bacteria in flask cultures on mixed hydrocarbon substrate was observed to precede growth of yeasts and fungi. In addition, oscillations were noted in the occurrence of several bacterial genera in mixed cultures growing on a mixed hydrocarbon substrate.

Combinations of bacteria, yeasts and filamentous fungi provide approximately two times greater degradation of the mixed hydrocarbon substrate, as compared with bacteria or yeasts and filamentous fungi grown individually (Walker and Colwell, 1974). It is surprising that these facts are not always emphasized when seeding or control of spills with microbial degradation is discussed. Furthermore, the populations of yeasts and fungi are sparse in open ocean waters (ZoBell, 1946). Hence the fate and extent of microbial degradation of petroleum in the deep ocean may be much less than in coastal or estuarine areas. Also, residual components may persist for much longer periods of time.

Microbial seeding of petroleum, as opposed to mechanical removal (Sebba, 1971), has been proposed by a number of investigators over the years. Seeding has been examined in laboratory experiments (Liu and Dutka, 1972; Robichaux and Myrick, 1972; LaRock and

Severence, 1973) and in simulated field conditions (Atlas and Bartha, 1973b; Miget, 1973), both of which demonstrate that bacteria enhance oil degradation. However, seeding must be undertaken only under certain circumstances and always with great precautions (Cobet, Guard and Chatigny, 1973). A single micro-organism will not possess the enzymatic capacity to metabolize all of the many varieties of compounds present in a spilled oil. A mixture of microorganisms, including fungi, yeasts and bacteria would be preferred. Either pure cultures or mixed, enriched cultures may serve as inocula.

There are several general areas of concern in applying microbial preparations to oil in nature. The problems include: selection of method of application, viz., wet slurry, dry powder, pelletization, or aerosol; potential pathogenicity or adverse human reaction, viz., presence of human pathogens in the microbial preparation or opportunistic pathogens; hypersensitivity reactions to the microbial preparation; increased density of micro-organisms in the environment and effects on the natural biota; greater potential toxicity of emulsified oil residues; and formation of extracellular metabolic products resulting from biodegradation. The pitfalls are many; studies by Cook, Massey and Ahearn (1973) suggest that microbial degradation of oil may increase the toxicity of oil and of components of oil. Metabolites or byproducts during yeast decomposition and emulsification of high asphalt crude oil by yeasts may be harmful to fish.

In lieu of in situ experiments designed for the marine environment, some actual in situ experiments were conducted, that is, oil-contaminated soil was seeded with bacteria by Jobson et al. (1974). It was found that, compared with addition of nitrogen and phosphorous sources (fertilizers), bacteria had little effect on removal of the oil. The most serious danger appears to be that the introduction of new bacterial species can produce effects on the natural populations, such as fish kills or loss of other biota.

Rates of petroleum biodegradation are dependent upon temperature, salinity, concentration of inorganic nutrients, extent of dispersion of oil in water, abundance and kinds of microorganisms, chemical composition of the oil, whether the system is open (in situ) or closed (laboratory), and a variety of other factors. ZoBell (1969), in summarizing work accomplished prior to 1969, has reported rates of degradation of crude oils, lubricating oils, cutting oil, and oil wastes, ranging from 0.02 to 2.0 g/m^2/day at 24 to $30^\circ C$.

Rates of oil biodegradation have been measured as g/m^2/day, g/m^3/day, mg/day/bacterial cell or percentage removed after a known number of days or weeks. In Table 2-1 oil degradation rates given by several investigators are summarized. It is difficult to compare the data from the relatively few studies that have been done. Differences in experimental conditions are additional problems when comparing studies in which biodegradation rates of petroleum were determined (Johnson, 1970; Kator, Oppenheimer and Miget, 1971; Kinney, Button and Schell, 1969; Robertson et al., 1973; Bridie and Bos, 1971). Although rates of biodegradation of petroleums have been estimated, the experiments, in general, involved measuring amount of oil at the beginning and end of the experiments, with the assumption that the rate was linear.

Investigations on the fate of saturated, aromatic, and other fractions of petroleum degraded by estuarine sediment bacteria (Walker, Colwell and Petrakis, 1976) showed that there was a marked difference in extent and rate of biodegradability among the fractions. The maximum amount of degradation of the total residue of South Louisiana crude oil occurred during the log phase (1.42 mg/day), levelling off at the stationary phase, as would be expected. Asphaltenes and aromatics increased after stationary phase was reached, but saturates were degraded throughout the seven-week growth phase.

The rates of biodegradation of South Louisiana crude oil using sediment from an oil-polluted estuary were determined by a computerized curve fitting to obtain graphs of first, second, or third degree polynomials (p less than or equal to 0.05). Saturates were the only components which showed a consistent decrease. Aromatics were observed to decrease initially but increased at the end of the period of degradation, i.e. at 3-7 weeks. Components comprised of pyridines, quinolenes, carbazoles, sulfoxides, and amides, decreased initially and increased thereafter. Asphaltenes appeared to be the most likely products of

TABLE 2-1

Oil degradation rates under varying conditions as shown by selected authors. It should be noted that sea water temperatures around the U.K. vary from about $2^\circ C$ to $16^\circ C$. The inorganic nitrogen concentration varies from <1 to around $500 \mu gN/l$

Bacteria involved	Kind of oil	Experimental conditions	Summary of results	Reference
Garden soil aerobes of several genera	Hydrocarbon mixtures in common use	Batch culture. Mineral salts media. Several temperatures between $20^\circ C$ and $37^\circ C$	0.4–0.75 g m^{-2} d^{-1} of some materials measured at $28^\circ C$	Sohngen, N.L. (1913) *Zentralbl. Bakt. Parasitkde. Ab. II*: 37, 595
Soil aerobes	Emba crude and lubricating oils	Batch culture. Nitrate or ammonia in mineral salts media. $23^\circ C$	1.2 g m^{-2} d^{-1} for crude oil (45% of added oil) 0.4 g m^{-2} d^{-1} for lubricating oil	Tausson, V.O. and Shapiro, S.L. (1934) *Mikrobiologiya* 3: 79
Enriched culture consisting predominantly of a marine bacterium	Clear refined mineral oil	Probably batch culture. $25^\circ C$ Aged sea water plus 0.5% KNO_3	The oxidation of the mineral oil was indicated by O_2 uptake, CO_2 output and bacterial growth. The Q_{10} is given as about 3.0 for temperatures between 0 and 40 C. The average amount of oil degraded at $25^\circ C$ is given as 1.2×10^{-10} mg per day per bacterial cell. Hence it is calculated that if the oil is uniformly distributed in the water and the population is constant at 8×10^5 organisms ml^{-1} then the rate of oil degradation will be about 350 g m^{-3} yr^{-1} at $25^\circ C$ and about 36.5 g m^{-3} yr^{-1} at $5^\circ C$.	ZoBell, C.E. (1964) *Advances in Water Poll. Res.* 3: 86. This paper also appears in *Air & Water Poll.* (1960) 7: 173

(Table 2-1 cont'd)

Bacteria involved	Kind of oil	Experimental conditions	Summary of results	Reference
Mixed culture of oil-oxidizing bacteria	American crudes	Batch culture, aerated by shaking 25°C sea water medium reinforced with 0.01% $(NH_4)_2HPO_4$. About 1 g placed in 100 ml medium. Oil dispersed on ignited asbestos	Between 17.8% and 98.8% by weight of oil removed in 30 days. Average around 45%.	ZoBell, C.E. and Prokop, J.V. (1966). Z. allg. Mikrobiol. 6: 143
Natural sea water population	Crudes and several refined products	Batch culture, 18°C sea water medium reinforced with NH_4Cl and phosphates	The influence of various physical and chemical factors on oil degradation is illustrated. The presence of nitrogen and phosphate was shown to increase markedly the breakdown of diesel oil in 8 weeks. The presence of easily degraded material 'spared' the oil. The effect of temperature is also shown.	Gunkel, W. (1967) Helgolander wiss. Meeresunters. 15: 210
Natural marine population	Atmospheric residue of Kuwait oil	Sea water percolated through columns of beach sand (median grain size 250μ) with natural meio and micro fauna. Sands were lightly or heavily oiled. 10°C.	Oxygen uptake used as indication of degradation. Using a 'B.O.D.' value of 5.0, the author calculates a loss of oil from 0.09 g oil $m^{-2} d^{-1}$ to 0.04 g oil $m^{-2} d^{-1}$ depending upon dosing. These rates applied for several months and accounted for 10% of the oil. Preliminary gas chromatograms suggested the main loss was of the alkane fraction. The remaining 90% decayed 'immeasurably slowly'.	Johnson, R. (1970) J. mar. biol. Ass. UK 50: 925

(Table 2-1 continued)

Bacteria involved	Kind of oil	Experimental conditions	Summary of results	Reference
Selected mixed cultures of oil-oxidizing organisms	Louisiana crude	Shaken flasks with sea water enriched with inorganic nitrogen, phosphates and yeast extract. Approx. 70 mg oil added to 200 ml medium, $20^\circ C$ and $30^\circ C$. Also simulated field studies of large tanks (900 l). Sea water enriched with (N H), 50-100 ml of crude	Initial oxidation attributed to breakdown of n-alkanes smaller than C. The initial rate was followed by a decrease and then another increase. Up to approx. 50% of the crude was lost. No evidence of utilization of aromatics was found. In the large tanks the bacteria...	Kator, H., Oppenheimer, C.H. and Miget, R.J. (1971) Prevention and control of oil spills. American Petroleum Institute Conference 1971, pp.287

From Floodgate (1972a).

weathering and biodegradation. Computerized mass spectrometry was used to monitor the biodegradation rates. It was obvious, from the data obtained by Walker, Colwell and Petrakis (1976a), that the components of petroleum, although degraded simultaneously, are degraded at different rates.

The presence of bacterial oxidation products is an ecological factor of some magnitude. Complete biological oxidation of petroleum by micro-organisms yields carbon dioxide, water, sulphates, and nitrates as major products. However, many of the biological reactions do not go to completion. Intermediate products of the reactions, for example, acids, aldehydes, ketones, alcohols, peroxides, sulfoxides, are soluble in water and will be removed from the floating sample.

Micro-organisms attacking films or thin layers of oil on water will bring about changes in colour, constituency, fluorescence, emulsification, and other visual properties. The underside of the oil will become stringy and gradually disintegrate into small droplets which slowly sink. Part of the oil will be emulsified in the water which, together with the great increase in the microbial population, will render the medium cloudy, turbid and sometimes coloured (Kator, 1973). The visual disappearance of oil does not necessarily indicate its complete destruction. While microbial attack may convert oil to carbon dioxide and water, part of the oil may be converted into other water-soluble substances, emulsified in water, or be deposited on the bottom of the container.

2.1.11 Tar balls

The occurrence of petroleum residues in the form of tar balls or tar lumps on beaches and on the oceans' surfaces is well documented. Tar balls or lumps represent products of different degrees of the physical, chemical and biological weathering of petroleum. Their physical appearance varies appreciably, ranging in size from a few millimetres in diameter up to several centimetres. Some samples are soft, and others are quite hard, almost brittle, and many incorporate much sand and small particles.

Koons and Monaghan (1973) report the collection of 34 samples of tar balls from the Gulf of Mexico. Overall, the samples showed a wide variety in chemical composition. For example, the saturated hydrocarbon composition ranged from 1.6 to 56.1 percent, and the asphaltene contents from 8.8 to 54.7 percent. The average composition of the 34 samples were compared with the Challenger Knoll seep oil and an average for crude oils produced from offshore Louisiana platforms. The soft, fresh-appearing tar ball samples appeared compositionally rather similar to the Challenger oil and the offshore Louisiana crudes. Evaporation and solution have stripped away the hydrocarbons up to about C_{15} to C_{17}. For these fresh-appearing samples, the ratios of saturated to aromatic hydrocarbons and of normal paraffins to isoparaffins plus naphthenes do not appear significantly different from those for the seep oil or produced oils. Apparently oxidative processes have not been operating long enough to significantly affect chemical composition. The hard, brittle tar ball samples are the ones that appear to have undergone appreciable chemical and biological oxidation in addition to physical weathering. They contain the low amounts of saturated hydrocarbons (1.6 to 10 percent) and greater amounts of nitrogen-sulphur-oxygen containing compounds and asphaltenes. The saturate to aromatic hydrocarbon ratio decreases significantly (about 2.0 for fresh samples to 0.3 or less for the weathered samples). The normal paraffins to isoparaffins and naphthenes ratio for the weathered samples approach zero (the n-paraffin spikes on the gas chromatograms practicalyy disappear). This last chemical observation supports severe biological degradation of these samples. The hard tar balls appear to have undergone some further evaporation and/or solution since the content of C_{15} to C_{25} hydrocarbons is decreased relative to the hydrocarbon fraction above C_{25}.

2.2 States of oil available for uptake

2.2.1 Uptake mechanisms

Petroleum hydrocarbons are presented to pelagic organisms as dissolved or dispersed materials, adsorbed onto particulate material, or as small floating tar balls.

Lee and Benson (1973) suggested that hydrocarbons enter the marine food web by several routes:

(1) Adsorption onto particles, both living and dead, followed by ingestion of these particles.

(2) Active uptake of dissolved or dispersed petroleum.

(3) Passage into gut of fish which gulp or drink water.

A considerable portion of petroleum is adsorbed onto or dissolved in particulate matter, because the partition coefficient should favour the solution of hydrocarbons in the lipids of detritus. Detritus would then carry the adsorbed or dissolved hydrocarbons to the sea floor to be consumed by benthic organisms. Holcomb (1969) states that, as early as 1950, Soviet microbiologists demonstrated that, after the lighter fractions of oil from a river spill dissipated, the remaining oil was adsorbed on suspended matter and sank. Bottom-dwelling micro-organisms produced a new mixture of organic compounds that were carried to the surface with bubbles of methane and other gases. The new compounds again were adsorbed, sank and the cycle was repeated.

Burns and Teal (1973) hypothesized, using organisms from the pelagic Sargassum community, that petroleum hydrocarbons could enter the food chain either in food or through body (mainly respiratory) surfaces, the latter being the main route of entry.

Stegeman and Teal (1973) consider that it is not yet established whether the primary mechanisms concerned in uptake of nonbiogenic hydrocarbons by marine bivalves involves consumption of contaminated oil or equilibration with hydrocarbons in the water. However, the work of Lee, Sauerheber and Benson (1972) indicates that partitioning across the outer membrane, particularly gill surfaces, is an important means of uptake.

Dissolved hydrocarbons are taken up by the gill tissue of the mussel Mytilus edulis, followed by transfer of the hydrocarbons to other tissues (Lee, Sauerheber and Benson, 1972). Subsequent work on the uptake of dissolved hydrocarbons by marine fish demonstrated the entrance of hydrocarbons through the gills (Lee, Sauerheber and Dobbs, 1972). It was also reported that phytoplankton appeared to absorb hydrocarbons but that there was no transfer inside the cell in the short-term.

It is presently believed that large floating aggregates (i.e., tar balls) are not taken up by marine organisms and are thus not an important introductory pathway into the marine ecosystem. There have been reports, however, of commercial fish having large tar balls in their stomachs (Tendron, 1968). One report, not a detailed study, indicated that tar was found in the stomach of the saury (Scomberesox saurus) in three of ten specimens collected by a research group at Woods Hole, Massachusetts (Horn, Teal and Backus, 1970). The role of the fish netting operation in causing the ingestion of the tar balls is not known.

2.2.2 Storage, metabolism, discharge

In some cases, the biota may be severely affected by the presence of oil in their habitat, but they are not a major reservoir for spilled oil. However, they may act as a temporary storage site or transfer point. It is apparent that marine animals probably are not major factors in influencing the distribution patterns of spilled oil. After petroleum hydrocarbons are taken up by an organism, they may be excreted unchanged, they may be metabolized, or they may be stored with possible elimination at a future date.

Morris (1973) investigated the hydrocarbon content of barnacles (Lepas fasiculari) living on tar balls from the northeast Atlantic. The barnacles had a hydrocarbon composition intermediate between those of the inside and outside of the tar ball. The total hydrocarbon content amounted to about 6 percent of the total lipid, of which the non-natural hydrocarbons were estimated to compose 80-90 percent; thus only 5 percent of the animal's total lipid was non-natural. In spite of the fact that the animals were in very close contact with a large volume of oil for an extended period of time, they did not show gross hydrocarbon pollution. This would appear to indicate that the pollutant hydrocarbons were being assimilated and then discharged, unmetabolized, quite rapidly.

Lee, Sauerheber and Benson (1972) found in particular that saturates and aromatics were taken up rapidly by mussels but were also discharged without metabolic breakdown after the mussels were returned to clean sea water. They found no evidence that mussels could not metabolize such aromatics as toluene, naphthalene and benzpyrene.

Following the ARROW incident in Chedabucto Bay, plankton were observed to ingest large quantities of small drops of Bunker C oil and eliminate them in the form of faecal matter (up to 7 percent Bunker C oil by weight) (Conover, 1971). The plankton always avoided the small oil particles within 24 hours and showed no sign of stress when viewed under a dissenting microscope. Conover (1971) went further to suggest that under the conditions observed in Chedabucto Bay, the plankton could graze as much as 20 percent of the oil particles (less than 1 mm in diameter) in the water column and sediment them in their denser than sea water faeces. Parker (1970) also demonstrated that copepods can ingest considerable quantities of oil and pass the oil, unchanged, into the faecal material.

The fact that oil can enter marine organisms and be discharged unchanged is of considerable interest. Oil from a slick can be grazed by plankton and the ingested oil precipitated in the faeces. Faecal matter is in most cases denser than sea water, thereby providing a mechanism by which oil can be made available through the water column and ultimately within the sediments, where it will be exposed to benthic organisms. It is also a well known fact that faecal pellets can be ingested by marine organisms, thus providing a possible mechanism for passage and/or concentration in the marine produce.

Morris (1974) has demonstrated that the surface zooplankton from the Mediterranean take up large quantities of petroleum hydrocarbons from the heavily polluted surface film. Hydrocarbons, having the same complex molecular weight distribution as those in the surface film, constituted some 17-33 percent of the total lipid material in the organisms. This was considered to be indicative of the storage and concentration of pollutant hydrocarbons by these organisms. Stegeman and Teal (1973) found that No. 2 fuel oil was accumulated rapidly by oysters and dropped to a stable level of approximately 30 $\mu g/g$ wet weight, where after two weeks of discharge it remained constant for at least $3\frac{1}{2}$ months. Blumer, Souza and Sass (1970) found that in some cases, shellfish had a similar concentration of No. 2 fuel oil eight months after a spill.

Studies by Lee and Benson (1973) have indicated some storage of hydrocarbons in the liver of marine fish and in the hepatopancreas of several invertebrates. Since the liver and hepatopancreas are generally high in lipid, this observation would be predicted. The gall bladder in fish was also a temporary storage site, but apparently serves mainly as an avenue for discharge. The complex lipoproteins of cell and organelle membranes of all tissues are also possible storage sites.

Metabolism of ingested petroleum hydrocarbons by marine organisms has also been shown to occur, followed by either incorporation into the body tissues or subsequent excretion. Marine fish tested by Lee, Sauerheber and Dobbs (1972) were found to take up aromatic hydrocarbons via the gills. Metabolism then occurred in the liver, followed by transfer of the hydrocarbons and metabolites to the bile, and finally excretion. This suggests an efficient detoxification mechanism in the fish which allowed for the removal of polycyclic aromatics from the body tissues.

Scarratt and Zitko (1972) found that 26 months after the ARROW spill, urchins, mussels and periwinkles showed a considerable reduction in Bunker C content. However, the absence of fluorescence emission maxima at 360 nm in species showing high background fluorescence strongly suggests that fluorescent compounds from ingested Bunker C or Bunker C-contaminated prey are either metabolized or bound up in the tissues with different fluorescent compounds being generated in the process.

Blumer (1967) showed that naturally occurring hydrocarbons contained in the basking shark's major food source, zooplankton, passed through the shark's digestive tract without fractionation or structural modification. The hydrocarbons were then deposited in the liver.

The metabolism of hydrocarbons by marine organisms other than bacteria is not well understood. The pathways involving oxidases and other enzymes, important in the degradation of aromatic and paraffinic hydrocarbons by mammalian systems, is well documented (Boyland and Soloman, 1955; Daly, Jernis and Witkop, 1972; Falk et al., 1962; Diamond and Clark, 1970; McCarthy, 1964). In the case of aromatic compounds, hydroxylation is followed by conjugation with sulphate or glucose and finally by excretion of the water-soluble product. Straight chain hydrocarbons are hydroxylated at the terminal end and further oxidized to the fatty acid which can be broken down by B-oxidation. Highly branched chain hydrocarbons, such as pristane and phytane, are probably oxidized to an acid (e.g., phytanic acid) which can be further oxidized by a combination of α- and B-oxidation (Mize et al., 1969). Whether all these pathways are available for use in marine organisms is still being examined.

Thus it can be seen that the work on metabolism of hydrocarbons within marine organisms is as yet very patchy, but some patterns are beginning to emerge. However, observations are that fish appear to have much the same metabolic pathways as mammals, but the bivalves have a diminished capacity for hydrocarbon metabolism and that capacity differs from one species to another.

Until the metabolic fate of the various classes of hydrocarbons present in petroleum, in the food web, and in aquatic organisms is examined, it is impossible to evaluate the fate of persistent low level petroleum pollution in the marine environment. Research on sublethal amounts of petroleum pollution must include experiments with water-soluble fractions, oil adsorbed on the food that goes through the digestive system, and oil metabolized by plants and animals lower in the food web. Two critical variables to be considered are the temperature of the environment and the concentration of the hydrocarbons. The former is important because it affects activity and even determines which enzyme systems are operative in cold-blooded animals such as fish, while the latter determines whether some of the components are inert, beneficial or harmful as well as whether the organisms utilize, store, or excrete the pollutant Finally, once hydrocarbons get into the lipid pool, an indication of turnover time of these compounds is needed.

2.3 Biomagnification

It has been previously stated that, although the biota may be severely affected by the presence of oil, it is not a major reservoir for spilled oil, nor is it a major factor in influencing the distribution patterns of spilled oil. One might therefore question the importance of discussing the biota, other than micro-organisms, in a chapter on fates. The reason is that once accumulation and/or concentration of components of petroleum are shown to occur within marine organisms, the pathways for the passage and magnification of oil within the marine food chain are opened up. In this section, the possible mechanisms for biomagnification will be discussed, including assimilation of oil by lower trophic level marine organisms, possible routes of transfer through the food chain, and actual evidence supporting the phenomenon of biomagnification. In Chapter 6 on Human Effects, the concept is discussed in that context and should be considered, while noting that the following definitions were not generally used by the authors.

In discussing the problem of biomagnification, it is highly desirable to distinguish between bioaccumulation, bioconcentration, and biomagnification. Most authors were noted as not making these distinctions and therefore the use in Chapter 6 is not as used here. Bioconcentration refers to the ability of an organism or a population of organisms of the same trophic level to concentrate a substance from an aquatic system. Bioaccumulation refers to the ability of an organism to not only concentrate, but to continue to concentrate essentially throughout its active metabolic lifetime, such that the ratio of tissue to ambient concentration, once calculated, would be continuously increasing during its lifetime. When the rate of intake of a given material is below the excretory and metabolic capacity of an organism, no accumulation of this material will occur. When the rate of intake of a given material exceeds this threshold, the material will be accumulated, the concentration of the material in the organism at any time reflecting the integrated rates of intake and excretion. thus bioaccumulation applies to individual organisms. Biomagnification, however, refers to an increase of a given material at successive trophic levels within an ecosystem, i.e. the predatory organism will have a higher concentration in its tissues of the material in question than those organisms upon which it feeds. This process has been demonstrated for such organochlorine pesticides as dieldrin and DDT.

Although bioconcentration has been shown to occur the process of biomagnification remains a theoretical possibility. There is no convincing evidence at this time for the food chain biomagnification of petroleum hydrocarbons. The bioconcentration of petroleum hydrocarbons that has been observed in organisms of higher trophic levels is more likely a function of the ability of these species to concentrate hydrocarbons from the water than a function of their position in the food web. A more detailed treatment of these phenomena can be found in Chapters 5 and 6.

Chapter 3: ANALYTICAL TECHNOLOGY

Definition: Analytical technology includes sampling, isolation, separation and methods of measurement of individual or groups of hydrocarbons (e.g., aromatics) in sediments, sea water and organisms.

Problem: Analytical technology involves:

(i) <u>collection</u> of <u>representative samples</u> without contamination and storage so as to prevent any alteration;

(ii) <u>work-up</u> of <u>the sample</u>, which involves clean-up of the sample and concentration since many of the methods in current use are not sufficiently selective and sensitive at the levels encountered in most marine samples;

(iii) <u>chemical analyses</u>, which involve differentiating between recently biosynthesized components present in the marine environment and the analysis of oils or petroleum hydrocarbons at varying levels.

There is a need for sophisticated modern analytical apparatus and for highly trained personnel. The vast area of the ocean leads to vast amounts of data and thereby a need for data-reducing systems. Interpretation of results and analytical data also requires highly trained personnel. For intercomparison of data on a worldwide basis, intercalibration is necessary.

Background: Insufficient attention has often been given to the collection of representative and uncontaminated samples of which analyses were carried out using insufficiently selective and sensitive analytical techniques. Current awareness of the problems involved has led to refined methods of sampling and the application of modern sophisticated analytical methods to environmental studies. A more detailed review of analytical techniques was prepared by the National Academy of Sciences (1975).

Further difficulties arise in studies of microbial degradation, photochemical oxidation and other processes because of the necessity to determine the nature as well as the concentration of metabolites (Walker, Colwell and Petrakis, 1975; Hansen, 1975). This is not a field that has been fully explored and further work on this topic is clearly required.

3.1 Collection of samples

The importance of ensuring that samples are representative and uncontaminated cannot be too highly stressed. A summary of types of equipment, methods of sampling and means of preservation in current use for sediments, sea water and biological material is given in Table 3-1. The references quoted should be consulted for further details.

Particular points to note include the need to ensure that subsurface samples do not come into contact with surface water, and that contamination from the vessel is avoided. It may be prudent to collect samples of all oils used aboard the vessels employed in the sampling programme. This enables a comparison to be made with the oil recovered from the samples, and thus establishes the freedom from this particular source of contamination. The samples should be frozen immediately to prevent microbial alteration.

3.2 Isolation and separation

A summary of the various methods currently in use for the isolation and separation of fractions containing the petroleum hydrocarbons from sediments, sea water and biological material is given in Table 3-2. The solvents used need to be of exceptional purity.

TABLE 3-1

Sample collection

	Water	Biological materials	Sediments	References
Equipment	Slurp bottles for surface samples Glass bottles for near surface samples Other samplers made of glass, teflon or stainless steel for deeper samples Aluminium or teflon-lined caps Normal sample volume 1-3 l	Various fishing and sampling gear	Coring device Dredges	United States, National Bureau of Standards (1974)
Method	Samples must be collected as far as possible away from the vessel Samplers must be lowered and retrieved closed	Samples should not come in contact with surface water	Hand-coring by divers down to 20-30 m of water Gravity, coring Dredging	Mackie, Whittle and Hardy (1974) Giger and Blumer (1974)
Preservation	$HgCl_2$, NaN_3 (not to be used with Al-lined caps), H_2SO_4, CCl_4 Freeze Extract immediately (as described below) and store extract	Freeze immediately wrapped in Al-foil and/or placed in glass jars	Freeze immediately	

TABLE 3-2

Isolation and separation

Water	Biological Materials	Sediments	References
Extraction Pentane Hexane Benzene CH_2Cl_2 CCl_4 10-100 ml/l sample 1-3 times Hand or mechanically shaken Magnetic stirring Organic layer separated in separatory funnel or by pipette Important to rinse the sample jar with the solvent (1) Gas equilibration for C_1-C_{10} (2) Vacuum degassing for C_1-C_4 (3)	Washing of sample Dissecting Homogenizing 1. Extraction by homogenizing: 15 g sample + 100 g Na_2SO_4 + 150 ml pentane 2. Extraction by soxhlet apparatus: 15 g sample, 300 ml 1:1 MeOH-C_6H_6 for 24 h + 24 h 3. Extraction by digestion: 50 g sample with 150 ml 6.7% KOH in MeOH. Reflux for 1 h. This extraction recommended by Farrington and Medeiros (4) Saponification: extracts obtained by method 1 and 2 must be saponified to breakdown the co-extracted fatty acid esters which will otherwise interfere with the isolation of alkene and aromatic hydrocarbons The extracts are refluxed with KOH in MeOH-H_2O (3:1) for 24 h Upon completed saponification (or extraction by digestion), the hydrocarbons are extracted into pentane or hexane washed with ether, dried and concentrated prior to fractionation	Extraction by soxhlet apparatus Wet sample: 100-150 g soxhlet extracted with 275 ml MeOH for 24 h, then 75 ml C_6H_6 added and extraction continued for another 24 h (5) Freeze-dried samples are extracted with C_6H_6-MeOH (1:1) for 16 h (6) Co-extracted S_8 will chromatograph together with the PNAH fraction and is removed by reaction with copper (5)	(1) Gordon, Keizer and Dale (1974) (2) McAuliffe (1974) (3) Fort, Prescott and Walters, U.S. Patent (4) Farrington and Medeiros (1975) (5) Giger and Blumer (1974) (6) Blaylook, Bean and Wildung (1974)

Once the isolation of the fraction containing the petroleum hydrocarbons from the sample is complete, the subsequent separation stages and measurement are independent of the different sample types. Non-hydrocarbons in the initial extract are removed by column or thin-layer chromatography, which also serves to fractionate the hydrocarbons. Silica alone (Brown, R.A. et al., 1973) or aluminia on silica (Blumer, Souza and Sass, 1970; Blumer et al., 1972; Ehrhardt, 1972; Farrington and Medeiros, 1975) can be used. Both aluminia and silica require to be activated by heating and thereafter partially deactivated with water prior to use. The sample is normally applied to the column in, for example, hexane or pentane solution. Further fractionation into alkanes, naphthanes and aromatics can be achieved by gradient elution involving increasing concentration of benzene in a pentane/benzene solvent mixture.

3.3 Chemical analysis

A number of methods have been used for the determination of "oil", individual hydrocarbons, hydrocarbon fractions. These include the following:

Gravimetric methods   (American Public Health Association, 1971; Environmental Protection Agency, 1971; American Society for Testing Materials, 1972). These give a measure of "total oil" content. They suffer from evaporative and dissolution losses;

Infra-red spectrometry   (e.g., Ahmed et al., 1974; Brown, R.A. et al., 1973) is based on the absorption of the C-H stretching band $3.4\,\mu$. The sample to be analysed must be in a suitable solvent such as carbon tetrachloride. This technique has sometimes been misused by an inappropriate choice of wavelength for measurement. Care must be taken that solvents containing C-H bonds are not used and that other solvents do not contain such compounds as impurities;

UV-fluorescence spectrometry   (Hornig, 1974; Gordon and Keizer, 1974, 1974a; Frank, 1975). Both excitation and emission spectra can be recorded. This technique suffers from the disadvantage that the measurement is a response to minor constituents which may be taken, erroneously, as equivalent to "oil". Difficulties also arise from the use of impure solvents and from the necessity of choosing an appropriate reference material for calibration;

High performance liquid chromatography. Although basically a separation technique, it is used in combination with mass-spectrometer and other sophisticated instrumental techniques to give a measure of individual oil compounds;

Gas chromatography. In a similar way, this is used in combination with instrumental measurement techniques for the determination of individual oil compounds. Retention time measurements, although valuable, should not be used without positive identification by some other means, e.g., mass spectrometry. In most cases, high resolution gas chromatography is required. This may be achieved by using Support Coated or Wall Coated Open Tubular Capillary Columns (SCOT or WCOT Columns). One of the most powerful and highly sensitive analytical tools is the fully computerized combination of gas chromatography with mass spectrometry using glass capillary columns.

3.4 Reporting analyses

The flow diagram reproduced as Figure 3-1 is useful to consider, in the light of the variety of oils listed in Chapter 1, the complex pathways by which the different oils reach the marine biota and their effects (see Chapters 2 and 4-6). All of these discussions depend upon a clear definition of the material being studied and therefore gives rise to the continuing controversy of what procedure or method is most appropriate.

```
                              SAMPLES

                             Extraction

                       HYDROCARBONS AND LIPIDS

                           Saponification              UV absorption and/or UV
                                                       Fluorescence Spectro-
                  HYDROCARBONS AND UNSAPONIFIABLE      metry - initial screening
                              LIPIDS                   for presence or absence
                                                       of aromatic hydrocarbons;
SAPONIFIABLE LIPIDS                                    measurements of relative
                                                       absorption at certain
    Discard                                            frequencies

                           Chromatography

                   ONE OR SEVERAL HYDROCARBON FRACTIONS

UV Absorption and/or UV    Gas Chromatography - presence    IR Spectrometry - deter-
Fluorescence Spectro-      or absence of complex mixture    mines if lipids were
metry - presence or        of hydrocarbons; homologous      separated from hydrocarbons
absence of aromatic        series
hydrocarbons: Measure-
ment of relative
absorption at certain
frequencies

AROMATIC HYDROCARBONS

Mass Spectrometry -                                    Gas Chromatograph-Mass
Complexity of molecular                                Spectrometer Computer
structure and molecular                                System Analysis
weight range
                                                       Compound type analysis
```

Figure 3-1. *Flow diagram for analytical techniques to detect and and estimate petroleum contamination in marine organisms (from Farrington, 1973)*

A number of authors have used methods to determine particular components of oil or specific features (e.g., fluorescence) of oil and then reported the results as "total oil". As oil is subjected to processes of degradation and differentiation within the marine environment, this approach is not necessarily valid and may be misleading.

Ideally, the following information should be included with the results:

(1) precision and accuracy of the method

(2) limits of detection of the compounds reported

(3) molecular weight (or carbon number) of compounds detected or measured.

3.5 Comments

It is widely recognized that there is a need both for more analyses and for more complete analyses of sample materials of all kinds. In particular, many workers have asked that existing data on the composition of crude and refined oils be made more widely available.

For further understanding of the degradative and other processes involving oil in the marine environment, the development of analytical techniques must include not only more refined methods for oil compounds, but also better methods for the isolation, identification and measurement of those compounds that are formed in these processes.

Because the analytical procedures used are frequently not specified or described in detail, it is often impossible to establish the validity of results quoted in many papers. Unless this information is given, it is impossible to make an independent assessment or interpretation of the data.

A certain number of reference oils are now available.[1] Where appropriate, these should be used to establish the validity of the method, especially the effectiveness of the isolation and separation stages and the suitability of the measurement technique.

In view of the size and complexity of the marine environment and the limited investigative resources usually available, it is important that the sampling strategy is designed to be appropriate for the task in hand and subsequent reporting.

The Working Group recognized the potential of instrumental methods such as the fully computerized combination of gas chromatography with mass spectrometry. Such methods provide more extensive, more accurate and more precise data than many of the older, simpler methods, and their use is to be encouraged.

---

[1] Enquiries for reference oils should be directed to: The Director, Warren Spring Laboratory, Department of Industry, Gunnels Wood Road, Stevenage, Herts. SG1 2BX, England, or to: Dr. Jerry M. Neff, Biology Department, Texas A&M University, College Station, Texas 77843, U.S.A.

Chapter 4: CHEMICAL AND PHYSICAL EFFECTS OF OIL DISCHARGES

The interactions between oil and the marine environment are complex. Physical, chemical and biological factors alter the composition of the oil immediately after the discharge and possibly for months, even years, afterwards. In addition, the oils themselves have a variety of effects on the marine environment, both physically and biologically. The biological effects of oil will be discussed in Chapter 5. It is the purpose of this chapter to examine the chemical and physical effects of oil upon the marine environment, which, in turn, may have significant effects upon marine organisms. Those particular physico/chemical factors to be discussed are deoxygenation, heating effects, pollutant sorption and carbon dioxide gas transfer. There is limited discussion of pollutant sorption in other chapters.

4.1 Gas transfer and deoxygenation

Definitions: The effects of oils upon gas transfer and deoxygenation include the effects of oil on the oxygen levels in water columns and sediments by interference with biological, physical or chemical activity, including the transfer rates across the air/sea boundary. Also included is interference with gas exchange involving the transfer of carbon dioxide across the air/sea interface.

The exchange coefficient is the mass of oxygen entering the water per unit time through a unit surface of water in which there is a unit oxygen deficiency, and in which there is no oxygen production or utilization.

Problem: From time to time, fears have been expressed in scientific journals and in papers of national and international meetings, that oil pollution in the form of a layer floating on the surface of the sea will result in appreciable if not total depletion of oxygen in the sea. The argument runs that if such depletion were to arise, the conditions necessary for the growth of marine organisms would be destroyed, and severe limitation would then be imposed upon sea life in general, particularly on the production of plankton and other marine organisms at the foot of the food chain that are essential for the survival of a marine biosystem, including sea food resources utilized by man.

Whilst the argument so far has related to open sea conditions, similar considerations apply with equal or greater emphasis in coastal areas where the biological activity, and hence oxygen utilization, is high.

Background: Unfortunately, it has not been possible to trace the origin of the fears concerning oxygen depletion, nor to examine any original work in which such a depletion was established as occurring in the marine environment. The topic of oxygen exchange across a sea/water atmosphere boundary, and the extent to which oil films limit or interfere with this exchange, does not appear to have received a great deal of attention from the scientific community. Data that are available and considered on transfer rates were regarded as poor and conflicting and, since much of it predates the present concern for oil pollution, have to be interpreted with present-day oil slicks in mind.

Data relating to the transfer of carbon dioxide across the air/sea interface are almost non-existent, and few papers have been found describing the effect of oil layers on the surface of sea water on marine sediments.

It is important to realize that there are a number of processes affecting the oxygen concentration in water. These will not affect the transfer rate of oxygen across the air/sea interface unless the concentration falls below the saturation level or rises above it. To measure adequately the transfer rate to the water it would be necessary to remove, or at least reduce considerably, the existing oxygen concentration, as the rate is determined by the oxygen deficiency. Work by Garrett (1972) on the effects of capillary wave dampening may be considered as more significant to reaeration limitation than an oxygen transfer barrier established by oil "membranes".

Data: Attempts to locate the data or observations on which some fears have been expressed have been unsuccessful. The Working Group, although aware that data may exist from authorities such as Prof. Simonov of the U.S.S.R., has been, however, unable to gather either published or unpublished data in support of these comments.

There is a scarcity of published data concerning the effects of oil on oxygen exchange across the air/sea interface. One paper describes the effects of polluting discharges on the Thames Estuary (HMSO, 1964). In this paper, the effects of an oil film, ranging in nominal thickness from 0.01 to 5 micron (10^{-6} to 5×10^{-4} cm), are described from laboratory experiments in a stirred vessel. In all cases, there was a gradual reduction in the exchange coefficient with increasing oil film thickness. This reduction was greatest with the highest initial exchange coefficient, falling to zero or near zero as the initial exchange coefficient decreases. From this it follows that, with increasing saturation, the effect of oil films will be progressively less in reducing oxygen transfer - presumably because oxygen transfer is both from air to water and from water to air, and transport in both directions is subjected to the same hindrance processes. These processes have their greatest effect with thick films, and reduce to zero or near zero with thin films.

The position at sea is complicated by natural changes such as temperature, air velocity and wave activity which affect the oxygen saturation value, and by oxidation processes and photosynthetic activity which affect oxygen levels.

Dowing and Truesdale (1955) examined earlier work by Roberts (1926) on the effects of four types of marine fuel oil on the reaeration of quiescent water, and concluded that the results were too variable to be used. The earlier work using Spindle oil, which is a lubricating oil, was then extended using a heavy marine fuel oil. As noted earlier, very little effect was noted with thin films, and little change in the rate of solution of oxygen was noted until the thickness of the marine fuel oil exceeded 10^{-4} cm. Because films of greater thickness than this are unlikely to persist for very long in natural waters, it was concluded by Downing and Truesdale that oil pollution was not a significant factor affecting reaeration in such waters.

The only other paper noted as describing experiments to determine the effects of crude oil films on gas exchange is that by Kinsey (1973) who used the coralback-reef at Heron Island. His measurements were made under calm night-time slack water conditions in an area of water isolated by a fence from the main body of sea water. His work confirms that, under the conditions prevailing, the oil films used in the experiments (up to 0.7 mm) did not interfere significantly or directly with oxygen transfer between the atmosphere and sea water.

Comment: Theoretical models indicate that the existence of an oil layer on the surface of the sea will interfere with oxygen exchange, particularly with increasing oil layer thickness. However, attempts to measure this interference under practical conditions have shown that such effects are not discernible. It was thought by the Working Group that its role was to report on the available scientific evidence and not to prepare calculations on water reaeration, hydrocarbon oxygen absorption, etc.

Conclusions: No definite studies have been done to investigate suggestions that under open sea conditions the presence of oil layers such as those that arise from pollution of the sea by oil, will give rise to a depletion of oxygen in sea water. No areas have yet been identified where the local conditions give rise to local problems of oxygen depletion, although it was recognized that such areas may exist. It is likely that in such areas other effects of oil such as acute toxicity, or smothering, would be important. Mangrove swamps and intertidal zones were noted in particular as areas where study is required. Consideration of the upper few centimetres of the sea was viewed as important; it should be examined further, although it was thought that the effects of hydrocarbons on this layer would not be significant in the open ocean.

4.2 Heating effects

Definition: The effects on marine organisms and substrates due to elevated temperatures produced by adsorption of solar radiation by oil films.

Problem: Fears have been expressed that when layers of crude oil of high optical density are exposed to solar radiation, the absorption of this radiation by the oil could give rise to elevated temperatures, and that these could give rise to, or be a contributing factor toward, the mortality of marine organisms.

Background: The presence of layers of dispersions of oil on ice reduces the reflection of solar heat and light from the surface (as well as absorbing heat as described above). This accelerates the melting process (studies Alaska, B. Scott, etc., referred to by Jeffery, pers.comm.), reducing the extent and time that the area is covered by ice. Spills in polar areas might, therefore, reduce local ice cover with possible effects on global climate.

The absorption of heat by beached oil has been observed to mobilize the oil, allowing it to repollute and thereby imparting a chronic aspect to an acute incident.

This fear relates to those marine organisms such as coral that live in an intertidal zone, and those areas of shallow, quiescent water such as intertidal pools. In these areas, the absorption of solar radiation by layers of oil could result in elevated temperatures to which the organisms are exposed. In these particular areas it is difficult to separate the mortality effects due to direct smothering, acute toxicity and temperature changes.

Data: Of the topics under scrutiny, this one relating to heating effects has been the most difficult to evaluate. This is because, although there is abundant reference to the supposed, possible, or likely heating effects that could be encountered, no published paper available to the Working Group records experimental data to support such an effect. The physical/chemical data available to make calculations of heat build-up, such as that due to water evaporation suppression, were thought to be beyond the scope of the Working Group's tasks.

Comment: Prof. Johannes, in a draft of a paper for publication made available to the Group, reports his own work in which 22 species of corals were exposed to Santa Maria crude for 1.5 hours. Oil adhered to portions of the surface of most species. Tissue death ensued within a few days in these areas, but not where the oil did not adhere. He concludes that tissue damage may have been accelerated by the high temperature ($32^{\circ}C$) reached by the black oil in the midday sun. No details are given to enable this possibility to be examined further.

Straughan reports that larvae can settle, survive and grow on the (weathered) tar, that larvae prefer a black surface to a light surface and that temperature stresses are greater on a black surface than on a light surface. The black body effects - which are not always detrimental to a given marine species - were likely to be reported only from the intertidal zone. Those organisms adversely affected are likely to be those most susceptible, especially those at or near their geographic limit. At these limits, the species are more likely to be susceptible to other factors, and some realignment of the exact species distribution and population numbers is likely to occur from time to time (Jeffery, pers.comm.).

The so-called "albedo effect", by which is meant the possibility that adsorption of solar radiation by an oil-covered ice cap, could give rise to a significant reduction of the area covered by ice, has also been difficult to substantiate. It is believed that calculations have been made to the likely magnitude of this effect, but no figures appear to have been published. Persons interested in the Arctic oil treaty and other instruments do not discredit the possibility of a catastrophe from this effect. The complete study and summarization of this phenomenon was determined to be beyond the resources of the present study.

In view of the extensive area over which temperatures are considerably below zero, the albedo effect appears unlikely to be observed at all, except in the marginal areas of the polar ice cap. No measurements appear to exist to indicate that such an effect, if present, will be discernible above the diurnal and other changes that naturally occur.

Conclusions: Until data or calculations emerge to indicate that heating effects are likely to be a problem, they should be considered as an area for scientific study.

4.3 Pollutant sorption

Definition: The sorption of pesticides such as DDT, and metals, by oil films and layers on the sea surface and the effect this might have on marine resources.

Problem: It has been reported that high concentrations of pesticides, in particular DDT, and heavy metals of which mercury has been particularly implicated, can arise by sorption from the atmosphere into surface films and layers of oil. If substantiated, this could give rise to bioaccumulation problems in certain marine organisms.

In a similar way, oil films and surface layers could serve as extractants for both pesticides and heavy metals from the body of the sea. This could result in exposing certain marine organisms to increased concentrations of pollutants.

Background: The basic fear underlying this topic of concern appears to relate to the general toxicity to marine organisms of such materials as pesticides and heavy metals. Any concentration of either in an oil layer may give rise to an appreciable decrease in microbial activity and hence to an appreciable decrease in the rate at which oil and oil products are removed from the marine environment by this activity. Further fears relate to the possibility that pesticides may bioaccumulate by this mechanism.

Data: Seba and Cocoran (1969) refer to the concentration of pesticides in surface slicks in the marine environment; it is clear from their paper that the larger part, if not essentially all, of the pesticides observed in the surface slicks, is derived from pesticide application on land.

Similarly, although the surface layer ("surface slick") is referred to many times in the paper, nowhere is it suggested that the slick is an oil slick. The word "oil" is used only once in the paper, and on that occasion refers apparently to an adventitious oil patch reported by Croker and Wilson (1965) as found eight days after a DDT application.

With regard to heavy metals, interest appears to be centered around the possibility that mercury may be concentrated in oil slicks. It is well known that, under certain circumstances, mercury can be mobilized in the marine environment, especially as the methyl derivative. From their work in Colgate Creek, Baltimore Harbor, Walker and Colwell (1974b, 1975) suggest that, in an area of heavy metal enrichment, mercury and other heavy metals may be concentrated in the oil phase. In their work, water was extracted with benzene to concentrate an oil phase, and the oil content was determined by low resolution mass spectrometry (Walker and Colwell, 1975; Walker et al., 1975a). The mercury in that oil phase was determined using tandom atomic absorption spetrometry and gas-liquid chromatography (Iverson and Brinckman, 1975; Blair, Iverson and Brinckman, 1974). The results indicated that this extract contained 300 000 times as much mercury as the water from Colgate Creek (Colwell and Nelson, 1975; Walker and Colwell, 1974b); the sediments were also found by Brinckman and others (Brinckman et al., 1975; Blair, Iverson and Brinckman, 1974; Iverson and Brinckman, 1975; Huey et al., 1974; Iverson and Blair, 1976) to contain enhanced concentrations of mercury.

A reference to crude oil later in the paper suggests that the oil phase is being considered as at least derived from petroleum because of GS-mass spectrometry work carried out during the study. The authors are particularly careful to describe their oil-rich extract as "benzene extract" in their description of what was done, but the implication is that this material is present in the water column as an oil phase, although this has not been specifically established in the paper. Colwell unequivocally established that the oil-rich extract in Colgate Creek of Baltimore Harbor is derived from petroleum (Thompson, pers.comm.).

There are a number of statements in this paper that need further elaboration, qualification or quantification:

1. Measured concentrations of mercury probably represent the amount present in the oil in Colgate Creek rather than that which is preferentially extracted with benzene. In subsequent publications (Walker and Colwell, 1976a; Sayler and Colwell, 1976) experiments are described as involving mixing oil, mercury, water and sediments. The mercury partitions into the oil phase and into the sediments. This is also shown for PCB's.

2. Preliminary results indicate that the mercury is associated with specific organic complexes in the oil.

3. The nature of the mercury compounds associated with oil is not known and is presently under investigation.

The transformation of inorganic mercury to oil-soluble compounds, followed by concentration in an oil phase could provide a mechanism for the uptake of mercury by higher organisms and into the food chain. The ultimate destination appears to be subsequent deposition and incorporation in the oil phase of the sediments of the creek.

It should however be borne in mind that at first examination, the sediments from Colgate Creek are by no means unique. Stock refers to paleozoic marine shales with 0.5 ppm Hg, and others in the range of 0.025 to 1.4 ppm Hg (Stock and Cucuel, 1934; Goldschmidt, 1954). If such levels are not due to diagenesis, the concentration effects noted by Walker and Colwell have been in action for a long time, and certainly pre-date the use of petroleum hydrocarbons by man. However, Colwell contends that the heavy petroleum contamination of these sediments is unique and that these levels are uncommonly high (Thompson, pers.comm.).

Comment: The evidence to-date shows that oil of petroleum origin can serve to concentrate pesticides from sea water into an oil slick, but further work must be done to prove this as a general phenomenon. Bioslicks that may be of land origin have been observed with high pesticide values, but there is little to indicate that any concentration mechanism is in action in the marine environment or that oil of petroleum origin is solely or significantly implicated.

Preliminary work has indicated that in an area of mercury enrichment, a mechanism exists for its conversion to a benzene-soluble form and this may be concentrated in a dispersed oil phase. Further work is required to support this.

Conclusions: The fear that sorption of pesticides by oil slicks could have a significant effect on marine resources in general appears to require additional work.

Similarly, the concentration of mercury in an oil phase as a widely occurring phenomenon remains to be convincingly substantiated. Clearly, any suggestion that a new mechanism exists for the translation of inorganic mercury to the food chain must be taken seriously, but the Working Group's concern is only whether or not petroleum is implicated. This is not yet generally established, and in those areas where high mercury values exist, shore discharges of inorganic matter may be important. Industrial and domestic waste discharges interacting with petroleum should also be evaluated.

4.4 Hydrocarbon interference with carbon dioxide transfer

References to this problem were most difficult to locate and the subject may be considered more in the immediate future based upon informational leads developed during this study.

Chapter 5: EFFECTS ON MARINE LIFE FORMS [1]

Definitions:

For the purpose of this report, the following definitions have been adopted.

<u>Higher Marine Life Forms</u> denotes those animals and plants that spend an important part of their lifetimes in the marine environment and includes mammals, birds, fish, benthic organisms, and both permanent and temporary planktonic forms. In addition to invidivual species of organisms, these broad terms of reference are understood to include the effects of oil on populations of species and communities of populations of various species as well.

<u>Lethal Effects</u> occur when components of the oil interfere with cellular and subcellular processes in the organism to such an extent that death follows directly. In severe cases, this may take the form of smothering and suffocation or interference with movements to obtain food or escape predators as a result of being coated with oil.

<u>Sublethal Effects</u> are those that disrupt physiological or behavioural activities but do not cause immediate mortality, although death may follow because of interference with feeding and reproductive activities, abnormal growth or behaviour, greater susceptibility to predation, lesser ability to colonize, or other indirect causes. These effects may not only lead to changes in populations of individual species but may also result in shifts in species composition and diversity. Uptake and discharge of oil, or certain fractions of it, by the organism are considered in relation to the initiation of a sequence where the pollutant is transferred to other members of the food web, while tainting and carcinogenesis are considered in chapter 6. The effects may be acute; that is, the responses to a single sudden infusion of oil, or chronic, if exposures recur sufficiently frequently over extended periods of time so that the biota do not have time to recover between doses.

The term <u>oil</u> includes crude oils, fuel oils, sludges, oil refuse and refined oils but excludes petrochemicals. Particular attention is given to the effects of light refined oils.

Problem:

The objective of this study is to review critically the recent scientific literature regarding the acute and chronic toxicological effects of crude and refined petroleum products on the marine environment and their effects on living resources. Specifically, the subjects covered by this review are, as laid out in GESAMP VI/10, Annex IV (GESAMP, 1974):

(i) "Lethal effects - direct and immediate effects of oil including, but not limited to, toxicity, smothering and clogging.

(ii) Sublethal effects - chronic biological effects of oil (other than those covered by tainting, carcinogenesis and microbial population effects) including, but not limited to, chronic toxicity and effects on behaviour, growth, reproduction, colonization and species distribution."

Although it is necessary to deal first with the impacts on individual species, an understanding of the effects of oil pollution on higher levels of biological organization including populations, communities and ecosystems is ultimately essential. This subject is, of course, infinitely more difficult and at present very little information is available in the literature.

---

[1] This chapter includes reports of work carried out in the USSR which has inevitably suffered in translation. It has not been possible in the time available to verify or correct all the texts. The readers' and the authors' indulgencies would be appreciated.

A great number of factors, acting both individually and in combination, govern the effects that an oil spill may have on marine life. The biological damage that ensues depends on the following: (i) The type of oil involved, particularly with respect to its content of aromatic compounds; (ii) the dosage of oil to which the organism is exposed and the duration of exposure; (iii) whether the oil is in a fresh, weathered or emulsified form; (iv) whether it is in solution, suspension, dispersion or adsorbed on particulate material; (v) whether plankton, pleuston, nekton or benthos are involved and within these categories the individual species affected; (vi) the season of the year with respect to the annual cycle of the organism and whether it is in a dormant state or actively feeding and reproducing; (vii) whether adult or juvenile forms are involved; (viii) the effects of the oil on competing biota; (ix) previous history of exposure of the organisms to oil or other pollutants, (x) whether a coastal, estuarine or open ocean area is involved and especially if it is a nesting or wintering ground for sea birds, a migration route for birds or fish, etc.; (xi) natural environmental stresses imposed by meteorological conditions or fluctuations in water temperature, salinity and other oceanographic parameters, particularly currents and wave action; (xii) the clean-up procedures, if any, that have been used and particularly whether chemical agents have been employed; (xiii) and all other stresses of both natural and pollutant origin, to which the organisms are subjected. In general, the biological damage is much more severe if the spill occurs in a coastal or estuarine environment, especially if the intertidal zone is affected, than if it occurs in the open ocean because there are generally many more types and numbers of organisms in these areas as well as the sensitive juvenile stages of many oceanic species (Evans and Rice, 1974).

The biological responses that might accompany or follow an oil spill include: (i) lethal toxic effects, where the components of the oil interfere with cellular and subcellular processes in the organism to such an extent that death follows directly; (ii) sublethal effects that disrupt physiological or behavioural activities but do not cause immediate mortality, although death may follow because of interference with feeding and reproductive activities or other abnormal behaviour, greater susceptibility to predation, or other indirect causes; (iii) up-take of the oil, or certain fractions of it, by the organism causing tainting, or in some cases carcinogenesis; (iv) possibly the initiation of a sequence whereby the pollutant is transferred to other members of the food web rendering them unfit for consumption by other animals including man; (v) direct smothering and suffocation or interference with movements to obtain food or escape predators as a result of being coated by oil; and (vi) alterations to the chemical and physical habitat which result in changes in the populations of individual species as well as shifts in species composition and diversity. Furthermore, the effects may be acute; that is, the responses to a single sudden infusion of oil, or chronic, if exposures recur sufficiently frequently over extended periods of time so that the biota do not have time to recover between doses. To make it even more difficult to determine which biological effects may be attributed to the spill, many marine organisms show marked seasonal variations in abundance and sensitivity to stress, and in only a few instances has the necessary pre-spill information regarding the affected area been available.

Different organisms react to oil pollution in different ways. What kills one species may have little or no effect on another. Individuals within a species may differ - eggs, larvae and newly molted individuals have different sensitivities to the same level of pollution. For example, the prelarval stages of the barnacle Balanus are 100 times more sensitive than adults (Mironov, 1968). However, the eggs and the fry of pink salmon were 10 times more tolerant than older fry (Evans and Rice, 1974). Renzoni (1973) found that in the oyster Crassostrea angulata and the mussel Mytilus galloprovincialis, the eggs and larvae showed a high degree of tolerance but the fertilizing capacity of sperm was markedly affected.

With such a complex situation it is obviously very difficult, if not impossible, to predict the impact that any particular oil spill might have on marine life (Straughan, 1972, 1972a). It is not surprising, therefore, that widely different effects are observed in different instances; for example the extreme biological damage from the TAMPICO MARU spill (North, Neushul and Clendenning, 1964) versus the relatively minor effects of the Santa Barbara incident (Straughan, 1969). These spills had vastly different impacts on the marine environment (Mitchell et al., 1970).

5.1 Effects of oil on marine organisms

Since the effects of the persistent oils on the marine environment have already received considerable attention by previous groups, the Working Group decided to focus its attention upon the effects of light refined oils on the marine environment with particular reference to their impact on the 'important' living resources. Since spills of light oils are generally much less spectacular than those of crude and residual fuel oils and because they seem to 'disappear' comparatively rapidly, the attitude has gradually emerged that these spills have only a superficial effect on living resources, and that what effects they do have are of a short-term nature and limited to the immediate vicinity of the spill. The Working Group recognized that light oils disappear, in part, by going into solution or becoming dispersed in the water column and thereby become more readily available to many species of marine organisms than if they remained floating on the surface of the sea. Since the more soluble components are often the more toxic, light oils may present a more serious threat to living marine resources than the persistent 'black' oils. Furthermore, light refined oils are very frequently spilled in highly sensitive estuarine and coastal areas where the impact on living resources from a single spill, or the accumulation of effects from repeated spills over extended periods of time, might have much more profound effects on living resources than has previously been anticipated.

5.1.1 Birds

Sea birds are perhaps the only group of marine organisms that have so far been affected by oil pollution to an extent sufficient to jeopardize local populations which in some cases constitute the total world population (Clark, 1973). The total mortality per year from chronic oil pollution in the North Sea and the North Atlantic has been estimated at 150,000 to 450,000 (Tanis and Mörzer Bruyns, 1968). However, of the many species of sea birds relatively few, notably the auks (Murres, Guillemots, Puffins, etc.) and the diving sea ducks (Eiders, Scoters, etc.), have suffered severe mortalities. These species are particularly susceptible to oil pollution since they spend almost their entire lives at sea, collect their food by diving, seem to be attracted to slick areas, and in most cases have a very low rate of reproduction. Furthermore, these birds are highly gregarious, particularly in their breeding and wintering areas, and what would seem to be a minor spill can inflict very large casualties. For example, heavy casualties resulted from a relatively small slick near breeding colonies in the Shetland Islands in 1971 when an estimated 10,000 birds perished (Bourne and Johnston, 1971). Included were 5,000 Guillemots (Uria aalge) or 10 percent of the population of these colonies. Even more severe losses occurred in an estuarine wintering area in the Netherlands in 1971 when 5,000 birds, including 100 percent of the wintering Mallard (Anas platyrhynchos), 75 percent of the Greylag Goose (Anser anser), 80 percent of the Bewick's Swan (Cygnus bewickii) and 100 percent of the Coot (Fulica atra) populations, were killed (Belterman, 1972). A spill of unknown origin off Jutland, Denmark, in December 1972 resulted in 5,000 dead birds and an additional 25,000 that were oiled but still alive at the time of the survey (Joensen, 1973); 30 percent of the wintering population of Eider ducks (Somateria mollisima) perished. An even more drastic reduction of Eiders occurred as a result of the spill which arose from the collision between the tankers FORT MERCER and PENDLETON off Chatham, Mass., in 1952 when the wintering population was reduced from 500,000 to 150,000 (Burnett and Snyder, 1954).

It was estimated that 2,000 birds were killed in Chedabucto Bay, N.S., during the ARROW spill. An additional 5,000 drifted onto Sable Island (Anon, 1970), a tiny island in the Atlantic some 200 miles from the spill. In actual fact, however, the number killed was probably very much greater, and Brown, R.G. et al., (1973) estimate that at least 12,000 birds were killed by the ARROW and the WHALE spills off southeast Newfoundland combined.

The effects on sea birds of some recent major oil spills in which heavy mortality occurred and in which an organized effort was made to determine the number of birds killed are summarized in Table 5-1 (Clark, 1973). At best, the data in this table are approximations since they are based on about twice the number of corpses per length of shoreline sampled and are therefore very conservative. Other workers have indicated that even under favourable conditions only 20 to 25 percent (Hope-Jones et al., 1970) and under less favourable conditions about 10 percent (Hardy, 1959; Clark, 1973) of the birds killed by oil

TABLE 5-1

Known and estimated seabird casualties
in recent oil spills (Clark, 1973)

Incident	Beached Birds	Total Casualties
TORREY CANYON March 1967 SW England U.K. 7851 Crude Oil, 60,000 tons	10,000+ France 2500 est. 98% Auks 78% Guillemot	20-30,000
Terschelling, Feb. 1969 Holland Crude oil		30-35,000 45% Eider 55% Common Scoter
N. Zeeland, Feb. 1969 Denmark	5000 71% Eider 16% Common Scoter	10,000
Santa Barbara, March 1969 California Crude oil	3600 Chiefly Western Grebe, Loons, Scoters, Cormorants	

Palva May 1969
  Kökar, Finland
  Crude oil, 150 tons
		3000-3500
		2400-3000 Eider;
		Long-tailed Duck

NE Britain Jan. 1970 Unknown ? Fuel oil	12,856	50,000
East Jutland, Feb. 1970 Denmark	3000 47% Eider 26% Common Scoter 21% Velvet Scoter	12,000
DELIAN APPOLON, Feb. 1970 Tampa, Florida Bunker C. 80-100 tons	4500	9000
South Kattegat Dec. 1970- Jan. 1971	6500 74% Eider	15,000
San Francisco Jan. 1971 Bay Bunker C. 300-350 tons		7000
HAMILTON TRADER Apr. 1971 Heavy fuel oil, 600-700 tons	4548 91% Guillemot	

at sea ever drift ashore. In the case of the ARROW, beach counts were unreliable because many corpses were buried in oil and snow, or the dying birds crawled off into the forest to die (Brown, R.G., 1973). It is not surprising that other estimates of the number of birds killed in the TORREY CANYON incident run as high as 40,000 to 100,000 and the HAMILTON TRADER between 6,000 and 10,000 (Nelson-Smith, 1972). Several other oil spills in which large numbers of birds were killed are summarized in Table 5-1.

The data in Table 5-1 and the incidents cited above pertain, for the most part, to the effects of crude and heavy fuel oils. Since the light refined oils are less persistent on the surface of the sea than the 'black oils', it might be predicted that the chances for a spill of light oil to encounter sea birds would be less and therefore the chance of extensive mortality would be lessened. On the other hand, however, should sea birds become fouled with light oils, mortality would probably be just as severe, or more so, since these oils would be as fully damaging to the bird's down and its water repelling and insulating properties (Clark and Croxall, 1972) while any oil ingested by the bird would be more toxic. Although very few accounts of the effects of light oils on sea birds are to be found in the scientific literature, the limited information available substantiates the hazards presented by these oils. For example, a rather small spill (750,000 l) of No. 2 diesel oil into Puget Sound at Anacortes, Wash., in 1971 killed some 30,000 Brant which represented about 25 percent of the entire population of this species in the Pacific Flyway of North America (Shiang-Chia, 1971). Further, Kirby (1970) reports that 86,000 birds, chiefly Common Murres, died within two days following the sighting of a slick of light diesel oil off the west coast of the Alaska peninsula in April 1971.

There are several theories as to why there is such high mortality among oiled birds. Such effects as loss of waterproofing and subsequent chilling and loss of buoyancy leading to death by drowning have been cited. In addition, there must be toxic effects leading to the high mortality among self-cleaned birds. A number of pathological conditions have been noted in contaminated birds and in those fed small doses of crude oil and derivatives. These include fatty degeneration of the liver, toxic nephrosis, enlargement of the spleen, adrenocortical hyperplasia, acinar atrophy of the pancreas, and lipid pneumonia (Hartung and Hunt, 1966).

In addition, Crocker, Cranshaw and Holmes (1974) have suggested that ingestion of crude oil disrupts the intestinal absorptive mechanisms impairing water and ion absorption and contributing significantly to the high mortality among oil-contaminated birds. Since the kidneys alone are not able to maintain homeostasis under all environmental conditions, ducks and gulls survive through activation of an extrarenal excretory mechanism which complements the limited excretory capacity of the kidney. This mechanism consists of paired nasal glands which are activated by a rise in plasma osmolarity following ingestion of hypertonic drinking water. The glands secrete a fluid containing Na^+, K^+, and Cl^- at concentrations higher than that found in sea water, thus enabling the bird to gain osmotically free water, excrete the electrolytes ingested in their food and drinking water and thus remain in a positive water balance. This extrarenal excretory mechanism is essential for all pelagic and many coastal species. Further, these authors (Crocker, Cranshaw and Holmes, 1974) found that crude oil interferes with the rate of intestinal mucosal absorption, which is vital to the function of the extrarenal excretory mechanism, and the bird enters a state of negative water balance, becomes dehydrated and ultimately dies.

Wildfowl, edible aquatic birds were used to be commercially exploited on a moderate scale in many parts of the world. Nowadays their main exploitation is for sporting purposes. Because of their aquatic habit which required the maintenance of a waterproof plumage, and their tendency to gather in particular sheltered areas for breeding or moulting, they have suffered heavily in oil pollution incidents. Gusey and Maturgo (1971) summarized the effects on wintering wildfowl in Louisiana over the period 1956-1969. Their data showed a reassuring increase which was maintained over the whole period despite short-term losses due to pollution. Straughan (1971) and Drinkwater et al. (1971) studied the short-term losses in wildfowl populations which resulted from the Santa Barbara incident. On their evidence, about 3,600 birds of all kinds were known to have been killed and of these about 18 percent were wildfowl. This loss, although serious, would not however, be sufficient to cause lasting harm to the local wildfowl population. On the whole, the effects of offshore exploitation on wildfowl would seem so far to have been small compared with the serious losses caused by oil pollution from other sources.

5.1.2 Mammals

As is the case with bird plumage, mammalian fur readily loses its waterproofing and insulating properties when fouled with oil and the animals suffer from exposure and restriction of mobility so they are unable to hunt and are more prone to predation. Although these effects are reasonably well documented for a variety of terrestrial mammals, very little information is available concerning the effects of oil pollution on marine mammals. Kirby (1970) noted that 400 Hair Seals were affected by a spill of light diesel oil off the west-coast of Alaska in April 1970, while Dickason (1970) reported that Sea Otters might also have been affected. Following the ARROW spill only 500 Grey Harbour Seals (13 of them dead) were sighted in Chedabucto Bay where normally several thousands are found (Anon, 1970). On Sable Island a group of 50 to 60 Harbour Seals and 100 Grey Seals (11 of them dead) were observed. Most of them were oiled, some heavily. The oil affected the eyes, ears, nose, mouth and throat, often completely plugging them and causing considerable pain and suffering (Anon, 1970). In this report it was stated that "The cause of death was suffocation rather than from any toxic effects of the oil" although no mention was made of an autopsy to actually determine the cause of death. Such unsupported remarks are commonplace in the literature dealing with the effects of oil pollution on marine life. However, regardless of the mechanism involved, it appears that oil was directly responsible for the death.

During the Santa Barbara incident, large numbers of Sea-lion pups (_Zalophus californianus_) were found dead when heavy slicks surrounded the breeding colonies on San Miguel Island, although Brownell and LeBoeuf (1971) point out that, since the normal mortality ranges from 12 to 15 percent, the mortalities observed in this case could not be attributed for certain to the pollution. This incident occurred during the seasonal migration of the Grey Whale (_Eschrichtius glaucus_) and, while it has been suggested that the whales took some pains to avoid the oil (Straughan 1972a), five whales and several porpoises were found dead (Brownell, 1971). This was considered to be a "rather high mortality" by Nelson-Smith (1970) yet others (Straughan, 1972, 1972a; LeBoef, 1971) considered that there was not sufficient proof that oil was the cause of death.

From the meagre literature, one might speculate that there have not been many instances in which extensive mortality of marine mammals resulted from oil pollution. One could also speculate that since it is not likely that whales, dolphins, seals and other marine mammals that die at sea ever drift ashore, the severity of the problem is not truly reflected by the existing literature. Therefore, more extensive monitoring of marine resources and reporting of incidents must be undertaken.

5.1.3 Fish

Oil may have a variety of effects on free-swimming fish, the most obvious being the lethal effects of sufficiently high concentrations, that the functioning of the gills is disrupted, or the ingestion of large quantities of toxic substances. In general, it would appear that sufficiently high concentrations for lethal effects in fish are only encountered in the vicinity of major oil spills or when a spill has occurred in a very restricted area. Less obvious, but perhaps more significant, are the effects on fish at sublethal concentrations which may bring about changes in the feeding, migration or reproduction of the species, or the losses of equilibrium of individuals.

For the most part, the large amounts of oil associated with major spills tend to remain floating on the surface of the water. Consequently, adult pelagic fish are only exposed to those components of the oil that dissolve in sea water or become dispersed through the water column as droplets either because of the use of dispersants or through wave action. The latter phenomenon, as observed by Forrester (1971) during the ARROW incident, may account for the fairly rapid 'disappearance' of oil during many spills when there is sufficient wave action. This process is environmentally significant since the oil is made available to subsurface organisms in a form that is readily ingested, as shown by Conover (1971) in the case of zooplankton, and thereby gains entry to the marine food web.

Despite the possibility of large amounts of oil entering the water column in major spills, there does not appear to be any report in the literature of large mortalities of pelagic fish as a result of spills of crude or heavy fuel oil. Unlike birds and mammals, the external surfaces of most fish are coated with a slimy mucous to which oil does not adhere readily, although dispersants tend to destroy this protection. Although there is some evidence that adult fish may make an effort to avoid heavily contaminated areas (North, Neushul, and Clendenning, 1964; Straughan, 1971, 1971a), a large mortality at sea would probably not be noticed. However, mortality in coastal areas, estuaries, etc. where most major spills occur, would be observed so it is possible that adult fish in fact avoid heavily contaminated areas.

During the TORREY CANYON incident there does not appear to have been any severe kill of fish (Spooner, 1969), and those actually killed were believed to be victims of the oil-emulsifier mixture rather than the oil itself (O'Sullivan and Richardson, 1967). Commercial fish were taken in normal numbers in the vicinity of oil slicks, although Tendron (1968) reported a decrease in one region and also the presence of nodules of oil in the gut of Whiting. Similar observations were made with fish in the vicinity of the WAFRA spill in South Africa (Day et al., 1971). While no adverse effect on fisheries was associated with the HAMILTON TRADER spill (O'Sullivan, 1971), fish landings fell considerably during the Santa Barbara Incident. However, Straughan (1971a, 1971b) considers that the lower catches were the result of fish avoiding the heavily polluted areas and the reluctance of the fishermen to foul their equipment. Fish catches also declined as a result of a spill of 100,000 barrels of Arabian light crude into Tarut Bay, Saudi Arabia, when a pipeline broke in 1970 (Spooner, 1970). As a final example, the yield of the commercial fishery in Chedabucto Bay does not appear to have suffered from the ARROW spill (Anon, 1970).

In comparison to the apparently minor effects of crude and heavy fuel oils on pelagic fish, spills of light refined oils appear to have considerably more detrimental effects. For example, a spill of 26,000,000 l (6 million gallons) of high octane aviation gasoline, aviation jet fuel, aviation turbine fuel, diesel oil and Bunker C from the tanker R.C. STONER on Wake Island killed an estimated 2,500 kg of inshore reef fishes (Gooding, 1971), while an 'intermediate' oil containing large amounts of aromatic hydrocarbons was identified as the cause of an extensive kill of Herring (Clupea harengus) in Nova Scotia in 1969 (Zitko and Tibbo, 1971). Other incidents in which fish mortalities resulted from spills of refined oils include spills of diesel oil in Alaska in 1970, the TAMPICO MARU incident in southern California in 1957 (North, Neushul and Clendenning, 1964) and the FLORIDA spill of No. 2 fuel oil in West Falmouth, Mass., in 1969 (Hampson and Sanders, 1969). The kill of fish associated with the TAMPICO MARU spill was not as extensive as that of other organisms, which were virtually eliminated from the cove where the oil was spilled (North, 1967), and is considered to reflect the ability of the fish to avoid the contaminated area. Such an ability, however, was not displayed in the case of the FLORIDA spill which resulted not only in high mortalities of fish but also the almost complete destruction of benthic organisms in a formerly productive area (Blumer, 1971, 1971a; Sanders, Grassle and Hampson, 1972).

The tainting of fish and shellfish by oil has been recognized for many years and may be indicative of up-take of oil by the animals. An expanded discussion of tainting of marine produce is given in chapter 6.

Since the eggs and larvae of many varieties of pelagic fish, many of which are commercially important, float on the surface or inhabit the upper layers of the sea, they are particularly exposed to the effects of oil pollution and suffer high mortalities. For example, Smith (1970) reported that 50 to 90 percent of the Pilchard (Sardina pilcardus) eggs were dead and that juvenile fish were scarce or absent in plankton samples collected in the vicinity of the TORREY CANYON spill. This was attributed to the toxic effects of the emulsifiers used to disperse the spill rather than the oil itself. It is unfortunate that dispersants were used so extensively and freely during this incident, since their use makes it almost impossible to assess the validity of much of the information that has been reported regarding this spill and also disguises the lessons that might otherwise have been learned. It is even more unfortunate that statements disclaiming the effects of the oil on the grounds that environmental damage was the consequence of the dispersants and not the oil imply that there would not have been any damage from the oil itself. Nevertheless, it cannot be denied that the presence of the oil was responsible for the use of the detergents and thereby also directly or indirectly responsible for any environmental damage.

Kühnhold (1972) carried out laboratory experiments on the effects of extracts of Venezuelan, Iranian and Libyan crude oils on cod eggs (Gadus morhua) and the larvae of Cod, Herring (Clupea harengus) and Plaice (Pleuronectes platessa), all of which are of great commercial importance. Cod eggs were most sensitive during the first few hours after fertilization, and after 10 hours mortalities were significant. In some cases, hatching was delayed or did not occur. The biological effects were considered to be more significant than indicated by the hatching data, since most of the larvae that hatched had deformed bodies or abnormal body movements and died within the first day. Normal larvae exposed to the oils initially showed increased activity followed by a reduction of swimming activity which finally stopped except for sporadic twitches. They then showed the beginning of narcosis which gradually deepened until a 'critical' point was reached and beyond which recovery did not occur. Herring larvae were the least and plaice larvae the most resistant while Venezuelan crude was the most toxic and Libyan the least. Kühnhold (1972) observed that the larvae were unable to avoid contaminated water, especially when the oil was present as a dispersion, and believed that the chemoreceptors were blocked or destroyed. He concluded that larvae have little chance of survival if they remain in oil dispersions.

Mironov (1972) found that fish eggs were highly sensitive to oil and oil products and generally died during the second day of exposure to concentrations of 10^{-4} to 10^{-3} ml/l. At concentrations of 10^{-4} to 10^{-5}, only 55-89 percent of the eggs hatched and hatching was irregular and occurred over a prolonged period of time. Larvae were abnormal and died soon after hatching. Mironov (1968) used fertilized eggs of the plaice and found they were highly sensitive to oil products in the water. He found 0.1-0.01 ppm caused injury to 40-100 percent of the hatched prelarvae and that 0.001 ml oil/l was toxic to eggs of the anchovy, scorpion fish and sea parrot (Mironov 1969). Lindén (1976) investigated the effects of a Venezuelan oil with or without the addition of oil spill dispersants on the embryonic development of Baltic Herring (Clupea harengus). The results showed that the toxicity of the oil increased several hundred fold if the oil was dispersed by a newly developed (non-toxic) dispersant, and by an additional power of tenth if an older dispersant was used. Lindén (1975) found that newly hatched larvae of Baltic herring were 50 to 100 times more sensitive to an oil dispersion that was obtained by mixing a crude oil and a dispersant, than a "natural" oil dispersion without a dispersant. Further it was found that the acute toxicity of self-dispersed crude oil decreased considerably in 24 and 72 hours, but if the oil was dispersed by a dispersant, the high toxicity remained almost unchanged in the same time.

Wilson (1972) reported high mortalities when larvae of Herring (Clupea harengus), Pilchard (Sardina pilcardus), Plaice (Pleuronectes platessa), Sole (Solea solea), Lemon Sole (Microstomus kitt), and Haddock (Melanogrammus aeglefinus) were exposed to a variety of dispersants with kerosene and aromatic solvents; embryos of herring, plaice and sole showed abnormities when the eggs were exposed. Swedmark, Granmo and Kollberg (1973) demonstrated that emulsions of diesel and crude oil had more pronounced effects on Cod (Gadus morhua) and Flounder (Pleuronectes flesus) than heavy fuel oil. Cod, in general, were more susceptible than flounder, but both species showed increased activity, followed by impaired activity and loss of equilibrium, and finally immobilization when exposed to sublethal concentrations.

Tagatz (1961) tested the toxicity of gasoline, diesel oil and bunker oil to juvenile American Shad (Alosa sapidissima), a common anadromous marine and estuarine fish, and found that mean tolerance limit (TLM) increased in the order given. Rice (1973) reported 96-hour TLM values to Prudhoe Bay crude oil of 213 mg/l and 110 mg/l for the fry of Pink Salmon (Onchorhynchus gorbuscha) held in seawater aquaria to which the oil was added. Older fry were considered to be more susceptible than younger to oil toxicity and also more sensitive in their ability to detect and avoid the oil. The older fry were able to avoid concentrations of 1.6 mg/l of crude oil in water and, since such a concentration might be encountered in the event of a spill in an estuarine area, it was suggested that the ability of the fry to detect oil might lead to an alteration in their migration routes. The toxicity of Prudhoe Bay crude to other species of Pacific salmon, the Coho (Oncorhynchus kisutch) and

Sockeye (Oncorhynchus nerka) was shown (Morrow, 1973) to depend upon temperature and concentration with the toxic effects becoming more severe at lower temperatures and higher concentrations. One hundred percent mortality resulted at 'equivalent' concentrations greater than 500 ppm in 96 hours. Since equivalent concentration is the concentration that would have resulted if the oil poured into the tank had dissolved rather than remained on the surface of the water, the sensitivity of salmon parr to water extracts of oil is greater than this value would indicate.

Similar behaviour was observed by Brocksen and Bailey (1973) when juvenile Chinook Salmon (Oncorhynchus tshawytscha and Striped Bass (Morone saxatilis) were exposed to sublethal concentrations of benzene. Increased respiratory rates were followed by narcosis which was reversible when the fish were placed in uncontaminated water.

Mironov (1973) investigated the effects on some Black Sea fishes: Mugil aliens, Sargus anmularis, and Crenilabrus tinca. The selection of the given species of fishes was governed by the following: on the one hand, experiments were conducted with a valuable commercial variety grey mullet (Mugil) and on the other hand, the inhabitants of shallow water regions (sea-carp, Sargus, and sea-parrot, Crenilabrus), where oil pollution is frequent and intense. The results of the experiments concerning the effects of oil pollution on sea carp and sea parrot are presented in Table 5-2.

Results indicate that this year's brood of sea carp and sea parrot remain viable in the course of several days when the oil content of the sea water is in the order of 0.25 ml/l. At similar concentrations of bunker oil, survival is shortened. The survival of fishes depends largely upon the method by which the oil is introduced into the sea water. Thus, in the case of emulsification of oil in marine waters, the effects are more harmful than in the case of oil films lying on the surface.

The survival rate of the young grey mullet did not differ from the control samples when exposed to an oil and bunker oil concentration of 0.25 ml/l. It is important to emphasize the relatively high resistance of the grey mullet to oil pollution. It was observed that the grey mullet swallowed oil from the surface of the water. Later, when the grey mullet was placed in clean water, the oil was quickly released from the anal opening of the fish, coating the top surface of the aquarium with a film. The fish remained viable for several months after this. Unfortunately, for technical reasons, it was not possible to follow their fate (originally the plan was to raise the grey mullet in the experiment until maturity in the hope of attaining their offspring).

TABLE 5-2

Survival of young sea carp and Crenilabrus tinca in sea water
containing petroleum and bunker oil (M-12)

Petroleum products	Concentration ml/l	Death time of fish, hour	
		50%	100%
Sea carp			
Petroleum	0.25	96	140
Bunker oil	0.25	32	92
Bunker oil	0.1	98	154
Bunker oil	0.05	120	190
Crenilabrus tinca (Petroleum products on surface)			
Petroleum	0.25	340	380
Bunker oil	0.25	120	280

5.1.4 Effects on fisheries

It is apparent that light refined oils are considerably more toxic to adult fish than crude and heavy fuel oils. While investigations of the effects of major incidents have concentrated largely on the damage to adult fish or the immediate reduction in fish catches, insufficient attention has been given to damage to the more delicate juvenile forms, effects on fish eggs or the food organisms on which commercial fish feed. All of these factors play a role in determining the long-term effect on fisheries. Damage from oil pollution, therefore, may not show up immediately nor necessarily at the location of the spill but may lead to a gradual reduction in productivity over a large area and on a long-term basis. Furthermore, a gradual and widespread reduction in a fishery would be more likely to result from a chronic pollution problem than from a single acute incident. In these instances, considering other natural and pollutant stresses, it may not be obvious what was the actual cause of the decline in the fishery. The specific topic of losses of fish as food is considered in chapter 6.

The major fisheries of the world are concentrated in areas of high biological productivity, i.e. areas of upwelling and continental shelf waters. In some parts of the world, the latter are shared with the offshore petroleum industry, a situation already becoming more widespread as interest in offshore petroleum deposits increases. One approach in gaining an understanding of the possible harm done to fish and shellfish stocks by the offshore petroleum industry is to examine the statistical records of fisheries in parts of the world, such as the Gulf of Mexico, where offshore petroleum production has been practised since April 1938. By July 1970 there were over 11,500 offshore wells in the Gulf of Mexico. The fisheries there have, therefore, been subjected to chronic oil pollution and intermittent major spillages for many years. About 90 percent of the catch from this area is made up of coastal and estuarine species - oysters, shrimps and menhaden, forms which live in comparatively shallow water where the effects of oil pollution could be expected to be greatest. Nevertheless, data presented by Gusey and Maturgo (1971) show that total catches have increased substantially since 1939. It should be recognized, however, that also the efficiency of fishing has increased because of better detection and catching equipment. Catch per unit of effort rose to a peak in 1961 since when it has shown a decline and a change was found in the species composition, e.g., replacement of brown shrimp by white shrimp. Caution must be exercised in this type of approach. Landings may not relate in the short term to abundance and all fisheries statistics are no better than the system used to collect them. The presence of the oil platforms has affected the fisheries in another way. The artificial environment created by the platforms has proved attractive to many species of fish, including two species of tuna, two species of marlin, swordfish and sailfish. As a result, a major big game fishing centre has been established 100 miles off New Orleans.

Another area where offshore petroleum production is well-established is Lake Maracaibo in Venezuela which is reputed to be one of the most heavily oil-polluted areas in the world. Although described as a "lake", Maracaibo is connected to the sea by a narrow entrance at Bahia del Tablazo. Its waters are saline with a gradient running from north to south. There has been no decline in the fisheries (Battelle Memorial Institute, 1974, 1974a,b). It is, perhaps, worth noting that toxic effects on the eggs and larvae of fish and shellfish which are not in a stock/recruitment relationship are only likely to affect recruitment if the mortality is massive in scale. This would demand firstly, that the oil spill was very large and that a substantial proportion of the toxic components of the oil had not evaporated off but were dissolved in the water, secondly that the eggs and larvae were concentrated in the area of the spill and thirdly that the oil components were present in toxic concentrations for the bulk of the period in which eggs and larvae were in the pelagic phase. Korringa (1968), Simpson (1968) all agree, as fisheries scientists, that oil floating on the open sea, even oil freshly-spilled, is not a hazard to fish and shelfish stocks. For those few stocks which are dependent upon restricted spawning grounds, the chances of spilled petroleum threatening subsequent stocks are greater. No evidence has been found in the literature that floating oil has ever affected the recruitment to any fish or shelfish stock.

5.1.5 Turtle fisheries

Turtles are exploited in many tropical and subtropical regions of the world. They spawn on beaches and are thus potentially vulnerable to oil spills from offshore structures. At the present, there are no firm data confirming or refuting that such fisheries have been affected by petroleum pollution.

5.1.6 Benthic and intertidal organisms

By far the greatest amount of information concerning the effects of petroleum pollution on marine organisms pertains to benthic and intertidal organisms; that is, those organisms which spend a major portion of their lives at the sea bottom. Included are a large number of species of molluscs, crustaceans, echinoderms, polychaetes, coelenterates, and hydroids. Many of these, notably lobsters, oysters, scallops, and clams, not only constitute an important fisheries resource but are also amenable to mariculture, an industry which will undoubtedly become much more important in the future. These creatures are very susceptible to oil pollution because many of them inhabit the intertidal zone where they become coated with oil and smothered in the event that heavy oil drifts onto the shore. Many of them are filter-feeders which indiscriminantly extract fine particles of a certain size range from the water and thereby ingest oil present as droplets or adsorbed on the other particulate material. Since the intertidal zone is the most accessible region of the marine world and since the consequences of pollution are most readily observed there, the effects of oil on intertidal invertebrates have been extensively studied.

Molluscs. In a large spill in the coastal area, molluscs frequently suffer heavy mortalities. For example, a diesel oil spill off the coast of California killed "enormous" numbers of Clams (Tivella stultorum) and Abalones (Haliotis) as well as practically all other animals inhabiting the cove at the time of the spill (North, Neushul and Clendenning, 1964). While recolonization of most species occurred during the next few years, abalone were still absent after 16 years and many species, while present, were not as abundant as before the spill (North, 1973). During the spill at West Falmouth, large numbers of shellfish perished, while contamination of the survivors resulted in closure of the industry. Blumer, Souza and Sass (1970) demonstrated that Oysters (Crassostrea virginica) took up petroleum hydrocarbons from the environment into their lipid pool and many of the petroleum hydrocarbons, particularly the more toxic cyclic compounds, were retained (Blumer, Souza and Sass, 1970). Hawkes (1961) reported that Quahogs (Mercenaria mercenaria) in Narragansett Bay, R.I., could readily withstand concentrations of oil that were intolerable to oysters.

Alyakrinskaya (1966) has examined the effect of water extracts of oil on the Black Sea mussel. According to her observations, mussels do not experience any harm at an extract concentration of 1-10 ml/l; filtering water in the same manner as the control samples. However, a visible reaction was noted at a concentration of 20 ml/l; the mussels open up only 10-12 hours after the weakening of the extract, the opening of the valve itself is smaller than in the control organisms, and the reaction for irritation is retarded. At higher concentrations of 50-100 ml/l, the valves of the mussels open after 1-5 days. In some mussels, these concentrations retard the opening of the valves for 1-2 days, and active filtration and reaction to irritation are absent; however, the opening up itself is evidence for the fact that they are alive. It is not clear what concentrations of oil hydrocarbons were contained in the water-extract used.

According to Milovidova (Mironov, 1972), the mussels were placed into a pond, where additional, more thorough cleaning was being carried out. The mussels were lowered (on a stone) near the coast at a depth of 30 cm. During the first day, the mussels began to filter the water; in calm weather a light spot of transparent water was seen around the stone, and it was partially covered with brown pseudofeces (the water of the pond has a light yellow turbid colour).

Mironov (1967a) has conducted experiments with adult forms of molluscs (Rissoa euxinica Mil., Bittium reticulatum Da Costa, and Gibbula divaricata G.), which live in the shore areas of the Black Sea and serve as food organisms for many fishes. Five types of oil

(Anastasiev oil, low-sulphur Archadin, low sulphur Malgobek, Romashkin and Urusin desalted naval bunker oil M-12, and kerosene) were tested for their effects upon these organisms. The results show that the sensitivity of the molluscs to oil pollution differs from species to species. From the three species tested, R. euxinica died the most rapidly. In addition, the toxicity of the oils and oil products was not the same for the organisms tested, the most toxic being low-sulphur Malgobek oil and bunker oil.

That petroleum-derived hydrocarbons were readily taken up by the common blue mussel (Mytilus edulis) was convincingly demonstrated by Clark and Finley (1973) using modern analytical procedures that have been described in detail (Clark and Finley, 1973a). Such a description of methods is almost unique in reports of the effects of oil on organisms, yet is necessary to account for differences in results of experimental work and to assess the validity of the conclusions drawn. Kanter, Straughan and Jessee (1971) have shown that the effect of oil on the mussel, Mytilus california, depends on the locality from which it was collected, with those taken from areas of natural seepage, and therefore having a history of chronic exposure to oil, having the greater tolerance for pollution. Gilfillan (1973) demonstrated that low concentrations (1 ppm) of sea water extracts of crude oil reduced both the feeding and assimilation of carbon by blue mussels (Mytilus edulis) and marsh mussels (Mytilus demissus) while increasing respiration, the net effect being a significant reduction in the net carbon balance for both species. Gilfillan also suggests that the low energy reserves of the mussels might preclude the development of gametes and that this might account for the fact that the mussels (Mytilus edulis) that survived the West Falmouth spill failed to reproduce the following year (Blumer, 1971). The water-soluble fraction of crude oil (Boylan and Tripp, 1971) and No. 2 fuel oil (Boehm and Quinn, 1974) contains many of the toxic aromatic compounds in the oil and it would be expected that similar but perhaps intensified response would have resulted from refined oils with correspondingly higher aromatic contents.

Crude oils, oil derivatives and oil-dispersant emulsions are harmful to the larvae some marine bivalve molluscs as shown by experiments conducted by Renzoni (1973) on the gametes, developing eggs and larvae of oysters (Crassostrea angulata and Crassostrea gigas) and mussels (Mytilus galloprovincialis). He reports that while the hydrocarbons showed no toxicity towards developing eggs, embryos or larvae at the concentrations used (1 to 1 000 ppm), there was a definite decrease in fertilization in the polluted water and the swimming activity of larvae was hampered.

Swedmark, Granmo and Kollberg (1973) demonstrated that scallops (Pecten opercularis) and cockles (Cardium edule) were considerably less tolerant to oil pollution than mussels (Mytilus edulis). At sublethal concentrations the ability of the bivalves to close their shells was greatly impaired on exposure to oil with mussels being the more resistant species and the effects of diesel oil the most severe. Griffith (1972) found that the marine snail, Littorina littorea, was more sensitive to Arabian light crude than was Mytilus edulis. Recently, Hargrave and Newcombe (1973) observed that the crawling and respiration rates of Littorina littorea were increased by the presence of Bunker C oil as a suspension of finely dispersed particles at concentrations of 750 to 800 μg/l.

There is laboratory evidence that outboard motor exhaust has an adverse biological effect. Using oysters (Ostrea lurida) and mussels (Mytilus edulis), Clark, Finley and Gibson (1974) found that after a 24-hour exposure to diluted effluent (10 percent) almost all mussels showed stress (gaping) while the oysters reacted by closing their shells and apparently not pumping water. The mussels showed such stress that they were removed and placed in clean water but even so they showed evidence of gill tissue degeneration and after 10 days there was 66 percent mortality. The oysters fared somewhat better, showing 14 percent mortality after being in the effluent for 10 days. These studies were conducted using an effluent which was probably much more concentrated than that which could be found in the field.

Crustaceans. The mobile crustaceans, lobsters, crabs, etc., generally inhabit the sublittoral zone and are therefore not as subject to direct contact with oil as are the intertidal molluscs and attached crustaceans. In addition, their mobility gives them the advantage of being able to avoid heavy contamination; or conversely, the disadvantage of being attracted to it (Blumer, 1970). Large mortalities occur, for example the TAMPICO MARU incident, when lobsters (Pamulirus interruptus) and crabs (Pachygrapsus crassipes) suffered heavy mortalities (North, Neushul and Clendenning, 1964). As is commonly the case with other juvenile forms of marine life, lobster larvae are more sensitive than adults to oil pollution. Wells (1972) indicates that emulsions of crude oil are lethal to larvae at concentrations of 100 ppm and appear to have sublethal effects at concentrations down to 1 ppm. The lethal threshold concentration (96 hr LC_{50}) ranged between 2 and 30 ppm. Lobster larvae were particularly sensitive immediately after moulting (Wells, 1972; Engel and Neat, 1971). Atema and Stein (1972) report that sublethal quantities (0.9 ml/l of sea water) of crude oil depress the appetite and chemical excitability of adult lobsters (Homarus americanus) and increase the delay period between noticing food and going after it.

In comparison with lobsters, the crab, Pachygrapsus marmoratus, is a very hardy species and thrives in the highly polluted waters of the Black Sea (Mironov, 1972). The material obtained in the laboratory by Milovidova (Mironov, 1972) regarding the effect of oil pollution on Gammarus olivii and Idothea baltica has shown that in the first 10 days, the survival of Idothea falls only at a concentration of 1 ml/l. However, between the 10th and 20th days, and especially between the 20th and 30th days, the survival falls at a concentration as low as 0.1 ml/l.

The Idothea lived longer in the lower winter temperatures than during the spring and summer. The intensity of feeding, judging by the quantity of feces, falls somewhat at 0.01 ml/l and more sharply at 0.1 and 1.0 ml/l oil. This fall is especially noticeable in the summer when they consume more food than they do in the wintertime. At a concentration of 1 ml/l, the Idothea almost completely stop feeding.

Young specimens of Idothea are more sensitive to oil pollution than are the mature specimens. A noticeable fall in the survival of young specimens obtained from a female kept in clean water was noticed at a concentration of 0.01 ml/l. At 0.1 and 1.0 ml/l, the young ones frequently died on the very first day. At 0.1 ml/l, the young specimens did not live for more than 8 days, and at 1 ml/l, not more than 3 days.

Oil appears to have a more harmful effect upon G. olivii than on I. baltica; at a concentration of 1 ml/l oil, not one sample of Gammarus lived until the 10th day, their survival was close to that of the control samples, but between the 10th and 30th days it fell strongly. In spring and summer at the same concentrations, the survival of Gammarus fell during the first 10 days (Table 5-4).

Table 5-3 presents data concerning the effect of oil pollution on the intensity of feeding of Gammarus.

TABLE 5-3

Oil Conc. (ml/l)	Mean Quantity of Faeces (mg)	Mean Food Consumption
Control	29	0.16
0.001	15	0.20
0.01	30	0.29
0.1	13	0.16
1.0	1	0.00

The survival of juvenile Gammarus borne by females in oil polluted water and later placed in clean water varies little from the survival of those retained in the concentrations in which they were born.

TABLE 5-4

Survival of mature Gammarus in sea water, polluted by oil

Month	Temperature of water C°		Oil conc., ml/l	Number of samples	Samples which survived		
	from-to	mean			After 10 days	After 20 days	After 30 days
March	3-10	7,0	Control	44	100	86	52
			0,01	35	69	57	34
			0,1	40	98	48	5
			1,0	30	0	-	-
	10-18	16,5	Control	12	75	-	-
			0,001	10	90	-	-
			0,01	12	66	-	-
			0,1	12	8	-	-
			1,0	12	0	-	-
July	18-19	18	Control	10	90	-	-
			0,001	10	70	-	-
			0,01	10	60	-	-
			0,1	10	20	-	-
			1,0	10	0	-	-

From: Mironov (1972)

In the lower temperatures, the juvenile Gammarus survives slightly better than at 16-18°C. In temperatures higher than 20°C, the mortality was extremely high, both in the experiments and in the control.

Thus, for G. olivii and for I. baltica, oil pollution in the order of 0.1 to 0.01 ml/l is harmful, but its effects on Gammarus appear sooner than on Idothea.

According to Mazmanidi, Diasamidze and Zambachidze (1973), oil at a concentration of 35 mg/l decreases glycogen content in muscles and hepatopancreases of oyster females in one day. Thus, glycogen has decreased from 1900 ppm to 690 ppm in hepatopancreases and from 970 ppm to 280 ppm in muscles, as compared with the control. During chronic exposure experiments, considerable decrease of glycogen content in this organism was observed (at 7.5 mg/l and 3.7 mg/l of oil concentration).

Below are presented data on the effect of oil on some other types of benthic organisms of the Caspian Sea (Table 5-5).

Larval brown shrimp, Crangon crangon, are susceptible to concentrations considerably lower than those affecting adults of the species (Portmann, 1972). Emulsions of crude oil and of diesel oil were toxic to prawns (Leander adspersus) at concentrations of 70 to 170 ppm (Swedmark, Granmo and Kollberg, 1973), while at lower concentrations their swimming activity was noticeably impaired. Again the most intense effects resulted from exposure to light marine diesel oils. Katz (1973) observed that the water-soluble fraction of crude oil had a retarding effect on the moulting of the larvae of the decapod crustacean, Neopanope texana. The effects of a crude oil and two refined products (light fuel oil and heavy fuel oil) on the amphipod Gammarus oceanicus was investigated by Lindén (1976a). He found that larvae (4 to 6 days old) were several hundred times more sensitive to the oils than the adults during acute exposure. A number of sublethal effects appeared during long-term bioassays. The adults showed impaired swimming performance, decreased tendency to precopulate, impaired light reaction and decreased production of larvae. Decreased growth was found among larvae during chronic exposure to crude oil. Delayed mortality occurred among adults after a short-term exposure to crude oil with a long recovery period. In another study (Lindén 1976b) concerning the effects of oil on Gammarus oceanicus it was found that sublethal levels of crude oil significantly decreased the brood numbers if the females were exposed to the oil during the incubation period. The frequency with which the male and the female enter the precopulation stage which is necessary for successful fertilization, was also found to decrease in the presence of low concentrations of oil in the water.

The sessile crustaceans, particularly those inhabiting the intertidal zone, frequently suffer extensive mortalities in the event of an oil spill. For example, an estimated 4.2 million intertidal animals, predominantly acorn barnacles, were killed by a spill of 840,000 gallons of Bunker C oil in San Franciso in 1971 (Chan, 1973). This is one of the few spill incidents in which long-term pre-spill data were available, and Chan is convinced that "the decrease in marine organisms in the transect areas was not due to storm conditions, natural predators, nor zealous collecting by man, but was attributable mainly to the contamination of the organisms by the oil." Similarly, extensive kills of barnacles (Chthamalus fissus and Balanus gladula) occurred in the TAMPICO MARU spill, but their population rapidly recovered and was back to normal within two years (North, Neushul and Clendenning, 1964). In the Santa Barbara spill, barnacles (Chthamalus fissus) were one of the most heavily affected species (Straughan and Abbott, 1971). On the other extreme, the oceanic barnacle, Lepas fascicularis, which attaches itself to floating debris, readily accepts tar balls as its substrate.

Echinoderms and Other Benthic Organisms. Echinoderms are considered to be extremely sensitive to any reduction in water quality (Nelson-Smith, 1970). For example, sea urchins (Strongylocentrotus franciscanus and S. purpuratus) were virtually eliminated from the vicinity of the TAMPICO MARU wreck (North, Neushul and Clendenning, 1964) and did not reappear until two years later nor become plentiful until four years had elapsed. With the removal of these grazing animals, the giant kelp Macrocystis pyrifera proliferated (North, 1973). Water-soluble extracts of a number of crude and fuel oils, diesel and jet fuels appeared to have little effect on fertilization of sea urchin (Strongylocentrotus purpuratus)

TABLE 5-5

Average survival (days) of bottom organisms under the effect of various oils

Concentration, ml/l	Bulla-Duvunin layer	Artemov layer	Tar balls "Oil Stones"
Coruntodurma lamarcki			
0,1	32,0	31,5	37,0
0,2	31,9	31,0	36,8
0,4	28,5	30,0	36,0
0,6	28,2	29,3	34,1
0,8	26,6	28,1	32,7
1,0	24,4	26,9	30,9
Control	38,6	38,2	38,5
Hytilaster lineatus			
0,1	49,4	63,0	49,0
0,2	48,8	59,4	48,7
0,4	46,1	54,0	47,8
0,6	45,2	49,1	47,1
0,8	41,0	42,0	46,0
1,0	39,4	40,5	44,8
Control	64,0	63,7	63,4
Abra ovata			
0,1	26,3	50,2	40,8
0,2	26,1	48,0	40,6
0,4	25,0	43,5	40,2
0,6	24,4	32,2	37,9
0,8	20,9	30,1	36,2
1,0	19,6	29,1	35,3
Control	50,8	50,6	51,0
Pontogammarus macoticus			
0,1	17,0	21,4	34,0
0,2	15,2	20,6	32,4
0,4	12,7	20,5	30,1
0,6	12,0	19,1	28,9
0,8	11,6	18,6	28,0
1,0	10,0	18,1	24,4
Control	–	–	–

From: Kasymov and Granovskii (1970)

eggs, but most of these extracts were toxic to the developing eggs (Allen, 1971). The most pronounced effects were obtained with the crude and heavy fuel oils and the least effect with the light refined products. The author suggests that "the process of refining removes many of the more toxic fractions." This, of course, is a direct contradiction to the general tendency for the effects of the lighter oils on marine life to be more severe than those of the black oils.

A wide variety of benthic invertebrates was killed by diesel oil spilled into Puget Sound in 1971, with mortalities for 48 species of intertidal invertebrates ranging from 30-100 percent, the hardest hit being brittle stars, polychaetes, nemertans, chitons, hermit crabs, and limpets (Chia, 1971; Woodin, Nyblade and Chia, 1972). Animals from the higher intertidal regions were more severely damaged than those from the lower regions. In January 1972, a continuous leak of fuel oil from the GEN.M.C. MEIGS in Washington not only killed sea urchins (Strongylocentrotus purpuratus) but petroleum hydrocarbons were taken up by the intertidal community in general. Otherwise healthy goose barnacles (Mitella polymerus) and crabs (Hemigrapsus nudus) displayed the same n-alcane hydrocarbon patterns as the fuel oil (Clark et al., 1973).

Massive destruction of a wide range of fish, crustaceans and invertebrates occurred during the spill of No. 2 fuel oil at West Falmouth. Roughly 95 percent of the animals collected in the vicinity of the spill were either dead or moribund (Ketchum, 1973). After the competing species were eliminated, however, the polychaete, Capitella capitata, flourished and built up huge populations (Sanders, Grassle and Hampson, 1972). The intertidal polychaetes, Cirriformia tentaculata and Cirratulus cirratus, apparently tolerate oil-polluted mudflats (George, 1971), and a spill of fuel oil apparently had no effect on the spawning, growth and mortality of these two species (George, 1970).

Johannes (1972) reported that there was no evidence of damage to reef corrals from oil floating over them, although they were seriously damaged when they came into direct contact with the oil (Johannes, Maragos and Coles, 1972). Lewis, J.B. (1971) demonstrated the sensitivity of four species of corals during laboratory experiments with crude oil and a dispersant Branching corals were more severely affected than the encrusting species and showed less ability to recover. Both pollutants had harmful effects at concentrations of 100 to 500 ppm.

5.1.7 Zooplankton

Most of the data on permanent zooplankton (holoplankton) pertain to copepods. Young Acartia clausii and Oithona nana die after 3-4 days immersion in sea water containing 10 μl/l of oil (Mironov, 1968, 1972a), while the adults of these and three other copepod species suffer an accelerated death after longer exposures to 10 μl/l or after 5-60 minutes in 1 ml/l. Acartia dies within 24 hours in sea water containing oil at 100 μl/l as does Calanus although Oithona is a little more resistant (Nelson-Smith, 1968).

Beneath a thick oil slick, penetration of light may be reduced by more than 90 percent (Nelson-Smith, 1968) drastically reducing the rate of photosynthesis in phytoplankton and also interfering with the daily vertical migration of zooplankton which is regulated by light intensity. Many of the large zooplankters locate their food by visual selection so that overshadowing by oil may exert an indirect effect on nutrition and behaviour in addition to its direct chemical or mechanical effects (Nelson-Smith, 1968).

During the first day in clean water, the percentage of live organisms Acartia clausii in crude oil and in diesel oil was practically equal, and was found to be independent of the time of contact of Acartia with the oil products. Thus, at the beginning of the first day all of the organisms which were in contact with oil for 5 and 30 minutes were alive and 90 percent were alive after contact with diesel oil and with oil for an hour. Bunker oil constituted a certain exception: on the first day in clean marine water, 40 percent of the organisms remained alive, which had previously been in contact with bunker oil for a one hour period.

In the following days, death of the hydrobionts was observed, which was proportional to the amounts of time in which they had earlier remained in polluted marine water. A similar situation was observed with another represesentative of the zooplankton, Centropages ponticus. However, in this case, death of the organisms was recorded during the first day, and reached 60-80 percent after a one hour contact with oil products. It should be emphasized that the harmful effects in both of the above cases was caused by dissolved and thinly emulsified components of the oils.

5.1.8 Phytoplankton

The literature prior to 1970, regarding the effects of oil on phytoplankton, has been reviewed by Nelson-Smith (1970, 1972a). The general impression emerging from this earlier work is that although natural plankton communities are adversely affected by oil, these organisms are sufficiently prolific that individual spills have only short-term effects on their overall populations. However, since phytoplankton is largely responsible for the fixation of the energy utilized by marine ecosystems, it is vital to know how oil in the water column affects it - both as large spills and as continuous additions in estuarine, coastal, and oceanic environments. Unfortunately, it is very difficult to detect the effects of oil on species abundance and composition of phytoplankton in the field, as these organisms show drastic natural seasonal variations with one species of diatom or dinoflagellate taking over the predominant position held by another species as a consequence of changes in temperature, light or the availability of nutrients.

Table 5-6 shows the differences in sensitivites of various algal species to oil, in the form of the concentration of oil necessary to cause death or retarded cell division. The concentration of oil necessary to cause death ranged widely, from 1.0 to 10^{-4} ml/l. An absence or retardation of cell division occurred, depending on the species, at 0.1 to 0.00001 ml/l. In these ranges of concentrations the rate of cell division of many algal species did not differ from the control. Out of the ten species tested, death or retardation of cell divison occurred in six species at an oil concentration of up to 10^{-2} ml/l, and in two species at a concentration of 1.0 to 0.1 ml/l. These results demonstrate the high sensitivity of the majority of microscopic algae from different sources to marine pollution by crude oil.

When Prorocentrum micans and Coscinodiscus sp. remained for half-an-hour in marine water containing 1.0 to 1.1 ml/l bunker oil and kerosene, retardation of multiplication of these algae was shown to occur. In this case a high sensitivity of P. micans to bunker oil pollution was observed. Thus, after a four hour contact with a bunker oil, this algae died on the third day in clean sea water, and on the first day after a six hour contact. At the same time, a six hour contact of this algae in sea water containing a similar quantity of kerosene merely caused retardation of cell division. This may also be observed in the earlier experiments with this algae (Mironov et al., 1975).

One of the best methods to determine the effect of oil on the growth of natural phytoplankton communities is the radiocarbon bicarbonate technique as employed by Dickman (1971) and by Strand et al. (1971) which is based on the standard radiocarbon method for measuring planktonic primary production (Strickland and Parsons, 1965). A known amount of oil in sea water is added to raw sea water samples after removing large zooplankton. Samples and controls are inoculated with ^{14}C-bicarbonate and incubated at ambient light intensity and in situ temperatures for six hours. The radiocarbon uptake of samples containing oil is compared with that of controls to determine effects of the oil.

Several studies have demonstrated the differences in the sensitivity of phytoplankton species to crude and other oils. Nuzzi (1973) carried out laboratory experiments to demonstrate the effects of No. 2 and No. 6 fuel oils and outboard motor oil on natural phytoplankton populations from Long Island, N.Y., area and also on axenic cultures of Phaeodactylum tricornutum, Skeletonima costatum, Chlorella, and Chlamydomonas. The water-soluble components of No. 2 fuel oil were toxic to both the natural cultures and axenic populations. On the other hand, water extractable substances from No. 6 fuel oil and outboard motor oil had little effect on growth of the axenic populations, while the growth of some species of

TABLE 5-6

Reaction of algae to various oil concentrations (ml/l)

TYPE	DEATH OF CELLS	THE ABSENCE OR RETARDED CELL DIVISION	DO NOT DIFFER FROM CONTROL
Glenodinium foliaceum	1.0-0.1 ml/l	0.1-0.01 ml/l	0.001-0.0001 ml/l
Chaetoceros curvisetus	1.0-0.01	0.01	0.001-0.0001
Gymnodinium wulffii	1.0-0.1	0.01-0.0001	--
Ditylum brightwellii	1.0-0.0001	--	--
G. kowolovskii	1.0-0.0001	0.001-0.0001	0.00001
Prorocentrum micans	1.0	0.1-0.00001	--
Peridinium trochoideum	1.0	1.0	0.1-0.00001
Licmophora chrenborgii	1.0	0.1-0.001	0.0001-0.00001
Platymonas viridis	1.0	0.01-0.001	0.0001
Coscinodiscus granii	1.0	1.0-0.1	0.1-0.0001
Molosira moniliformis	10.0-1.0	10.0-0.1	0.1-0.01

natural populations was stimulated by the motor oil. Lacaze (1967) reported a depression in the growth rate of cultures of Phaeodactylum tricornutum exposed to one percent extracts of Kuwait crude. Strand et al. (1971) found that the uptake of labelled bicarbonate by Monochrysis lutheri was significantly reduced by crude oil emulsions at concentrations of 50 ppm. Similarly, Mommaerts-Billiet (1973) observed diminished growth rates when marine planktonic algae were exposed to crude oils and emulsifiers. Some relative data on crude and fuel oils are given in Table 5-7.

Gordon and Prouse (1973) studied the effect of Venezuelan crude oil and No. 2 and No. 6 fuel oils on natural populations of marine phytoplankton from Bedford Basin, Nova Scotia, and the North Atlantic between Nova Scotia and Bermuda. Although the water-soluble fractions of all three oils inhibited the uptake of labelled bicarbonate, the effects of No. 2 fuel oil were the most pronounced with a 40 percent reduction in carbon fixation at 100-200 ppb. This inhibition was assumed to be indicative of a decrease in the rate of photosynthesis. At a low concentration there was some indication that the extractable substances from crude oil had a stimulatory effect. It was concluded (Gordon and Prouse, 1973) on the basis of these observations, that the present levels of oil contamination in the areas studied are not having any serious detrimental effect on photosynthesis by the natural phytoplankton communities. However, it may be wise to add a note of caution in interpreting these observations. The fact that phytoplankton photosynthesis was not seriously affected by oil concentrations normally found in the areas studied cannot be used to draw the general conclusion that oil pollution in the marine waters studied causes no ecological damage.

In tests with naphthalene, a concentration of 3 ppm caused almost complete elimination of bicarbonate uptake by Chlamydomonas angulosa (Kauss et al., 1973). Mironov and Lanskaya (1967) found that the diatoms Ditylum brightwelli, Coscinodiscus granii and Chaetocaros curvisetus were killed within 24 hours by 100 μl/l or less of kerosene and fuel oil, although Melosira moniliformis and Grammatophora marina tolerated concentrations up to one percent. Much lower concentrations (as little as 0.1 μl/l for the more sensitive species) retarded the rate of cell division and thus the growth of the culture.

TABLE 5-7

Relative effects of crude and fuel oils on primary productivity (IMCO MEPC II/INF. 13).
(Concentrations in μg/ml of contaminant that result in 50% reduction in photosynthetic activity).

Test Organism	No. 2 Fuel Oil		South Louisiana Crude Oil		Kuwait Crude Oil		Bunker C	
	Dispersed	Undispersed	Dispersed	Undispersed	Dispersed	Undispersed	Dispersed	Undispersed
Morchrysis lutheri	8.4	0.9	75	250	87			
Chlamydomonas spp.	50	7.2			250	550	575	575

In the Arctic environment a major portion of the energy retained must be fixed by aquatic plants. At the low mean temperatures in the Arctic, evaporation rates of the highly toxic, lighter fractions are reduced thereby prolonging the time that sensitive aquatic organisms are exposed to the toxic influence of the lighter hydrocarbons. Dickman, M. (1971) found that although phytoplankton density and productivity was low to begin with, the addition of crude oil reduced the primary productivity by nearly tenfold. It is impossible, of course, to predict the consequences of such a reduction without understanding the relative importance of algae in the Arctic energy pyramid.

5.1.9 Macroscopic algae

Seaweeds, like most plants, can suffer damage over a considerable area and retain their capacity to recover. Many of the larger algae on exposed shores normally produce new growth near the base during the growth season and lose the distal parts every winter. Thus the ill effects of an oil spill are usually not as severe amongst intertidal seaweeds as they are on animals. The large brown seaweeds on temperate zone shores are covered by a protective mucilaginous slime that is not readily penetrated by oil. After the Santa Barbara spill the offshore beds of giant kelp Macrocystis pyrifera prevented the approaching oil from coming in contact with the plants and animals of the lower intertidal zone until after the tide rose to cover them. At that time the oil washed readily off the kelp blades (Straughan, 1971a). The kelp beds were relatively unharmed with no abnormal decay or damage and no oil beneath the kelp canopy (Foster, Neushul and Zingmark, 1971). On the other hand, North, Neushul and Clendenning (1964) reported that Macrocystis was seriously affected by the TAMPICO MARU spill and found that a 0.1 percent emulsion of diesel oil almost completely inhibited the photosynthetic activity of young blades while irreversible damage was caused by exposure to the suspension of oil for 6-12 hours. Fucus spiralis, an intertidal algae, was eliminated from rocky shores in Chedabucto Bay, N.S., where oiling with Bunker C was heavy following the ARROW incident and there was no subsequent recolonization (Thomas, 1973). In contrast, the mid-tide seaweeds (Fucus visiculosus) did not appear to be severely affected in Brittany after the TORREY CANYON spill, presumably because the plants were protected by a mucilage layer which prevented adhesion of the oil (Stebbings, 1970). Crapp (1969) reported that light fuel oil clung to Ascophyllum nodosum, and topshore species like Pelvetia canaliculata strongly retain oil when dry during neap tides.

Emulsified oil seemed to cling more readily to Porphyra umbilicalis than to other red algae. Red algae seem to be the most sensitive and suffered the greatest losses in both the TORREY CANYON disaster and the TAMPICO MARU spill of diesel oil (Nelson-Smith, 1968). Although red algae retain crude oil for long periods of time, the heaviest pollution associated with the Santa Barbara spill occurred before the spring bloom of Porphyra and mortality was low except for one location (Foster, Neushul and Zingmark, 1971).

The green algae, Enteromorpha intestinalis, Chaetomorpha aerea and Ulva angusta in the upper-mid and high-intertidal zones were only slightly damaged by the Santa Barbara spill unless they were completely coated or hot water cleaned as the oil did not seem to stick to them but washed on and off with the tide. Any damage was generally in the high intertidal area (Foster, Neushul and Zingmark, 1971).

Pathological morphological modifications under the oil effects are observed in marine algae, juvenile forms having the most considerable modifications. In the study (Mironov and Tsimbal, 1975) of the effect of oil in the concentration of 1.0; 0.1; 0.01 ml/l on the development of sporelings of Polysiphonia breviarticulata, Polysiphonia opaca, Dilophus fasciola, the slowing down of their growth was marked.

The destruction of the sporelings of Polysiphonia breviarticulata and Polysiphonia opaca when oil is in the concentration of 1.0 ml/l comes on the fifth day, as a rule in the phase of a spore or a few cells. When oil concentration is 0.1 ml/l the sporelings of P. breviarticulata die on the fifth day of the experiment, and those of P. opaca and D. fasciola correspondingly on the ninth and fifteenth days.

Most brown algae in the Santa Barbara area occur in the low intertidal and subtidal areas and escape from much of the large initial dose. Oil that did get onto them seemed to wash off and caused little damage (Foster, Neushul and Zingmark, 1971).

Some lower forms, particularly the blue-green algae, appear to be resistant to oil pollution and may even obtain nutrients from it, as reported by Spooner (1970), Baker (1971), and Cabioch (1971), all of whom observed blue-greens in patches around refinery outfalls which were otherwise bare of vegetation. Oscillatoria - a typical blue-green algae - grows abundantly on trickling filters used to remove oil from refinery effluents (Crosby, Rudolfs and Heukelkian, 1954).

5.1.10 Marine grasses

Much of the information regarding the effects of oil pollution on marine grasses and marsh plants has been gathered by workers at the Orielton Field Station in Pembroke, Wales, where extensive studies have been made of the effects of oil on salt marsh plants. Baker (1971), for example, reported that a single spill of oil does not cause long-term damage to marsh vegetation although short-term effects include death of oiled shoots. This is followed, however, by new growth from plant bases and the area soon recovers. During the recovery period, there may be reduced germination and flowering, a reduced population of annuals, and growth stimulation of some species. Recovery times vary from two to three years depending on the severity of the incident. To determine the effect of chronic pollution, experimental plots of Spartina angelica, Pucinnella maritima and Juncus maritimus were sprayed with Kuwait crude oil at monthly intervals (Baker, 1971, 1973a, 1973b). Recovery from four oilings was good, but more than this resulted in a rapid decline of the vegetation. Species varied considerably in their tolerance to successive spillages, from the very susceptible annuals, Sualda maritima and Salicornia, and seedlings of all species through the grasses and rosette perennials to the very tolerant Umbellifer Oenanthe lachenalii. The continuous discharge of oily waste water from a refinery resulted in the destruction of an area of the marsh grass, Spartina angelica (Baker, 1973a). Death of the plants was attributed to their becoming coated with thin films of oil during high spring tides, and this also prevented recolonization. The effects of oil on marsh grasses were shown to have a strong seasonal component with a marked reduction in flowering if the plants were oiled when flower buds were developing. Flowers, if oiled, rarely produced seeds, although oiling of seeds during the winter did not prevent their germination in the spring (Baker, 1971). No difference was observed in the damage to salt marsh grasses whether the area was cleaned with emulsifiers or if the grasses were burned or cut, and it was, therefore, concluded that it is best to leave an oiled salt marsh to recover naturally (Baker, 1971). If the soil of a salt marsh is coated with oil, the diffusion of oxygen to the rhizosphere may be hindered and toxic conditions may be created (Baker, 1971). Experimental evidence was obtained that indicated that the growth of some marsh grasses (Pucinella maritima and Festuca rubra) may be stimulated by oil pollution and, while the reason for this was not clear, it may be related to an increase in nutrients from other oil-killed organisms, the presence of growth-regulating compounds in the oil, or an increase in nitrogen fixation (Baker, 1971). The low-boiling fractions of crude oil were found to be the most toxic to salt marsh plants, with fresh crude being more toxic than after it weathered (Baker, 1971). Studies of plant physiology suggest that oils penetrate into the plants, where they may travel in the intercellular spaces and possibly also in the vascular system (Baker, 1971). Cell membranes are damaged by penetration of hydrocarbon molecules leading to leakage of cell contents. Oils reduce transpiration rates, probably by blocking stomata and intercellular spaces, and this may also be the reason for the reduction of photosynthesis which occurs. An increase in respiration rate often occurs, and oils inhibit translocation. The severity of these effects depends upon the constituents and amounts of oil, environmental conditions and species of plant (Baker, 1971).

Several species of marine plants and shore grasses were affected by the ARROW spill of Bunker C oil in February, 1970 (Thomas, 1973). Spartina alterniflora, the salt-marsh cord grass, was also eliminated from heavily oiled areas and failed to grow during the 1970 season. In less heavily oiled areas, spikes of leaves penetrated the oil and oily sediment and grew well in 1970 but failed to do so in 1971 suggesting that the effects of the oil were more severe during the second year after the spill. In contrast to this delayed effect, Stebbings (1970) considered that the salt marsh grasses in Brittany had completely recovered from the effects of the TORREY CANYON spill within 16 months. The initial heavy dose of oil from the Santa Barbara spill had deleterious effects on the surf grass Phyllospadix torreyi when the oiled blades stuck together and later turned brown and disintegrated (Foster, Neushul and Zingmark, 1971). Plants in the lower intertidal and subtidal zones, however, were protected by water and escaped severe damage. Straughan (1971a) listed Phyllospadix as one of the few organisms seriously damaged by this incident.

The results of experimental studies of emulsifier treatment, burning and cutting as possible cleaning methods for oiled salt marsh vegetation show that none of these methods decreases the damage due to oil and may increase it (Baker, 1971). Following a spill in the northern Baltic the burning and emulsifying that was done caused much damage to the flora and fauna in the littoral zone (Leppakoski, 1973). However, three months later there was found to be very little difference between polluted and clean areas - probably because the spill occurred in early spring so that the annual algae crop had not yet begun to grow and the sublittoral aquatic vegetation was not damaged. As a general rule it appears better to leave an oiled salt marsh to recover naturally rather than to attempt to clean it.

There have been a number of incidents where oil has actually stimulated growth (Stebbings 1970). Although the reason is not yet known, it could have to do with an increase in nitrogen fixation, the nutrients supplied by oil-killed animals, or bacteria could also have an effect. The possibility of oil containing growth regulatory compounds has not even been ruled out (Baker, 1971).

5.1.11 Other marine plants

Some maritime plants occur on cliffs and are generally quite sensitive unless it is one of the deeply-rooted species like Armenia or Plantago which recovered from both oil and emulsifier contamination (Crapp, 1971). Although most lichens are extremely sensitive to pollution, Xanthoria parietina, the familiar orange encrusting lichen, has been reported to survive beneath and eventually grow through a hardening layer of oil on a sea wall (Nelson-Smith, 1968).

5.2 Effects of oil on marine populations

5.2.1 Birds

It would appear from the literature that only sea bird populations have suffered from oil pollution to the extent that certain species or subspecies are threatened with extinction. For example, there is little doubt that the more southern colonies of puffins (Fratercula artica) (Bourne, 1971), razorbills (Alca torda) and guillemots (Urea aalge) are declining rapidly on both sides of the Atlantic (Tuck, 1960; Clark, R.B., 1973). Those that nest on the islands to the north of Scotland have already suffered seriously and are considered to be in a particularly precarious position (Bourne and Johnston, 1971; Bourne, 1971). It is almost certain that the primary cause of these declines is oil pollution (Nelson-Smith, 1972).

Other species with higher reproductive rates, on the other hand, seem to be able to withstand repeated and heavy losses from oil pollution and are not as seriously threatened. However, certain subspecies may be vulnerable. For example, during the winter of 1971, the borealis subspecies of eider were more seriously affected than the number killed would indicate (Brown, R.G. et al., 1973), since 98 percent of those killed were adults and since 83 percent of these were adult males, a large number of females might have gone unmated. Since 90 percent of the total borealis population winters in Newfoundland, it is particularly vulnerable to losses from oil while the birds are in their winter habitat.

The only satisfactory approach to the conservation of sea birds is to reduce the number of spills and the amount of oil on the surface of the ocean. Clark (1973) strongly favours the use of chemical dispersants to remove spilled oil from the surface as soon as possible and considers this the only technique presently available that offers any hope for the future of sea birds. As an illustration he cites the successes at Milford Haven where the use of dispersants has virtually eliminated sea bird casualties without damaging the rich marine fauna and flora of the area and, by way of contrast, the unnecessary kill of 900 birds, mostly scoters, from a spill of 80 to 100 tons of oil from the DELIAN APOLLON in Tampa Bay, 1970, because of a ban on the use of dispersants.

Cleaning and rehabilitation of sea birds has, for the most part, proven to be unsuccessful and is unlikely to make a material contribution to the conservation of the declining populations of those species seriously affected by oil pollution, except possibly in small, isolated breeding colonies (Clark, 1973).

5.2.2 Mammals and other higher marine life forms

In comparison with the rather bleak situation regarding certain sea birds, there is no evidence in the literature to suggest that existing levels of oil pollution present any serious threat of extinction to marine mammals, fish, benthos or plants, except in the vicinity of a spill or an area of chronic pollution where the effects on all forms of marine life may be disastrous.

Biological observers of the TORREY CANYON disaster concluded that spilling a large quantity of crude oil leads to deleterious effects of a local character. When a productive shellfish area is contaminated by an oil spill, this natural resource will be adversely affected. However, such local events will probably not lead to extermination of any given species because the geographical range of virtually all marine organisms living in shallow water is much larger than a single affected area.

From this point of view, the change in biocenoses in Novorossiiskaya Bay under conditions of pollution is of great interest, as crude oil is the main polluting component. In comparison to materials obtained by Morozova-Vodianitskaja (1927, 1930), Kalugina et al. (1967) showed that during a period of 30 years, the distribution of algae in some areas of Novorossiiskaya Bay changed, due to a shifting of flora boundaries and minimizing of some alga thickets. Changes occurred in the animal distribution as well. Meretrix rudis and Venerupis lineatus, formerly predominant in the port, was transferred to the middle part of the bay. The most resistant to oil pollution of the benthic animals, Nassa reticulata, Rapana thomasiana, Capitella capitata, and Platynereis dumerilii, appeared now at the port and other polluted areas (pier, the Galatskaya Bay). Modiola adriatica originally predominant in the middle part of the bay, was shifted to the bay outlet.

5.2.3 Plankton

There does not appear to be any data to support or refute any adverse effects of oil pollution on planktonic communities.

Mironov (1961) was not able to identify any qualitative or quantitative differences among the planktonic diatomic algae in relation to the pollution of Black Sea water, except that the bentho-planktonic species M. moniliformis var. subglobosa occurred exclusively in polluted areas of the sea.

5.2.4 Microbial populations

Effects of a mixed hydrocarbon substrate (Walker and Colwell, 1975b), a No. 2 fuel oil, and a South Louisiana crude oil on water and sediment microorganisms from Chesapeake Bay and the Atlantic Ocean, in the latter case from 9 to 5,000 m, were determined. It was found that addition of mixed hydrocarbon substrate limited growth of sediment bacteria from the oil-free estuarine site. Similar results were also observed for ocean sediment bacteria cultured in sea water with a mixed hydrocarbon substrate added. In a subestuary marsh site of about 0.81×10^6 m^2 wetlands supporting abundant microbial populations, petroleum either had no significant effect on the sediment and water microorganism total populations or, in some cases, provided a degradable carbon source supporting microbial growth. Growth-limiting effects of fuel oil No. 2 and South Louisiana crude oil were, however, detected for certain ecological important components of the marshland populations, namely the proteolytic, lipolytic, and chitonolytic populations (Walker, Seesman and Colwell, 1975, 1975a) A slight inhibiting effect on yeasts was also noted (Walker and Colwell, 1975b). Thus, there are measurable effects of oil on ecologically important bacterial groups, caused by exposure to crude and refined oils. Addition of petroleum hydrocarbons to water collected from an oil polluted environment may promote growth of the petroleum degrading bacteria already present in the water (Walker and Colwell, 1975a).

In general, the microbial response to an oil spill is of two types, an inhibition of some of the naturally occurring groups of bacteria that may be critical for maintaining the ecosystem balance and/or the development of petroleum degrading populations that contribute to the removal of the oil. This is, of course, a highly simplified statement. In areas already exposed to oil, the numbers of petroleum-degraders are higher than in areas never before exposed to petroleum hydrocarbons of an allochthonous nature. Mironov (1971) observed that there are significant increases in petroleum degrading microbial population along established tanker routes in the Indian Ocean and elsewhere.

5.3 Effects of oil on the behaviour of marine organisms

In addition to toxicological effects on organisms at lethal and sublethal concentrations, oil and other pollutants may have subtle effects on marine life at concentrations that are not high enough to cause obvious negative effects on the environment, by interfering with the chemical signals that control many important biological processes. The detection of food, feeding impulses, escape from predators, territory definition, moulting, homing of migratory species, and other biological processes that are crucial to the survival of the species are regulated by very low concentrations of substances in sea water. Natural chemical signals, such as pheromones, that trigger these responses may be masked or mimicked by the presence of low concentrations of pollutants (Blumer, 1972a; Todd, Atema and Boylan, 1972). Blumer (1972a) considers that the aromatic fractions of crude oil and the full range of olefinic hydrocarbons in refined products are likely to interfere with chemically mediated behaviour.

Atema and Stein (1972) report that sublethal quantities of crude oil (0.9 ml/l) interfere with the feeding behaviour of lobsters (Homarus americanus) by increasing the delay period between noticing food and going after it as well as by depressing their appetite and chemical excitability. Rice (1973) demonstrated that the older fry of pink salmon (Onchorhynchus gorbusche) were able to detect and avoid water polluted with oil, and suggested that oil pollution might change the normal migration behaviour of salmon fry. On the other hand, Kühnhold (1972a) reported that the larvae of cod, plaice and herring were unable to avoid well-defined milky clouds of oil dispersions after they had been exposed to oil contaminated water.

Kittredge (1972, 1973) investigated the effects of the water-soluble component of crude oil on the chemoreception and sex pheromone response of lined shore crabs (Pachygrapsus crassipes) and found that these substances completely inhibited the chemoreceptors of the crabs, apparently by destroying the sensitive neuronal dendrites of the chemoreceptor organs. Recovery of the chemoreceptive sense did not occur until 8 to 11 days after exposure. Studies with individual hydrocarbons revealed that straight chain compounds were innocuous, monoaromatic compounds were effective inhibitors but the inhibition was transient, while polynuclear aromatic hydrocarbons produced an effect similar to that of the water-soluble fraction from the crude oil.

Jacobson and Boylan (1973) demonstrated that 1 to 4 ppb of kerosene in sea water is sufficient to interfere with the chemically mediated attraction of the marine snail, Nassarius obsoletus, to food extracts. These authors suspect that naphthalenes extracted from the kerosene were responsible for the interference with the normal feeding behaviour of the snail. Chemical communication is one of the most ubiquitous and powerful controlling factors in the marine environment and is not restricted to higher organisms. Fogel, Chet and Mitchell (1971) have shown that the majority of the motile forms of marine bacteria have chemoreceptors and are attracted to their food sources by chemotaxis. Mitchell, Fogel and Chet (1972) subsequently demonstrated that 0.6 percent phenol, 0.6 percent toluene and an unspecified amount of crude oil totally inhibit the chemotactic response of marine bacteria, apparently by blocking the chemoreceptors. The bacteria were not immobilized, but their activity although at a normal rate, was random. Blockage of the chemoreceptors was reversed when the bacteria were washed free of hydrocarbons. Clearly, one of the major consequences of interference with microbial chemotaxis by petroleum is that the decomposition of organic materials in the sea would be affected, since this is not only dependent on the ability of the microorganisms to degrade the substrate enzymatically, but also on their ability to detect and become attached to the substrate. Another aspect of chemotaxis inhibition to be considered is that slime-forming bacteria "prepare" surfaces for attachment of higher organisms (ZoBell, 1946a). In coastal waters and in estuaries, the intermediate stages of shellfish and other biota require attachment for completion of their life cycle.

The total inhibition of bacterial chemoreception by low concentrations of petroleum hydrocarbons would be of tremendous environmental importance since the marine microflora are responsible for the degradation and turnover of the majority of the organic matter in the sea.

5.4 Miscellaneous topics

5.4.1 Bioassay methods of determining the toxicity of oils

It is very difficult to summarize and evaluate existing data pertaining to the toxicity of oil to marine organisms because it is virtually impossible to intercompare the results of toxicity tests obtained by different workers using different methods and different organisms. Bioassays to measure lethality within a specified period of time seldom permit a meaningful evaluation of the toxicity of an oil in the field. (There are, of course, exceptions; for example, the work of Swedmark, Granmo and Kollberg, 1973; LaRoche, Eisler and Tarzwell, 1970; Pickering and Henderson, 1966, and others). In many experiments oil is floated on the surface of the water in the test container and the concentration reported is calculated from the total volumes of oil and water used without regard for the amount that has actually become dissolved or dispersed in the water. In comparatively few cases attempts have been made to measure the concentrations of oil to which the organisms are exposed. In other instances, a fine emulsion of oil in water is used or water extracts of oil are added to the test container without identification of the fraction being tested. It is not surprising, therefore, that the ranges of toxicity reported in the literature by different workers using different methods often differ by several orders of magnitude.

In addition to the problems associated with the concentration of oil used, experimental conditions are seldom described in enough detail to interpret and compare the results. Factors such as degree of agitation, static or flow-through system, aeration, temperature, pH, salinity and health of the organism including stress all play a role in determining what effect a given pollutant may have on an organism. Furthermore, it is difficult to extrapolate from laboratory experiments on single species to the environmental situation where many species may be present and where the stresses to which the organism is subjected are different.

Although a properly designed and carried out bioassay provides information on the relative toxicities of various substances, it generally does not provide any information regarding the effects on behaviour, reproduction, migration and other long-term sublethal effects that may influence the survival of a species within an ecosystem.

Despite their shortcomings, however, there is no ready alternative to bioassay tests, and if properly carried out using a standardized method or methods which are validly comparable, toxicity tests yield meaningful information regarding relative lethal effects. One of the standardizations that is possible is the use of standard animals and Reish (1973) advocates the use of a laboratory-bred population of test animals so that specimens are always available when required, there is no need for a conditioning or acclimation phase, they do not deplete a limited wild population, their diet is known and controlled, and specimens could be transported to other laboratories making cooperative studies and intercomparison experiments. However, cultured organisms are so adapted to life in the laboratory that they can hardly be compared with a natural population because of genetic drift from the wild condition with time. More research needs to be done to determine which is actually the better type of organism although a compromise between wild and cultured organisms is probably the most practical. Nadeau and Roush (1973) consider that ecologically meaningful data can only be obtained by controlled field experiments and point out the advantages of salt marsh communities as microcosms for studying the effects of pollutants on marine ecosystems. Notini and Hagström (1974) have used a physical littoral model system as a link between the field and laboratory studies. The main floral and faunal elements of the Baltic Sea littoral community were introduced to outdoor pools provided with a constant flow-through of water. Oil spills were simulated in the pools. The effects on the model community were studied with regard to the fate of the oil and environmental factors. The choice of a standard animal would allow comparability between laboratories and methods. However, a standard animal would not always be suitable for evaluation of particular discharge effects in particular localities.

Although there are several methods of determining toxicity, there are very few cases in the literature where the methods have been adequately described. The work by Swedmark, Granmo and Kollberg (1973) in which a 96-hour flow-through system was used to determine the effects of oil dispersants and oil emulsions on a variety of fish, bivalves and decapod crustaceans is the type of reports which are needed. The description of experimental techniques and equipment was remarkably complete making it possible to reproduce the experiments and to compare the results with anyone else who has the foresight to carry out work of similar quality. In addition, observations on sublethal effects were made at the same time as the acute toxicity work. Differences in bioassay technique and the condition of the test animals may account for the differences between the LC 50 values of Swedmark, Granmo and Kollberg (1973), those of Portmann (1970), and Sprague and Carson (1970). The continuous flow system presumably eliminates several of the disadvantages of static systems and gives more accurate results.

LaRoche, Eisler and Tarzwell (1970), for example, used a 96-h static test but did not consider the changes that might have occurred in the oil mixtures during that time. Mummichogs (Fundulus) and worms (Nereis), and grass shrimps (Palaemonetes vulgaris) were chosen because of their local abundance, wide distribution, ease of collection, adaptability to laboratory conditions and representation of taxonomically discrete phyla with divergent ecological habitat. Each species was acclimated at $20^{\circ}C$ under reduced light for 10-14 days in synthetic sea water. Groups of organisms showing $\geq$ 20 percent mortality during the first 48 hours of acclimation or $>$ 5 percent after 48 hours were discarded. The animals were fed twice daily but not for 48 hours prior to or during the experiment. The physiological parameters that were maintained during acclimation included temperature, $19.5 \pm 0.5^{\circ}C$; pH, 8.0 ± 0.1; salinity, 20 ± 0.5 ppt and dissolved oxygen ≥ 4.0 mg/l, conditions which are naturally encountered by test species during most of the year. For these tests 4-litre glass jars with screw caps containing 2 litres of fresh synthetic sea water and measured amounts of test material were used with a minimum of six concentrations and five replicates for each concentration. After addition of the test material the jars were tightly capped and shaken for five minutes to ensure reproducible results. Homogenization of a crude oil-water mixture was found to produce a product which was five times more toxic to mummichogs than identical samples that were merely agitated. Dead animals were recorded and removed daily. After 96 hours the assay was terminated and TL 50 values were determined. Values for TL 100 (highest concentration tested with 100 percent survival) and TL_o at 24, 48 and 96 hours were determined. The TL 25, TL 50, TL 75 were estimated for 24-, 48-, and 96-hour periods (Table 5-8).

The work of Pickering and Henderson (1966) exemplifies some of the shortcomings of bioassays. They applied a 96-hour static bioassay test on fathead minnow, bluegills, goldfish and guppies. Their data cannot be compared with that of other authors since several of the test components (all important petrochemicals) had low water solubility, they were homogenized with water. The TLM values were calculated in terms of the initial concentration of toxicant, and since several of the compounds used are very volatile while others are susceptible to biological breakdown, this forms a major source of uncertainty in evaluating their data (Table 5-8).

TABLE 5-8

Some static bioassay data

Organism	Toxicant	Mean Tolerance Level			Reference
		24 hr mg/l	48 hr mg/l	96 hr mg/l	
Fathead Minnow	benzene	35.6	35.1	33.5	Pickering and Henderson (1966)
Bluegill	benzene	22.5	22.5	22.5	"
Goldfish	benzene	34.4	34.4	34.4	"
Guppy	benzene	36.6	36.6	36.6	"
Fathead Minnow	chlorobenzene	29.1	29.1	29.1	"
Bluegill	chlorobenzene	24.0	24.0	24.0	"
Goldfish	chlorobenzene	73.0	56.0	81.8	"
Guppy	chlorobenzene	45.5	45.5	45.5	"
Fathead Minnow	o-chlorophenol	22.0	19.1	11.6	"
Bluegill	o-chlorophenol	11.3	10.6	10.0	"
Goldfish	o-chlorophenol	14.5	12.4	12.4	"
Guppy	o-chlorophenol	22.2	20.8	20.2	"
Fathead Minnow	cyclohexane	35.1	35.1	31.7	"
Bluegill	cyclohexane	42.3	40.0	34.7	"
Goldfish	cyclohexane	42.3	42.3	42.2	"
Guppy	cyclohexane	57.5	57.7	57.7	"

Table 5-8 continued

Organism	Toxicant	Mean Tolerance Level			Reference
		24 hr mg/l	48 hr mg/l	96 hr mg/l	
Fathead Minnow	isoprene	86.5	80.5	86.5	Pickering and Henderson (1966)
Bluegill	isoprene	42.5	42.5	42.6	"
Goldfish	isoprene	180.0	180.0	180.0	"
Guppy	isoprene	240.0	240.0	240.0	"
Fathead Minnow	toluene	46.3	46.3	34.3	"
Bluegill	toluene	24.0	24.0	24.0	"
Goldfish	toluene	57.9	57.8	57.7	"
Guppy	toluene	62.8	60.9	59.3	"
Sandworm					
Nereis virens	crude oil	–	–	0.1 ml/l	LaRoche, Eisler and Tarzwell (1970)
Mummichog	crude oil	–	–	8.2 ml/l	"
Grass Shrimp					
Palaemonetes vulgaris	crude oil	–	–	<0.050 ml/l	"

Table 5-9 illustrates a wide variety of experimental results recorded by various authors and the impossibility of intercomparing either the toxicities of various oils because of the various techniques and organisms used. From these few examples it is apparent that a major shortcoming of bioassay methods lies in variance in technique and incomparability of results amongst researchers.

Once the experimental work has been completed the treatment of the data should also follow a standard type of pattern and in this there is more uniformity. Most current authors base their calculations on the theories of bioassays as expressed by Sprague, J.B. (1969, 1971) who considers that since the least and most resistant individuals in a group show a greater variability in response than those near the median, it is the average response which is desired. Since concentration within a test tank is fixed, exposure time is the variable stimulus while percentage mortality is the response. The effect on each animal is all or none (i.e. quantal) so that the test in each container is essentially a quantal bioassay. When the cumulated percentage mortality from a given test is plotted against exposure time, a skewed sigmoid curve is produced. The skew is eliminated by using a log scale on the time axis. The sigmoid nature results from the variable individual response and that can usually be straightened by plotting percentage response on a probability (probit) scale instead of an arithmetic scale. Probits express mortality in terms of standard deviations above and below the near response with the value 5.0 added for the convenience of eliminating negative numbers (Bliss and Cattell, 1943). Plotting results on log-probit paper also makes it apparent whether there are two or more different modes of toxic action indicated by changes of slope or grouping of lines.

There are two approaches to constructing a toxicity curve; either median lethal times may be estimated for a series of concentrations, or median lethal concentrations may be estimated for a series of inspection times. If response is reasonably regular and mortality checks are regular, either method should give essentially the same curve. However, some workers consider that, just as the choice of test method and organism should relate to the particular situation under investigation, the method of treatment of the data obtained should be related to the methods of observation and measurement used. No standardization is required if both the test method and the method of data treatment are clearly and fully stated.

The ultimate objective of bioassays of oils with marine organisms is to develop the ability to predict levels of oils in sea water that are safe to the survival, growth, reproduction, the general well-being of the species. The 'safe level', the concentration of a pollutant which does not have an adverse sublethal or chronic effect on fish is a measure empirically determined from the median lethal concentration. The LC 50 is multiplied by an application factor, such as 0.1, to obtain a concentration which presumably has no sublethal or chronic effects. The value of the application factor is assigned on the basis of the judgement of scientists. Safe concentrations are generally expressed as fractions of the acute lethal level called the toxic unit. Most estimates of safe levels are between the limits of 0.1 to 0.4 toxic units (0.1-0.4 x LC 50). Estimates for some pollutants, especially cumulative and persistent poisons with important chronic or sublethal effects, are as low as 0.01 toxic units. Sprague (1971, 1971a) feels that the higher values approaching 0.4 come rather close to harmful levels.

The bioassay has traditionally employed lethality as the measurement of effect but for algal cells, for instance, lethality is not the only useful end-point. Since photosynthesis is the most vital and most characteristic function of plant cells, it seems a valid parameter of effect. There have been a number of publications on standard methods measuring photosynthetic rates using the C^{14} technique. Strickland (1960) and Ryther (1954) compared the C^{14} and oxygen production methods for measuring photosynthesis and found the former 50 to 100 times more sensitive. The <u>in situ</u> algal bioassay utilizing the C^{14} techniques represents a rapid and sensitive way to assess the effects of discharged oil and chemical treatment on phytoplankton. It is especially suitable for assessing the effect of oil in the open sea. Photosynthetic rate of samples collected just below the surface of an oil contaminated area can be compared with photosynthetic rates of samples from adjacent uncontaminated water. However, the patchy distribution of algal cells in the sea necessitates the use of several replicate samples. A minimum of four light and one dark are recommended. The incubation time varies with the estimated productivity of the area; for example, eutrophic coastal regions require 2-6 hours incubation

TABLE 5-9

Variability of toxicity data for a variety of marine organisms and the difficulty of intercomparison (from Nelson-Smith, 1972a) (Note: the units are not common, but as authors have reported)

Organism	Toxicant	Lethal Level (unless otherwise specified)	Time	Reference
Diatoms *Ditylum brightwellii* *Coscinodiscus granii* *Chaetoceros ourvisetus*	Kerosene and fuel oil	100 µl/l	24 hr	Mironov and Lanskaya (1967)
Marine Micro-algae	Kerosene	188 µl/l	several days	Aubert, Chaira and Malara (1969)
Copepods *Acartia clausii* *Oithona nana*	Oil	1 µl/l	3-4 days	Mironov (1968)
Acartia Catanus	Oil	100 µl/l	24 hr	Mironov (1972a)
Herring eggs and fry	Dissolved hydrocarbons from crude oil	5 ppm	3-4 days	Kühnhold (1969)
Rhombus maeoticus eggs	Crude oil Fuel oil	0.1 µl/l	50% died in 2 days	Mironov (1972a)
		Mean Tolerance Level		
Shad	Gasoline Diesel oil Bunker oil	91 mg/l 167 mg/l 1952 mg/l	48 hr 48 hr 96 hr	Tagatz (1961)
Trout	Gasoline Diesel oil Heating Oil Lubricating oil	60-180 mg/l 300-4000 mg/l 1000-150,000 mg/l 3000-180,000 mg/l		Zahner (1962)

Table 5-9 continued

Organism	Toxicant	Mean Tolerance Level	Time	Reference
Sunfish *Lepomis macrochirus*	Kerosene	2820-2990 ppm		Turnbull, Demara and Weston (1954)
Toadfish *Opsanus tau*	Crude oil	10 ml/l 1 ml/l	1 day 10 days	Chipman and Galtsoff (1949)
Mummichog	Crude oil	LC50 = 18 ml/l 16.5 ml/l	24 hr 48 or 96 hr	LaRoche, Eisler and Tarzwell (1970)
Sunfish	Naphthenic acids	4.6-7.2 ppm	96 hr	Cairns and Scheier (1962)
Minnows	Naphthenic acids	5.0 ppm	72 hr	U.S. Public Health Service (1939)
Minnow *Phoxinus phoxinus*	Naphthenic acid leached from crude	29-36 ppm	48 hr	Malacea, Cure and Weiner (1964)
Bitterling *Rhodeus sericeus*	Naphthenic acid leached from crude	92-118 ppm	24 hr	Malacea, Cure and Weiner (1964)
Oyster *Ostrea edulis* Clam *Crossostrea angulata* Mussel *Mytilus galloprovincialis*	Fuel oil	30-40 ml/l	<7 days	Leenhardt (1925)

Table 5-9 continued

Organism	Toxicant	Mean Tolerance Level	Time	Reference
Black Sea Topshell *Gibbula divaricata*	Various Russian crude oils	1.0 ml/l	6-18 days	Mironov (1967, 1970)
Needleshell *Brittium reticulatum*	Crude oils	1.0 ml/l	6-18 days	Mironov (1967, 1972a)
Rissoa euxinica	Crude oils	1.0 ml/l	3-5 days	Mironov (1967, 1972a)
Pond Snail *Physa heterostropha*	Naphthenic acid	LC50 6.6-15.6 ppm	?	Cairns and Scheier (1962)
Hydroid *Tubularia crocea*	Crude oil	50 ml/l	24 hr	Chipman and Galtsoff (1949)

Organism	Toxicant	Concentration Applied	Effects	Reference
Kelp *Macrocystis*	Diesel oil	0.1%	Inhibited photosynthesis in young blades	North, Neushul and Clendenning (1964)
Macrocystis	Fuel oil	10-100 ppm	5% inactivation of photosynthesis in 4 days	Clendenning and North (1960)
March grass *Puccinellia Fastuca rubra*	Low boiling fractions of Kuwait crude	40% by volume sprayed on plants	Penetration commenced in 2 minutes	Baker (1971)

whereas oligotrophic areas in the open sea may need 6-10 hours incubation. For routine laboratory bioassays, purified cultures of marine phytoplankters are recommended, and for the prediction of environmental effects, cultures of indigenous algal species isolated from the region of concern are recommended (Strand et al., 1971).

Another algal bioassay method involves using the concentration of Chlorophyll-a as a measurement of productivity. The methods have been used extensively since 1955 (Creitz and Richards in Strickland and Parsons, 1965). While a reduction in Chlorophyl-a may not always represent a similar reduction in productivity it does represent a loss in the potential productivity of the system. During periods of low temperature chlorophyll concentration may not be the factor limiting photosynthesis. Instead, the amount and activity rate of cellular enzymes may be limiting. However, as environmental conditions change and chlorophyll becomes limiting any reduction in amount will be reflected in lowered productivity. Similarly, any behavioural or physiological response leading to known mortality rates from field experiments could be used as an end-point in laboratory bioassays.

5.4.2 Effects of oil on early stages of growth of marine organisms

Material was taken from a plankton net sample for experiments concerning the influence of oil pollution on the larvae stages of benthic and planktonic crustacea. Nauplii, which were most active and did not have any noticeable damages, were used in the experiment. During the course of the work, the experimental and control organisms were fed on pure cultures of Exuviaella cordata and Prorocentrum micans (Mironov, 1969).

As in the experiments with mature samples, rapid death of the Nauplii occurred at an oil-product content of 0.1 ml/l. In smaller oil concentrations, the death rate of the organisms was approximately twice that of the control. It is interesting to note that the effect of bunker oil concentrations of 0.01 and 0.001 ml/l were approximately the same. Death of Acartia clausi occurred in the time limit of 4-4½ days and in 3 days for Oithona nana.

Mironov (1967, 1969) conducted experiments on a series of Black Sea fishes, including some commercially valuable species such as flounder-turbot, anchovy, and also ruff and sea-parrot, in order to observe the survival of fertilized eggs. The organisms were chosen in such a way as to follow the growing pelagic and bottom eggs under conditions of oil pollution. Death of the developing fish eggs and/or the appearance of non-viable larvae occurred in all concentrations of crude, bunker and diesel oil products, ranging from 0.1 to 0.00001 ml/l. Similar data is presented by Krishtan (1968) concerning crude oil from different oil fields.

Lower concentrations of oil products were not employed in the present work, but it may be supposed that they, too, will be damaging to these species. The consequences of such action may not be manifested in the immediate death of the growing eggs or prolarvae during the course of the experiment, but may show up at later stages of ontogenesis. It is possible that these consequences may become apparent in future generations. However, a definitive answer can be arrived at only by conducting specific experiments, and these types of experiments are difficult to set up. In order to conduct these experiments with fishes, tens and hundreds of years will be needed. Thus, for example, in order to obtain similar information for beluga sturgeon, 250-300 years would be needed, for sturgeon 150-300 years, for carp 50-70 years (Stroganov and Kolosova, 1968). Therefore, it becomes necessary to seek out organisms with a short growth cycle to make prognoses possible, with a certain amount of caution for the far reaching consequences of hydrocarbon contamination for not only the specimens but for other hydrobionts as well.

Mironov (1970) has conducted such observations on two diatomic algae, Ditylum brightwellii and Coscinodiscus granii. In contrast to the above experiments, the cells of the species being studied were placed individually in dishes, where the required concentration of oil or oil products in marine water was prepared. A control was set up alongside. Based on the earlier data, the concentration of oil products was chosen at 0.01 ml/l, which is frequently found under natural conditions and begins to have a harmful effect on the selected species of phytoplankton. Some species of oil and oil products had no effect upon C. granii.

After the division of the mother cell, one of the daughter cells was removed and placed in clean marine water. The other was left in the original concentration of oil products until the next division. Observations were conducted through ten generations. The marine water with the assigned oil product concentration in the experiment and control was changed daily. Every experiment was set up ten times.

Considerable difficulties were encountered in setting up the experiment. In several cases, both in the experiment and in the control, cell division was either entirely absent or the cells died after one or two divisions. This occurred in spite of the fact that the original cells, those which were taken from culture and those isolated from the sea, were without visible pathological changes. It therefore became necessary to conduct the experiments many times in order to observe cells which had ten or more generations.

Addition of 0.01 ml/l concentration of crude, bunker and diesel oil to marine water leads to a decrease in the number of generations of D. brightwellii. When 30 percent of the control cells gave ten generations, then the cells in the crude oil and bunker oil gave 7 and 6 generations respectively. Beginning in the second generation, the amount of divided cells decreased to 30-40 percent. Cell division proceeded slightly differently in the presence of diesel oil. The number of divided cells in the experiment and control were initially equal, and towards the end of the experiment, i.e., by the tenth generation, it was twice as much in diesel oil, as compared with the control. Apparently, the differences in the chemical composition of the oil products play an important role.

Previous studies indicated the possibility of an increase in the rate of division of microscopic algae under the influence of small concentrations of oil, but how useful such a "stimulation" is for the species or for the planktonic community as a whole, remains hard to determine.

The division of the cells of C. granii proceeded along the same lines as the first species. However, since C. granii is more highly resistant, some of the cells continued to divide in marine water containing oil and bunker oil until the tenth generation.

In addition to the observations on the number of generations, calculations of the maximum number of cells formed as a result of the division of daughter cells placed in clean water was also made. Due to the higher rate of division, the number of cells of D. brightwellii was about ten times greater than C. granii within the same period of time. No correlation was noticed between the intensity of multiplication of daughter cells and the amount of time that the mother cells were in polluted water. On occasion, in the experiment and in the control, daughter cells of the first generation placed in clean marine water totally lacked division, but those daughter cells which did divide had the highest number in the second and third generations. On the contrary, in some cases an intensive division of the daughter cells of the first generation was observed, followed by an absence of division in some succeeding generation, in turn followed by a sharp rise in the number of divisions of those daughter cells which continued to divide.

The results show that bunker oil has a high toxicity as compared to crude oil and diesel oil. as measured by the growth of the daughter cells. C. granii gave the least divisions in bunker oil but in D. brightwellii the total number of cells was counted, and it is therefore difficult to judge whether the greater numbers occurred as a result of a greater proportional division or only a fraction of them. Judging from the data, however, the latter is the more probable case.

Thus, pollution of marine water by crude oil and bunker oil leads to a decrease in the number of generations of D. brightwellii and C. granii. In the given case, we were unable to determine any clear peculiarities in the future growth of daughter cells in clean marine water after contact of the mother cells with oil products at a concentration of 0.01 ml/l.

These observations on the productivity of phytoplankton are supplemented with the investigations carried out on a molecular level. Experiments concerning the influence of oil on the nucleic acids of three species: the green algae Ulva lactura, and the red algae Grateloupia dichotoma and Polysiphonia opaca were performed in aquaria where they were exposed to three different concentrations of Romashkinskaya oil (0.1, 1.0 and 10 ml/l) (Davavin, Mironov and Tsimbal, 1975). The oil was emulsified in sea water by stirring at 6 000 rev/min for 20 min. After 24 hours exposure, the algae were incubated for 24 hours in unpolluted (oil free) sea water and ^{14}C labelled $Na_2 CO_3$ solution. The algae were then washed until ^{14}C activity disappeared from the washings. The nucleic acid content of the algae was estimated by the method of Spirin and Belosersky (1956) and Nechaeva (1966).

DNA separation to determine the degree of polymerization was carried out by the Konarev method (Konarev et al., 1966) for labile and stable DNA, and by the Sulimova and Slyusarenko (1972) method for total DNA. Polymeric spectra were determined by the methods of ion exchange chromatography (Kuzin et al., 1960). ^{14}C activity was measured by an SBT-13 meter and specific activity expressed in counts/min per mg of DNA or RNA.

Metabolic heterogenecity has been demonstrated in various test animals such as mice (Morin et al., 1957) and sprouts or embryos of pea seeds (Tokarskaya, 1967). The aim of this study was to determine the differences in carbon-14 activity between labile and stable DNA as well as within various size fractions of each. For this purpose, fractions of labile and stable DNA of Ulva obtained by the Konarev method were fractionated on columns with DEAS-Sephadex-25.

There is no substantial difference between the degree of polymerization of stable and labile DNA of Ulva. The lower content of the fourth (high molecular weight) fraction and the higher content of low polymeric fragments of stable DNA may be due to the method of isolation since stable DNA is exposed to the more prolonged treatment which may result in breakdown of the larger DNA polymers.

The activity distribution in the fractions is variable. In comparison with stable DNA, much more activity is associated with the first and third fractions of labile DNA. The greatest activity (63.9 percent) of stable DNA occurs in the second fraction, but this fraction has the lowest specific activity in labile DNA, for which much of the labile is included in the first fraction. The third fraction also shows high specific activity for labile DNA. The high polymeric fractions for both labile and stable DNA have the least specific activity. It was thought important to study the influence of oil on this part of the molecule, particularly because the molecules of labile DNA undergo substantial changes through ionizing radiation (Ibragimov et al., 1971). It seemed likely that these DNA areas are the least stable to oil. Changes in the degree of polymerization of DNA may be explained by an accumulation of individual chain breaks, resulting in double breaks. Therefore the increase of low polymer fractions indicates the process of DNA depolymerization. Changes in the degrees of polymerization of total DNA in Ulva do not occur even at very high concentrations of oil indicating the considerable resistance of this alga to oil pollution.

The experimental procedure was slightly modified for the red algae. In experiments with total DNA of Ulva, four fractions were obtained with eluents 0.5 M NaCl, 1 M NaCl, 1 N NH_3 in 2 M NaCl and 0.5 N NaOH. Because of the insignificance of the first fraction, it was omitted in the case of the red algae.

The fraction of labile DNA of Grateloupia is the most sensitive, resulting in decreasing content of the second and third fractions, as compared to the decrease of the high polymeric fraction of Polysiphonia. This indicates also that the oil influences lesser polymerized fractions of labile DNA of Grateloupia. Stable DNA is more resistant to oil, but the nature of the changes are comparable to those of labile DNA.

Fractions of labile and stable DNA have their own functions. Stable DNA transmits genetic information to further cellular generations, and labile DNA is characterized by metabolic activity, while RNA synthesis occurs on the labile DNA, determining the synthesis of specific proteins. Labile DNA was found to be more sensitive to oil pollution; this may result in the destruction of the mechanism of the expression of heredity, the synthesis of associated proteins, and, in turn, the death of the organism.

The effect of length of exposure and oil concentration on the incorporation of carbon-14 activity on Ulva indicated that increasing concentration causes a decrease in specific activity: at 10 ml/l it is reduced to 3.7 percent of that for the control after 2 days, and 2.6 percent after 3 days. It should be noted that specific activity in the control and at all oil concentrations increases sharply at 2 days and decreases in the third day, though in the control and oil at 0.1 ml/l concentration, it remains above the level of the first day.

Specific activity of RNA of Ulva increases at 0.1 ml/l concentration after 24 hours, but higher concentrations cause a sharp decrease. On later days, the situation is comparable to that for DNA.

Thus there appears to be nearly complete inhibition of biosynthesis of nucleic acids in Ulva at 10 ml/l concentrations of oil after 48-72 hours.

Experiments with the two red algae and Ulva, showed that oil does not selectively influence some nucleic acids, there is simultaneous inhibition of biosynthesis of both DNA and RNA (Davavin, Mironov and Tsimbal, 1975). The greatest reduction of specific activity of red algae occurs after 24 hours. RNA biosynthesis of both species being inhibited to a greater extent than that of DNA during the first day. Experimental specific activities of DNA and RNA are much greater for Polysiphonia than Grateloupia. During the following 2 days the specific activity of RNA and DNA increased. It should be noted that RNA synthesis of Grateloupia is more depressed than DNA synthesis after 2 days. (RNA reduction in Polysiphonia occurs at a higher rate. By the end of the test, the same reduction ratio of specific activity of RNA remains. Complete reduction of RNA biosynthesis of Grateloupia (22.4 percent more than the control) occurs and there is a slight decrease in the specific activity of DNA. These variations may be explained by differences in the systematic positions of these algae and therefore to individual peculiarities of these organisms).

At the end of 3 days exposure to 1 ml/l concentration of oil in water, quantitative determinations were made of nucleic acids. Oil pollution appears to decrease the RNA and DNA content of red algae by inhibition of their biosynthesis. The reduction of both nucleic acid contents of Polysiphonia appears to be equal and represents 75 percent of that in the control. The DNA content of Grateloupia is greater and there is a 6 percent reduction of RNA content as compared with the control.

In view of the data on ^{14}C laballed carbonate incorporation into nucleic acids, it can be concluded that the decrease in RNA and DNA content is due to the inhibition of their biosynthesis.

Thus we may generally conclude that the influence of oil on these algae results in a modification in the degree of polymerization of deoxyribonucleic acids, and inhibition of biosynthesis of DNA and RNA. Specific changes, however, depend on the specific characteristics of the organisms in question.

5.4.3 Habitat alteration and destruction

Another aspect of oil in the marine environment is the adverse effect it may have on marine habitats. In many areas habitats have been severely damaged and, in some cases, destroyed for certain forms of marine life. Such destruction may result from the effects of a major spill or from chronic exposures such as occur in the vicinity of refineries, sewer outfalls, etc.

Kuwait crude oil from the TORREY CANYON and the aromatic dispersants used in its cleanup affected the marine habitat along the coast of Cornwall to an extent that species abundances were significantly changed; for example, a prolific growth of algae resulted when grazing and browsing animals were killed thereby altering the habitat to such an extent that settlement of the area by the usual fauna of mussels and barnacles was altered for a time (Spooner, 1969). In due course, however, the habitat recovered. Similarly, following the TAMPICO MARU spill of diesel oil, habitats were modified to the extent that even after 10 years, complete recolonization of some species that were abundant before the spill had not taken place (North, Neushul and Clendenning, 1964).

In other incidents, for example the ARROW spill, oil becomes buried in the sediments where it alters the habitat of burrowing animals (Thomas, 1973). In this incident, oil was often found buried to a depth of 5 feet in shingle beaches (Owens, 1971). Beaches are constantly changing in response to a variety of forces which vary with the season, tidal cycle, weather conditions, etc., as manifested by seasonal changes in slope, grain size distribution, etc. The presence of oil must affect these processes and in cases of heavy contamination and insufficient wave action the beach may become 'paralyzed' (Drapeau, 1970). On high energy beaches, on the other hand, this movement provides a rather effective self-cleaning process.

Finally, restoration of a contaminated shoreline may drastically alter the habitat. For example, the use of dispersants, burning, absorption, stabilization or physical removal must all affect the area as a habitat for marine life regardless of whether it is a marsh, mud flat, sand, shingle or rocky shoreline.

Any offshore structure which remains in position long enough to become fouled by epi-flora and fauna is itself an artificial littoral and sublittoral environment which attracts to it at least a proportion of the attached and free-swimming marine organisms associated with steep rocky shores. Similarly, the submerged parts of rigs, platforms and exposed pipelines are all potential habitats for the range of organisms normally associated with a rocky sea bed, wrecks, etc.

Chapter 6: HUMAN EFFECTS FROM OIL DISCHARGES

6.1 Carcinogenic

Definitions: Evidence relating to oil discharges of:

(i) Accumulation of compounds known to be carcinogenic
(ii) transfer of carcinogens up the food chain
(iii) induction of carcinomas in marine biota
(iv) the significance to man of ingestion of cancerous tissue
(v) the significance of increased contents of carcinomas in marine produce to the induction of carcinomas in man as a consumer.

6.1.1 Problem - General

Oils contain carcinogens such as the polynuclear aromatic hydrocarbons (PNAH). It has been alleged that these PNAH's will persist in tissues of marine biota exposed to oil discharges and will be accumulated up the food chain, reaching concentrations presenting a hazard of cancer-induction in man as a consumer. It has also been alleged that oil discharges have induced carcinomas in marine produce.

6.1.1.1 Occurrence in oils

Crude and refined oils contain many compounds of known or suspected carcinogenicity to mammals or man, including nitrogen and sulfur bearing heterocyclic compounds, methyl derivatives of polycyclic and heterocyclic compounds, and the polynuclear aromatic hydrocarbons. Weathered and partially degraded oils may further contain oxidation products of potential carcinogenicity (Feldmann, 1973). Comprehensive surveys are given by Medical Research Council (1968) and IARC (1973), and technical information was received from G.J. Brockis of BP. The literature on oil-derived carcinogens other than the PNAH's is relatively sparse, and it is on the latter that public, political and scientific attention has been focussed. Estimates of the relative potency of the carcinogenic PNAH's vary, but perhaps the most documented from crude oils are 3,4-benz-pyrene (BaP) and 1,2 benz-anthracene (BaAnth) (Carruthers et al., 1961; Graf and Winter, 1968). Since it is thought that a particular molecular configuration renders a compound potentially carcinogenic, few of the many other PNAH's occurring at detectable levels in mineral oils are, at present, suspected of carcinogenicity. However, they may act as accelerators or cocarcinogens in promoting the activity of carcinogenic constituents, as may constituents from the other oil fractions mentioned above, and even some aliphatic compounds such as dodecane (quoted in IARC, 1973). Thus, the established tumorigenicity to mammals of whole mineral oils or of particular extracts and fractions may greatly exceed that to be expected from their content of specific carcinogenic compounds (Hueper, 1963; Medical Research Council, 1968).

6.1.1.2 Levels of PNAH's in oils

There is little data available on the concentrations of known or suspected carcinogenic PNAH's in crude and refined oils, and such as exists may have been misrepresented. It must be remembered that the accuracy and specificity of the various analytical methods differ widely, and that some figures are estimates based on the tumorigenic potency of a sample known to contain a given compound relative to solutions of that pure carcinogen. Pancirov and Brown (1975) found 1 ppm BaP and 1.7 ppm BaAnth in a South Louisiana crude oil, and 2.8 ppm BaP and 2.3 ppm BaAnth in a Kuwait crude oil, while several less strongly carcinogenic PNAH's were present in quantities from less than 0.5 to 17.5 ppm.

Graf and Winter (1968) reported 40, 132 and 166 µg/100 ml BaP in crude oils from the Persian Gulf, Libya and Venezuela, respectively. Figures of 450-1800 g/t BaP in crude oils (Blumer, 1971a) have been widely quoted, but these are miscalculations from Graf and Winter's figures above, which are more nearly equivalent to 450-1800 mg/t. Sullivan (1974) quotes Dooley et al. (1973) that a Gach Saran crude oil contained 3.54% PNAH's from the fraction boiling between 370° and 535°C, but some 75% of this consisted of sulfur compounds and only traces of 4-6 condensed aromatic ring compounds were reported. Recent analyses by the U.K. Laboratory of the Government Chemist found 100-200 µg/l BaP and 200-300 µg/l BaAnth in a Kuwait crude oil. Lower boiling point direct distillation products may contain lower concentrations of PNAH's than crude oils, while catalytic cracking of distillates increases the PNAH content.

Pancirov and Brown (1975) found 0.6 ppm BaP, 1.2 ppm BaAnth, and 44 ppm BaP, 90 ppm BaAnth in a No.2 heating oil and a Bunker C fuel oil, respectively. Masek (1961) reported that a 22% residue from a Soviet crude oil contained 0.001% BaP by straight distillation, but 0.1% BaP after cracking. The U.K. Government Chemist found 400 µg/l BaP in a high aromatic heating oil (63% total aromatics), but even a sample of commercial grade hexane has been reported to contain 23 µg/kg BaP and 280 µg/kg BaAnth (Lijinsky and Raha, 1961). Graf and Winter (1968) found 2.6 µg/100 ml BaP in both a diesel and a heating oil. Catchpole, MacMillan and Powell (1971) found 14-26 ppb BaP in an unrefined Iraq gas oil and 0.001-0.6 mg/kg BaP in lubricant distillates from various crude oils. They used a similar analytical technique to that of Graf and Winter above, but found only a 20-50% efficiency in extraction and analysis of spiked samples. A recent Stichting Concawe Report (1974) gives levels for three automobile gasolines of 34-44% total aromatics as 0.005-1.29 ppm BaP and 0.01-2.05 ppm BaAnth while a total of 13 PNAH compounds (not all suspected carcinogens) varied from 0.16-37.4 ppm in the three fuels. Sullivan (1974) quotes analyses of commercial grade gasolines by the Esso Research and Engineering Company (1970) which found 212-3120 µg/kg BaP. These latter figures illustrate the wide variation in reported PNAH content within even a single class of petroleum product. Petroleum bitumens from Soviet crude oils have a reported BaP content of only 0.00006-0.0272% (Ianysheva, 1963), while petroleum residues from Western fluid catalysis-cracking processing gave residues containing 0.4% BaP, 0.4% BaAnth and alkyl derivatives, 1.5% phenanthrene, and other identified PNAH's (Tye et al., 1966). Pancirov and Brown (1975) quote levels of 3-5 ppm BaP in petroleum bitumens and levels up to 27 ppm BaP and 35 ppm BaAnth in asphalts.

6.1.1.3 Levels in other products and wastes

These concentrations should be compared with data available on the concentrations of PNAH's in some other products which may enter the marine environment. Coal tar and coal tar pitch have been reported to contain 1.25-1.5% BaP and 0.70-1.25% BaAnth. Creosote may contain 2.75 g/l BaAnth, but is estimated to contain less than 100 mg/l BaP. Graf and Winter (1968) found 2.6 µg/100 ml BaP in unused motor lubricating oils, but 580 µg/100 ml in a sump oil after 2 200 km of motoring. However, Pancirov and Brown (1975) quote a level of 0.28 ppm BaP in unused motor oil. Palmork, Wilhelmsen and Neppelberg (1973) have recently found high levels of PNAH's in the scrubber sludges from aluminium smelters using Soderberg electrodes, including pyrene, anthracene, phenanthrene, methylphenanthrene and fluoranthrene. Concentrations in the dried sludge were 240 mg/kg anthracene and phenanthrene, 250 mg/kg fluoranthene, 220 mg/kg pyrene.

6.1.1.4 Relative contribution of oil-derived PNAH's

Concentrations of PNAH's in crude and refined oils must be seen in the perspective of the probable quantities contributed to the marine environment. For instance, from the figures above, the loss of 10^6 t of crude oil to the sea would contribute no more BaP than the loss of 10^2 t of coal tar pitch. However, without reliable data on the quantities of other products entering the sea, their relative contribution of PNAH's cannot be determined. Estimates of the quantity of mineral oils entering the sea from petroleum production, transport and refining, vary from 10^6 t.p.a. (Jeffery, 1972) to 10^7 t.p.a. (Blumer et al., 1972).

Natural oil seeps have been estimated to contribute 0.6×10^6 t.p.a. (Wilson et al., 1974). Duce, Quinn and Wade (1974) estimate a marine atmospheric burden of 22×10^6 t of hydrocarbons in the range nC_3-nC_{32} for the northern hemisphere, with a residence time of 0.5-2.3 years. The input to the sea of mineral oils and biogenic hydrocarbons from terrestrial sources has been largely neglected, but urban sewage and sewage sludges may contain considerable quantities of PNAH's (Harrison, Perry and Wellings, 1975), e.g., 31.4 mg/m^3 BaAnth in sewage water (IARC, 1973) and of mineral oil, e.g., 9 000 ppm by wet weight in undigested sludge, as has been reported in the U.K. (Bennett, Dee and Harkness, 1973). Hydrocarbon inputs from nontidal waters may also be appreciable (Cooper, Harris and Thompson, 1974; Mallet, 1965, 1965a). Hites and Biemann (1972) found a correlation between increased naphthalene contents of the Charles River, Boston, and the quantities of storm-water runoff from its drainage area. Farrington and Quinn (1973, 1973a) estimated that in Narragansett Bay, over the three years 1968-71, while 200 t of oil were lost in major spills adjacent to the Bay and 4.3 t actually entered the Bay from minor spills, 435-2190 t entered the Bay in sewage discharges. Calculations of the relative inputs of mineral oils from some industrialized coastal areas of the U.K. in 1972 indicated that sewage contributed 10-20 times the volume from refineries and that these in turn contributed, on the average, 20-50 times that from marine spillages. Stich (1975) has suggested that motor boat marinas and creosoted wharf piles represent major sources of PNAH to the (coastal) marine environment, having found an average of 570 mg/kg wet weight BaP in the outer 4 mm of creosoted wood from such pilings (Blackman, personal communication).

ZoBell (1971) has compiled evidence for the biosynthesis of PNAH's by marine bacteria and phytoplankton, freshwater algae and terrestrial higher plants. Farrington and Quinn (1973) doubt the ability of marine organisms other than bacteria or yeasts to synthesize such complex mixtures of aromatic hydrocarbons as may be found in marine organisms exposed to oil pollution, and Farrington (Blackman, pers.comm.) reports that Hites has been unable to remove contamination sufficiently to demonstrate any biosynthesis of PNAH by bacteria grown in PNAH-free culture. However, ZoBell's documentation of the presence and levels of PNAH's in river waters, terrestrial soils, humus, plants and vegetable oils leave little doubt that biogenic PNAH's, whether of marine origin or derived by runoff from the land, are a primary source of the PNAH's found in marine organisms and sediments, on a global scale. Oppenheimer (1974) reports that, even in an area of Louisiana waters that has been subjected to pollution from oil-production activities for many years, the pattern of hydrocarbons in water, plankton and benthos appears still to be largely determined by that in the terrestrial debris reaching the sea. ZoBell points out that over the Arctic, North Atlantic and Mediterranean, high levels of PNAH's in samples of marine algae and grasses, invertebrates and fish are not necessarily correlated with oil or other terrestrial pollution. Highest concentrations in plankton (mainly phytoplankton) seem to be correlated with known pollution, but care must be exerzised in interpreting many reports, where few or no precautions have been taken against the contamination of samples with surface films or particulate oil residues (e.g. Lima Zanghi, 1968; Horn, Teal and Backus, 1970). On a local scale, ZoBell notes that bottom-living animals in areas of high productivity and terrestrial pollution are generally reported to have high BaP levels (e.g. Lalou, 1964).

All that it seems reasonable to state from this evidence is that _globally_, the input of PNAH's to the seas from mineral oils, as a result of human activity, can be only a minor proportion of the total input. On a _local_ scale, sites receiving industrial, domestic and shipping wastes, may have significantly increased PNAH levels, but, with the exception of areas suffering major oil spillage, the contribution of PNAH's from shipping is likely to be greatly outweighed by those from other inputs.

6.1.1.5 Levels in marine produce and other foods

The few reported levels of PNAH's, notably BaP and BaAnth, in marine produce from both polluted and unpolluted sites, vary greatly as shown in Table 6-1. This is no doubt due to the differing analytical techniques employed and to the inclusion of varying amounts of sediments, etc., in gut contents of whole tissue samples. In some cases published results

TABLE 6-1

Reported levels of PNAH's

	Relatively Unpolluted		Relatively Polluted	
	BaP	Ba Anth	BaP	Ba Anth
SHELLFISH				
Mytilus	ND-28 (55)	3.8-4.3	(11-750) 0.06-2.1 (tainted)	-
Mercenaria	0.38-1.1	4.4	8.2-16 (tainted)	-
Ostrea	ND-4	5.8	ND-6	-
Crassostrea	ND-1.4	8.7	26	<10
Cardium	ND-5	<2	ND-(25-780) 11.7 (tainted)	6 ND
Pecten	4-6.7	5.3	-	-
Chlamys	(90)	-	4	-
Solen	ND-(12.5)	-	-	-
Tapes	ND	-	(100-550)	-
CRUSTACEA				
Crangon	ND-0.5	0.7	ND-(90)	-
"Crabs"	ND-5	-	(ND-30)	-
FINFISH				
Pleuronectes	ND	-	<0.05 (tainted)	<0.05 (tainted)
Gadus	ND-3(15)	-	ND	ND
Clupea	ND-0.1	-	0.4-13	-
Sardina	-	-	(60-65)	-
Scomber	-	-	ND	ND
Solea	2	-	(38)	-
Anguilla	-	-	ND-(30)	-
Mugil	3.5	-	ND	-
Osmerus and Anchovy	-	-	200-1 000	-

These figures should be compared with a compilation of levels in some foodstuffs quoted in IARC (1973) and from the U.K. Laboratory of the Government Chemist, shown in Table 6-2.

TABLE 6-2

Levels in foodstuffs of PNAH's
(figures in µg/kg wet)

BaP		Ba Anth
0.17-0.63	Cooked meats, sausage	0.2-1.1
1.6-4.2	Cooked bacon	-
2.6-11.2 (50.4 recorded)	Charcoal-broiled meats	1.4-31
0.02-14.6	Smoked ham, sausage	0.4-9.6
up to 23 (107 recorded)	Heavily smoked ham (Iceland)	up to 12
0.9	Cooked fish	up to 2.9
0.3-60 (up to 37 in Japan)	Smoked fish	0.02-2.8 (up to 189 in Japan)
0.2-4.1	Cereal grains	0.4-6.8
0.1-4.1	Flour and bread	0.4-6.8
1.8-40.4	Bakers' dry yeast (yeasts grown on mineral oils are lower)	2.9-93.5
3.1	Soya bean	-
0.4-36	Refined vegetable oils, fats	0.8-1.1
0.2-6.8	Margarine, mayonnaise	1.4-29.5
2.8-12.8	Salad	4.6-15.4
0.2	Tomatoes	0.3
7.4	Spinach	16.1
12.6-48.1	Kale (only 10% removed by washing)	43.6-230
0.1-0.5	Apples	-
2-8	Fruits (not apples)	-
0.2-1.5	Dried prunes	-
0.1-4	Roast coffee and solubles	0.5-14.2
up to 15	Malt coffee	up to 43
3.7-21.3	Tea	-
0.04 µg/l	Whisky	0.04-0.08 µg/l
ND	Beer	ND
	Roast peanuts	up to 0.95
ND	Milk	-

do not even indicate a wet or dry weight basis, or may give the contents per organism. Some results have been compiled from: Clark (1971, 1971a), ZoBell (1971), Stich (1975), Cahnman and Karatsune (1957), Greffard and Meury (1967), Mallet (1961), Mallet and Priou (1967), Perdriau (1964) and unpublished results from the U.K. Government Chemist. Figures are in µg/kg wet tissue, and dry weight results have been divided by 5 to give wet weight figures. Figures in brackets are those possibly on a dry weight basis, and ND means Not Detected. Reported levels have been divided into those from sites not known to be polluted and those known to have been subjected to oil or terrestrial pollution likely to contain oil.

It is reported that a standard of 1 µg/kg BaP has been adopted by the Federal Republic of Germany for smoked foods (Stich, 1975). Sullivan (1974) gives a list of BaP levels in marine produce which includes figures of 1-112 µg/kg for French oysters and 24 µg/kg for Alabama oysters, but the basis for these figures has not been established. While qualified support can be given for Sullivan's conclusion that the quantities of PNAH in sea foods is appreciably higher than in other non-smoked foods, it is particularly the filter-feeding molluscan shellfish that contain the highest levels, and his assumption that higher values are undoubtedly due to petroleum contamination is contradicted by those cases where samples from relatively unpolluted sites such as the Arctic Ocean contain higher levels of PNAH than those from sites of known pollution. Such an attribution to oil-containing pollution can be drawn only with difficulty even for a local situation as described by Stich (1975) for mussels in the Vancouver region whose BaP content increases from 0.02 µg/kg wet weight, through 2-4, 33-50, and up to 60 µg/kg from uninhabited areas to the centre of the marine and wharf areas. An attribution solely to petroleum contamination is always difficult, since, as Stich also reports, elevated levels of BaP occur in sediments near sewage outfalls and mussels growing on the creosoted pilings used for such wharfs show BaP levels as high as 214 µg/kg.

Similarly, Shimkin, Koe and Zechmeister (1951) recovered a BaP-containing fraction of 5 mg from 1 kg of thatched barnacles (Tetraclita squamosa) which was shown to have a tumorigenic potency when injected in mice equivalent to a content of 10-40% pure BaP. The wooden pilings from which the material was collected had been given a surface creosoting 10 years previously. As the authors themselves point out, goose barnacles (Mitella polymerus) from pier pilings at La Jolla also showed the presence of BaP, but Tetraclita from rocks near La Jolla were found to be free of BaP or similar compounds.

6.1.1.6 Summary

Polynuclear aromatic hydrocarbons of known mammalian carcinogenicity occur in crude, and particularly in refined oils. Reported levels of named compounds, such as BaP and BaAnth, vary widely in both, from 0.005 ppm to over 3 ppm, but the residues from catalytic cracking and pyrolysis may contain over 10^3 ppm.

Compared with biosynthesis and terrestrial runoff, it is not yet clear whether oil provides a significant proportion of the PNAH input to the marine environment on a global scale, but it can be a major contributor on a local scale, particularly in sewage discharges and refinery effluents.

The reported incidence and levels of PNAH's in marine produce show that high levels can occur, particularly in molluscan shellfish, and that these high levels are frequently, but not necessarily, associated with known sources of terrestrial pollution, including oil. On the basis of samples from sites not known to be so polluted, the background level of PNAH in sea foods appears to be of the same order of magnitude as that in other non-smoked human foods, and only the notably polluted molluscan levels greatly exceed the levels in smoked foods.

6.1.2 Problems: What is the increased health hazard to man from oil-derived PNAH's

Given that oil-derived PNAH's may be a major local input and can occur at elevated concentrations in the tissues of certain marine produce consumed by man, what increased health hazards, if any, does such an occurrence present to the human consumer? The following discussion is intended to build upon the information available in the context of Chapter 2.

Allegations of potential hazard depend basically on the three arguments contained under <u>Definitions</u>, viz:

(a) that oil-derived aromatic hydrocarbons, including PNAH's, are bioconcentrated in the tissues of marine produce exposed to even low ambient oil levels until significantly elevated concentrations are reached, and that such bioaccumulated contents are transferred to higher members in the food chain leading, by a process of biomagnification, to significantly elevated concentrations in the tissues of produce not directly exposed to oil.

(b) either that there is a threshold concentration of PNAH carcinogens to man, above which the carcinogen takes effect, and that this threshold can be exceeded in marine produce consumed by man, due to these processes of bioaccumulation and biomagnification; or that there is no threshold concentration to man for carcinogens, i.e. that there is no safe level.

(c) that oil-derived PNAH's induce carcinomas in marine produce and that ingestion of such cancerous tissues would present a significant health hazard to man.

6.1.2.1 Bioaccumulation and biomagnification

Much of the contention about argument (a) above has centred on the work of Blumer and his colleagues as a result of the West Falmouth oil spill and the thesis for bioaccumulation of PNAH's that they have developed.

Blumer proposed that biologically synthesized hydrocarbons spread unaltered through the marine environment because of their relatively great stability in the food chain, and that oil-derived aromatic hydrocarbons, including PNAH's, will behave as biogenic hydrocarbons, passing without structural modification into tissue lipids and accumulating up the food chain.

The evidence for the accumulation of natural hydrocarbons is as follows (Blumer, 1967; Blumer, Mullin and Guillard, 1970). An olefinic hydrocarbon heneicosahexane ($C_{21}H_{32}$), "HEH", occurs in many marine planktonic algae. The copepod <u>Rhincalanus nasutus</u> also contains this olefin and accumulates it when maintained in cultures of appropriate algae in the laboratory. The same compound has been isolated in oysters, herring and the liver of basking shark. Since it is rapidly auto-oxidized in the laboratory, it appears to be protected once it enters the lipid pool of marine organisms and thus may be transmitted to higher members of the food web. Similarly, isomeric C_{19} olefins have been isolated from mixed zooplankton, and the livers of pelagic and predatory fish, the basking shark, and a sperm whale (Blumer and Thomas, 1965), while pristane, occurring in zooplankton and at very high levels in <u>Calanus</u> spp., is also found in the lipids of herring, other pelagic fish, basking shark, lobster, various sharks, and sperm whale (Blumer, Mullin and Thomas, 1963).

HEH is only a minor constituent of the lipids of the mixed wild zooplankton including <u>R. nasutus</u>, and several copepods related to this species are incapable of accumulating HEH when maintained in the appropriate algal cultures. Many of the wild algae contain very little or no HEH and some appear to metabolize it themselves. Blumer, Mullin and Guillard (1970) suggest that <u>R. nasutus</u> may be rather unique in its ability to assimilate and non-selectively accumulate this olefin. The isomeric C_{19} olefins and pristane appear not to be derived as such from the phytoplankton, but by metabolism of their phytol content when ingested by the zooplankton.

Blumer's thesis now appears in the more particular form that while organisms may specifically <u>reject</u> alien hydrocarbons from their food sources or pollution, the converse may also occur that an organism may <u>selectively</u> accumulate a hydrocarbon. They consider that, because of the relative inertness of hydrocarbons and the great similarity between members of homologous series, distinction between natural hydrocarbons and pollutants is difficult and may lead to the incorporation of both into the lipids. Since, in these authors' views, marine organisms other than bacteria are generally unable to metabolize hydrocarbons, the process of biomagnification up the food chain will occur even if the initial introduction into the food web is made by only a very few species concentrating a pollutant, as shown by R. nasutus and Calanus spp.

As Blumer and his colleagues point out, such a selective accumulation of natural hydrocarbons by only a very few species must have a function. In the cases of R. nasutus and Calanus spp., it is suggested that HEH and pristane are both accumulated as means of maintaining buoyancy during periods of lipid utilization, but pristane levels in the former copepod are very low, while the latter does not accumulate HEH. That marine organisms are capable of so selectively accumulating minor constituents of the food source hydrocarbon pool, presumably for very specific purposes, and that the factors controlling this accumulation are extremely complex, ill accords with the suggestion that pollutant hydrocarbons could be assimilated and accumulated, as it were by mistake, because of the similarity between members of homologous series. Admittedly, such a mechanism might expose the organism to a massive uptake if this minor constituent or its homologues suddenly become abundant. Whether such selective accumulation as an introduction to the food web would apply at all to the PNAH' remains a matter for speculation, as no arguments have been forwarded for any functional accumulation of the relatively harmless PNAH's, such as 1,2 benz-pyrene (BeP), which might lead to an accumulation of their much more carcinogenic homologues. However, the point most amenable to experimental evaluation is that the scheme for biomagnification presented depends on the inability of organisms higher than phyto-phagous zooplankton to metabolize or reject hydrocarbons in general, and the PNAH's in particular. The evidence presented by Blumer and his colleagues that this is the case must now be examined. Blumer, Souza and Sass (1970) published evidence that the total shell contents of oysters (Crassostrea virginica) and the adductor muscles of scallops (Aequipecten irradians) taken from polluted sites 57 and 58 days after the West Falmouth oil spill contained fuel oil components with much the same features, when analysed by gas chromatography, as that recovered from the sediments.

As MacKin (1973) has stressed in his review of these papers, this provides no evidence that the oil fractions have persisted in the tissues for this length of time, because, as Blumer has agreed (personal correspondence to Dr. K. Whittle, Torry Research Station, Aberdeen), the continuous recontamination of shellfish by oil from polluted sediments is probable in this field situation.

Blumer, Souza and Sass (1970) suggest that the greater relative height of the unresolved background envelope in the chromatographs of total oil hydrocarbons in oyster tissues compared with those from sediment samples is due to the greater resistance to leaching by water of the cyclic-aromatic fraction. However, the data and figures given do not allow a sufficiently detailed comparison to be made and, in the absence of identification and quantification of the aromatic hydrocarbons, inferences of their relative persistence within the tissues cannot be sustained.

The gas chromatographs of total hydrocarbons from the adductor muscles of scallops taken from a polluted area are interpreted by Blumer, Souza and Sass (1970) to show increased dissolution of the lower molecular weight hydrocarbons including the isoprenoid alkanes, over those in oyster tissues. Inspection of the published figures suggests that the broad unresolved envelope assumed to contain the higher boiling aromatic hydrocarbons is also lower in the scallop chromatographs and would not support the interpretation that, following ingestion, these PNAH-containing fractions are more resistant to breakdown.

In a second paper, Blumer et al. (1970) reported on the transfer of contaminated oysters to clean laboratory conditions. (Note: the doubts previously cast on the provision of clean running water in these transfers have been effectively answered by Blumer in personal correspondence to Dr. K. Whittle. However, Teal (Blackman, pers.comm.) reports that the sea water supply to the Woods Hole Aquaria was contaminated by low levels of oil pollution in 1974, and maybe previously). Oysters were removed on 12 November from a site polluted by the West Falmouth spill 57 days previous and at the time of removal the total hydrocarbon content from 7 animals was 6.9 mg/100 g. One oyster was frozen after 72 days in clean running sea water, when its total hydrocarbon content was 12.6 mg/100 g, and two additional specimens were kept in clean water for a total of 180 days, when their hydrocarbon content was 3.8 mg/100 g. On the basis of these samples, the authors state that the gas chromatographs of total hydrocarbons demonstrate that little, if any, further biodegradation of the hydrocarbons took place during this (180 day) period in clean water. However, not enough data are presented in the report for this to be confirmed from the published chromatographs. They also state that the hydrocarbon quantities remaining within the shellfish agree well with those present in November, especially if allowance is made for the apparent dilution of the oil by the growth of new tissues between November and May (180 days). This is not borne out by the data shown in Table 6-3.

TABLE 6-3

<u>Hydrocarbon residuals after washout</u>

Sample	Number of animals	Wet weight (g)	Average wet weight/ animal	Hydrocarbon content (mg/100 g)
57 days in polluted area	7	110	15.7	6.9
72 days in clean water	1	17	17	12.6
180 days in clean water	2	33	16.5	3.8

Thus the change in average tissue weight over the 180 day period is negligible, and even assuming no loss of hydrocarbons, this change would only dilute the original content of 6.9 mg/100 g to 6.57 mg/100 g.

The data may be further criticized on the grounds that analyses of oysters from control areas are missing, allowing no evaluation of the final hydrocarbon content. Conclusions like that drawn by Blumer <u>et al.</u> (1970) from their results that "Thus, once contaminated, shellfish cannot cleanse themselves of oil pollution" needs further evidence to be generally supported.

The thesis of persistence and biomagnification of oil-derived PNAH's has come to rely heavily on the behaviour of particular olefinic hydrocarbons in the food chain. It infers from the relative persistence in tissues of the "unresolved envelope" compared with the olefinic fractions that marine organisms are unable to reject or metabolize the PNAH-containing higher aromatic fractions of oils, to which they are not normally exposed (Blumer <u>et al.</u>, 1972). The data of Walker, Colwell and Petrakis (1976) show that at the microbial level degradation of these aromatic components found in oils is slow, if it occurs at all. There is other evidence that this "unresolved envelope", and more specifically, the higher aromatics and PNAH's appear to be rejected or metabolized at a slower rate than the lower molecular weight olefins or the saturated straight-chain hydrocarbons (Ehrhardt, 1972; Morris, 1973; Vaughan, 1973). This has been at least partly attributed to the greater water solubility of the aromatics leading to the exposure of organisms to solutions or accommodated dispersions containing relatively greater concentrations of these compounds than the original oil (McAuliffe, 1966; Vaughan, 1973; Anderson <u>et al.</u>, 1974). However, the effects of relative solubility are complicated by the behaviour of individual hydrocarbons in salt water (Sutton and Calder, 1974), and by the behaviour of various fractions of different whole oils when extracted by, or dispersed in sea water. Anderson <u>et al.</u> (1974) found that the ratio of aromatics to n-alkanes from two crude oils and a fuel oil varied greatly between the water soluble fractions of a 10% extraction of oil in water and the aqueous phase of a 0.1% dispersion of oil in water. Lee, Craig and Smith (1974) similarly found a wide range of concentrations of individual aromatic hydrocarbons in sea water extracts of a series of crude oils. Dissolved organic matter in sea water can solubilize n-alkane and isoprenoid hydrocarbons but not aromatic compounds such as phenanthrene and anthracene. However, the PNAH's can be accommodated in colloidal micelles and their solubility increased by the presence of short chain fatty acids (Andelman and Suess, 1970; Corner et al., 1976). It is therefore questionable whether the effective field exposure of the PNAH's or any one hydrocarbon fraction to an organism can be deduced from the composition of the parent oil, especially since most marine organisms will be exposed to adsorbed or particulate oil, from which more readily soluble components may have been extracted, and to these water solubles themselves in the water column. However, Lee (Blackman, pers.comm.) reports that 90% of 3H-labelled BaP and a water soluble fraction of a fuel oil were absorbed to clay and detrital material and deposited from solution. The effective exposure to any organism will therefore depend largely on the main routes of uptake. Another explanation that has been advanced for the "increased aromaticity" of hydrocarbon spectra from contaminated tissues relative to sediments or parent oils is that of selective uptake of particular hydrocarbons or fractions. However, this has also been suggested to explain the uptake or presence of greater quantities of low boiling point hydrocarbons in tissues than in the hydrocarbon source (e.g.,Farrington and Quinn, 1973). Evidence that the higher aromatics cannot be rejected or metabolized at the same rate as lower molecular weight hydrocarbons does not mean that they cannot be rejected or metabolized at all, and there is considerable evidence that a range of marine organisms are capable of so dealing with the PNAH's. Corner (1975) considers from mammalian studies that the metabolism of naphthalene and of the PNAH's are sufficiently similar to use the former to predict the behaviour and metabolism of the latter in marine organisms. However, the reported differences, in rates of uptake and elimination between PNAH's and n-alkanes, and between PNAH's themselves (Vaughan, 1973; Lee, Sauerheber and Dobbs, 1972; Lee, Sauerheber and Benson, 1972) and consideration of the difference in chemical structure between the higher aromatic fraction and the remaining hydrocarbon fractions of oils leads us to the conclusions that the mode of presentation and uptake, physiological effects and metabolic pathways are likely to be very different for the PNAH-containing higher aromatics and for the other fractions. The evidence that marine organisms do or do not possess the requisite mechanisms for rejection and/or metabolism of PNAH's therefore cannot be derived from their capabilities towards other classes of compounds. This rules out much recent work on the uptake and elimination of total oil hydrocarbons in general, and in particular of the n-alkane spectrum as an "indicator" of oil pollution (e.g., Clark et al., 1973). The problem of the uptake and elimination of the carcinogenic fractions of oil is even more difficult, since, as explained above, the tumorigenic potential of an oil fraction may be much greater or less than the sum of its known constituents. Evidence relating to the PNAH fraction alone at present falls into the two categories of experimental evidence of uptake and elimination rates of known compounds in isolation, or of the aromatic fraction as a whole, neither of which offers direct evidence of the carcinogenic potential to a predator of the body burden of an organism before, during or after contamination and depuration. Nevertheless, with these provisos in mind, we may examine the evidence relevant to the depuration of PNAH's from marine organisms.

6.1.2.2 Correlation of oil-derived hydrocarbons with position in the food chain

General

It would follow from the thesis of persistence of oil hydrocarbons in tissues that in areas of known oil pollution, oil-derived aromatic hydrocarbons and PNAH's should accumulate up the food chain. Specific evidence for or against this is lacking. However, Burns and Teal (1973) reported no correlation between the flesh levels of petroleum components in the "unresolved envelope" by their GLC method and the animals' supposed position in the food chain of a polluted Sargassum community. From their reported figures, despite the fact that a trigger fish (Canthidermis) of 7.4 g and a Sargassum fish (H. histrio) of 0.5 g would ingest some 10.2 µg and 0.64 µg of such hydrocarbons respectively in devouring 1/20 of its body weight of crabs (Portunus sayi), the body burdens of these fish were only 10.46 µg and

0.71 μg respectively. By contrast, the crab of 0.3 g or a pipe fish (Syngnathus pelagicus) of 0.2 g devouring 1/20 of its body weight of shrimp (Leander tenuiformis) would only ingest some 0.005 μg and 0.003 μg of such hydrocarbons respectively, but their body burdens were 1.39 μg and 7.66 μg respectively.

Scarrat and Zitko (1972) surveyed the sediments and fauna of Chedabucto Bay over 26 months following a spillage of Bunker C oil. Fluorescence spectroscopy at 360 nm against a standard, which will have quantified almost entirely aromatic compounds, showed no overall change in sedimentary oil contents over this period. In contrast, levels in the fauna all fell considerably, and by the end of the 26 months several samples no longer showed the characteristic emission peak of the oil. Browsing and filter-feeding species generally showed higher oil contents than carnivorous species. Scallops (Placopecten magellanicus) showed positive results for oil in the mantle, digestive gland, adductor muscle and gonad, showing some assimilation of the oil, while the browsing periwinkle (Littorina littorea) showed fluorescence only in gut contents and faeces, and not in digestive gland, foot or mantle, suggesting no assimilation at least of the fluorescing components. Lobsters (Homarus americanus) showed little fluorescence over 5 months in the field and lacked the characteristic emission peak. Lobsters exposed in the laboratory to 10 000 mg/l of oil for $6\frac{1}{2}$ days followed by 7 days in clean running water showed uptake of the fluorescing components but levels in the stomach and intestine were 10-100 x those in the abdominal and claw muscle.

Mackie, Whittle and Hardy (1974) isolated the PNAH fraction from sediments in the Clyde estuary, whose GLC trace revealed a very complex mixture of compounds. A crude assessment of the quantities of total PNAH against a BaP standard by spectrophotometry (tlc and uv light) indicated that the sediments and water surface film contained the most fluorescent hydrocarbons with a descending order of content from benthos, fish liver and fish flesh to plankton. The fish included bottom living, demersal and pelagic forms, while the benthos comprised predatory and omnivorous species. Thompson and Eglington (1976) have found that, whereas intertidal estuarine surface muds contained up to 140 μg/g dry weight total PNAH fraction, the diatom populations living on this surface contained no detectable PNAH's. Teal and Farrington (in pres report that at the site of the West Falmouth oil spill, after 1 year the territorial detritus-feeding crab Uca, the territorial bottom-living omnivorous fish Fundulus and the filter-feeding mussel and clam contained a hydrocarbon pattern resembling that of the degraded fuel oil in the sediments. After 5 years, this pattern was retained in Uca (except for some n-alkanes), while the bivalves now show a pattern appreciably lower in retention time than that of the sediments, suggesting either preferential release from the sediments and assimilation of the lighter ends, or preferential metabolism or release from the bivalve tissues of the higher hydrocarbons. The fish hydrocarbon content and pattern has returned to that of uncontaminated animals from before the spill.

Working with mudsuckers (Gillichthys mirabilis), tidepool sculpins (Oligocottus masculosus) and sand dabs (Citharichthys stigmaeus), Lee, Sauerheber and Dobbs (1972) found a rapid uptake of 3H-BaP and 14C-naphthalene from solutions of 1-6 μg/l and 32 μg-29 mg/l respectively. In the first hour of uptake, the liver, gut and gill tissues contained most hydrocarbon. Longer periods resulted in accumulation of hydrocarbon by flesh and especially the gall bladder. The liver of Gillichthys reached a steady state of uptake and elimination of BaP within 1 hour and increasing the water concentration from 1-6 μg/l did not result in further uptake by tissues. However, the uptake of naphthalene by Gillichthys liver continued to increase after 2 hours exposure, and increasing the water concentration from 32 μg - 29 mg/l resulted in a large increase in total naphthalene uptake. All three fish took up more naphthalene than BaP, but then the concentrations of naphthalene used were much greater. Uptake of BaP by the various tissues did not increase after 24 hours exposure, except for the gall bladder. After various periods of exposure, fish were transferred to hydrocarbon-free water. A rapid discharge of labelled hydrocarbon followed such that by 24 hours most tissues had lost over 90% of the naphthalene taken up, while the liver, gut, gill and flesh lost 50, 50, 90 and 20% of the BaP taken up respectively. However, there was a large increase in radioactivity of the gall bladder after transfer. After 24 hours discharge, gall bladder radioactivity due to naphthalene was also reduced, but not that due to BaP derivatives. After 30 minutes exposure hydroxylated derivatives of both BaP and naphthalene were detected in the liver and gut, but only the original hydrocarbon in the gall bladder. 24 hours exposure resulted in the accumulation of hydroxylated derivatives in the gall bladder also. After 96 hours exposure to BaP, only 10% of the original hydrocarbon remained in the various tissues.

When not feeding, sea fish accumulate bile in the gall bladder, so that the apparent storage in the gall bladder observed by Lee, Sauerheber and Dobbs (1972) may have been an artefact since the experimental fish were not fed. Pentreath, at the U.K. Fisheries Radiobiological Laboratory (pers.comm.) maintained plaice (Pleuronectes platessa) in water containing 0.3 mg/l 14C-BaP. Over 16 days the level of radioactivity in the muscle rose to, and was maintained at, less than 0.6 mg/kg, with the gall bladder counts remaining high and providing most of the whole body count. When these animals were then allowed to feed, the gall bladder counts fell rapidly. When fish were maintained in clean water, but fed worms containing 14C-BaP the muscle counts were much higher and less radioactivity appeared in the gall bladder. Hardy et al. at the U.K. Torry Research Station, Aberdeen, fed single doses of 14C-BaP in squid flesh to codling (Gadus morhua). Ninety six hours after feeding, a high proportion of the recorded activity remained in the lipid fraction of the stomach contents. Of the recovered activity released from the stomach, most was found in the intestinal contents, gall fluid and aquarium water, very little in the liver and none in the muscle. By contrast, Whittle (pers.comm.) has since carried out experiments hand-feeding herring (Clupea harengus) of some 10 cm length artificially reared on a diet of squid with pieces of squid containing 14C-BaAnth and 14C-BaP. After 42-43 hours, most of the countable activity was already present in the lipid fraction of the flesh, but 87% of the BaP activity was found in the stomach lipid and only 0.08% in the muscle while for BaAnth 11% was found in the stomach and 56% in the muscle; only 24-31% of the fed activity was recovered from the fish. Such differences in patterns of uptake and tissue distribution of PNAH's are perhaps to be expected between fish species with greatly varying muscle lipid contents and differing rates of lipid turnover and storage, following earlier work at this laboratory on n-alkane assimilation (Hardy et al., 1974; Mackie, Whittle and Hardy, 1974). This serves as an indication that the patterns of PNAH uptake, distribution and elimination are very likely to vary between different groups of fish. Shelton at the MAFF Fisheries Research Laboratory of the U.K. (unpublished) force-fed plaice (Pleuronectes platessa) with an artificial diet containing 330 mg Kuwait crude oil or a high aromatic heating oil per day for four days. The oils were found to contain 100-200 μg/l and 400 μg/l BaP respectively but, at the end of the feeding period and after 3 and 10 days replacement on a clean diet in running water, no BaP (i.e. less than 0.05 μg/kg) could be detected in the flesh. However, the artificial nature of the diet and feeding probably interfered with normal digestion and excretion, and no analyses were made of liver, faeces, or the fatty deposits below the skin.

Reported work on fish neural tissues (Roubal and Collier, 1975) states that, whereas paraffinic hydrocarbons show a greater affinity for neural than other body tissues, withdrawal from exposure leads to rapid losses over 15-20 days from all tissues to a level of some 10% of maximum build-up, and that this persists for at least 90 days. By contrast, low molecular-weight aromatic hydrocarbons drop to undetectable values after only 2-3 days withdrawal. Spin-labelling studies indicated that the aromatic hydrocarbons interact at cell-membrane surfaces, whereas the paraffins invade and interact at levels deep within the cell-membrane.

Teal and Farrington (in press) have recently shown at the site of the West Falmouth oil spill that, after 1 year, the territorial bottom-living omnivorous fish Fundulus contained a pattern of hydrocarbons resembling that of the degraded fuel oil. However, 4 years later, the content and pattern has returned to that of uncontaminated fish before the spill, despite the persistence of oil in the sediments and high levels in other biota. The initial ratio of total hydrocarbons between fish and sediments was 0.03-0.047 just after the spill, but 5 years later was 0.0032. Burns (Blackman, pers.comm.) has shown that the major part of this adaptation by Fundulus is by the induction of microsomal enzyme systems.

Vaughan (1973) exposed shiner perch (<u>Cymatogaster aggregata</u>) to a metered input of 50 μg/ml of a No. 2 fuel oil (40% aromatics) in either an undispersed form or chemically dispersed with 10% Corexit 7664. After 4 days, the entire tissue levels of 1- and 2-methyl naphthalenes were 8.3 μg/g wet weight for undispersed oil and 8.1 μg/g for dispersed oil and the levels of selected dimethyl naphthalenes were 11 μg/g and 12 μg/g respectively. These figures compare with initial entire tissue concentrations of < 0.1 μg/g mono-methyl naphthalenes and < 0.2 μg/g dimethyl naphthalenes, and with the 2.3 μg/g and 2.4 μg/g, respectively, of nC_{12}-nC_{20} alkanes taken up from an initial concentration of < 0.5 μg/g. After exposure to 50 μg/ml of Kuwait crude oil under the same conditions for 14 days, n-alkane levels were only 1.0 μg/g and methyl substituted naphthalenes did not exceed the background levels.

Shellfish

Working with shellfish, Shelton (1971) showed that drained shell contents of clams (<u>M. mercenaria</u>) from an area polluted by discharge of oils from a shipyard, contained 16 μg/kg wet weight of BaP. When transferred to relatively clean estuarine conditions this level fell to 8.2 μg/kg over 7 weeks and to 0.9-1.1 μg/kg over 4 months. Lee, Sauerheber and Benson (1972) working with mussels (<u>Mytilus edulis</u>) found that 20 μg of the PNAH tetralin was taken up from a solution of 10 mg/l per mussel of some 1.5 g wet tissue weight and that 80% of this was lost in 24 hours when replaced in clean water. Similarly, 2 μg of naphthalene per mussel was taken up from a solution of 32 μg/l over 4 hours and 60% of this was discharged over 72 hours in clean water. 0.02 μg of BaP was taken up from a solution of 3 μg/l, and a threshold seemed to be approached with time (over 16 days, pers.comm.), but no discharge rates in clean water are known. No evidence of metabolism of these PNAH's was found. However, Corner (1975) has pointed out that some marine molluscs can form sulphate esters of phenols and might therefore be expected to metabolize naphthalene to 1-naphthyl sulphuric acid, as in <u>Maia squinado</u>. The ability of several gastropod molluscs to synthesize sterols, which involves a microsomal NADPH-dependent oxidase, suggests that the apparent inability of several workers to detect mixed function oxidases in various marine invertebrates, including bivalve molluscs, does not preclude the possibility of their metabolism of PNAH's. Lee has also reported (Blackman, pers.comm.) that background levels of total hydrocarbons in mussels at the Scripps Institution averaged 0.7 mg/mussel. After a nearby oil spill levels rose to 10 mg/mussel with a typical petroleum profile and a high aromatic content. Within 2 weeks of the spill the hydrocarbon content of the mussels had returned to normal in both quantity and absence of aromatic content.

DiSalvo, Guard and Hunter (1975) performed a series of transfers for 4-10 weeks of mussels (<u>Mytilus edulis</u> and <u>M. californianus</u>) between presumed clean water sites and industrially polluted sites in San Francisco Bay. Insufficient data on ambient water hydrocarbon content and on variations in background hydrocarbon levels of mussels maintained on site make the results difficult to evaluate. Background tissue levels of total aromatics, total alkanes and the alkane:aromatic ratio at any one site varied greatly with time. Transfer from one clean water site to a polluted site resulted in a greater tissue uptake of aromatic than alkane hydrocarbons. Re-transfer to a site of lower aromatic but higher alkane background resulted in a continued rise in alkanes but no significant fall in aromatics. Transfer from another clean water site to the polluted site resulted in a greater initial tissue uptake of alkanes than aromatics, after which aromatic concentrations ceased to rise and began to fall while alkane concentrations continued to rise. Re-transfer to a further clean water site brought about a rapid fall in aromatics while alkane levels continued to rise. Re-transfer to the polluted site brought about a rise in aromatics to background level but a fall in alkanes to twice background levels. Mussels maintained at, or transferred to the polluted site showed ratios of aromatics in gonadal to somatic tissue of near unity, with fluorescent PNAH's appearing in the aromatic fraction. Mussels transferred to one of the clean water sites showed a great reduction in this ratio, the aromatic fraction in one case becoming undetectable in the gonads. This was attributed to discharge of gonadal material during spawning although spawning was not confirmed. However, analysis of eggs from mussels transferred to the polluted site showed 332 ppm dry weight total hydrocarbons with an alkane:aromatic ratio of 0.32, compared to a whole body concentration of 525 with a ratio of 0.91 in the incompletely spawned animals. This paper illustrates the difficulty of making any predictive models of the behaviour of aromatic hydrocarbons in tissues from field results and further illustrates the impossibility of predicting aromatic behaviour from alkane analyses.

Neff and Anderson (1975) exposed two batches of 30-48 estuarine clams Rangia cuneata to 12 litres of a 0.0305 ppm solution of 14C-BaP for 24 hours. Average total tissue concentrations reached 5.7 and 7.2 ppm BaP. The majority of the radioactivity (65-75%) resided in the viscera (including the gonads), only 3-5% each residing in the adductor muscle and foot. When returned to isotope-free water, the depuration rates were very different, 94% of the 5.7 ppm being eliminated in 6 days, but only 10% of the 7.2 ppm in the same period. However, after 13 days depuration, only 19% of the 7.2 ppm remained and after 30 days 1.2% of the 5.7 ppm. Depuration of the latter for 58 days reduced the levels to <0.01 ppm, the limits of detection. Discharge for 6-13 days altered the distribution of residual radioactivity between the various tissues very little. After this period, the viscera discharged at a greater rate than the other tissues but retained the majority of the total body burden. There was considerable individual variation both in total concentration and tissue distribution of BaP. The authors attribute this to the ability of bivalves to close the valves for variable periods. However, the clams (of unstated size) were maintained in "large" aquaria (i.e. volume > 12 l with water recirculated through activated charcoal filters). Since more than 22 and less than 45 of the clams began depuration, the volume of water available per clam may have led to initial recontamination of some individuals.

Anderson (1975) exposed oysters (Crassostrea virginica) to 1% dispersions of four oils. After 96 hours exposure, mono- and di-aromatic fractions were heavily accumulated from each of the oils (Venezuela Bunker C and No. 2 fuel, S. Louisiana and Kuwait crude), but "tri-aromatics" (including some di-aromatics) were accumulated in relatively small amounts (c.f. Vaughan, 1973). Oysters exposed to Kuwait crude oil took up high amounts of aromatics, particularly tri-methyl naphthalenes. There were wide individual differences in uptake of aromatics. Oysters exposed to 1% No. 2 fuel oil for 1-7 days were analysed by GLC when the greatest accumulation of all fractions occurred over the first day, with declining rates of uptake over days 2, 3 and 4, followed by rising rates of uptake over days 5, 6 and 7. Anderson comments that the water phase concentrations of oil in dispersion have been found to decrease with time, often followed by a rapid rise which might be due to microbial activity. However, analysis apparently included oil droplets in the gut contents. The stated methods imply that 30 oysters were exposed to 7.5 l of oil dispersion with no fluid exchange. Therefore, initial uptake may well have reduced the ambient oil concentrations to such an extent that, in subsequent days, animals were effectively undergoing depuration into cleaned water. Subsequent release of oil in faeces and pseudo-faeces would then have allowed further uptake. Dispersed oil may increase the gill filtration rate (Anderson, 1972) and animals were used 7-28 days after collection and were held in artificial sea water with no food. Oysters exposed to 5% S. Louisiana crude for 4 and 7 days had completely eliminated aromatic hydrocarbons from their tissues (i.e., <0.1 ppm) within 27 days and 24 days of depuration. Oysters exposed to 1% of each of the four oils for 4 days had completely eliminated aromatics within 52 days. All animals were held under constantly flowing sea water during depuration, and Anderson states that the most significant depuration occurred within the first 14 days. UV spectrophotometry of a series of single oysters during depuration showed considerable elimination of naphthalene, 2-methyl- and 1,2 dimethyl-naphthalenes over days 19-33, but with some further uptake of aromatics in the 210-220 nm wavelengths over days 5-19 before further elimination. No short or long-term effects of oil exposure on shell growth were noted. (More detailed results from these experiments are set out in section 6.2).

Stegeman and Teal (1973) worked with two populations of oysters (Crassostrea virginica) with different fat contents of 0.93 and 1.62% lipid weight/wet weight which remained constant throughout the experiment. Animals were maintained in running aquarium water containing 106 µg/l of total hydrocarbons from a No. 2 fuel oil, for a period of 7 weeks for those of higher fat content, and for a period of 5 weeks for those of lower fat content. The former were then returned to running uncontaminated water containing 11 $\mu g/l$ total hydrocarbons for a further period of 4 weeks.

On a wet weight basis, initial rates of hydrocarbon uptake were very similar for the two populations, but then diverged. Both groups showed a significant reduction in the rate of uptake at 10-14 days, and those of high fat content approached equilibrium after 5-6 weeks with a hydrocarbon content of 334 $\mu g/g$. By extrapolation, those of low fat content would have reached a content of some 161 $\mu g/g$ by this time. On a lipid weight basis, those of low fat content showed a greater initial rate of uptake, but the rates then converged, being very similar at the point of significant reduction in uptake rates at 10-14 days, and by extrapolation, reaching similar contents of 20-22 mg/g after 5-6 weeks.

When replaced in running uncontaminated water, oysters of high fat content showed a rapid elimination of hydrocarbons, 90% being lost of the first 2 weeks, after which the rate of loss slowed dramatically, the concentration remaining at 34 $\mu g/g$ wet weight (approximately 2 mg/g lipid) after 4 weeks' discharge. The contents of both groups of oysters before the experiment were approximately 1.0 $\mu g/g$ wet weight (100 $\mu g/g$ lipid).

Initial rates of hydrocarbon uptake over a 48 hour period of exposure was approximately 90 $\mu g/day/g$ wet tissue and showed a direct relationship with the hydrocarbon concentration in the water up to concentrations of 450 $\mu g/l$. At this level, a progressive reduction in the rate of production of faeces plus pseudo-faeces to approximately 35% of control levels was noted. At 900 $\mu g/l$, the oysters remained tightly closed and production of biodeposits was virtually zero, but the uptake rate was still approximately 45 $\mu g/day/g$ wet tissue. At the concentration of the long-term experiment, initial rates of uptake were approximately 20 $\mu g/day/g$ wet tissue, and biodeposit production was approximately 74% of control levels.

The authors state that fractions B and C from their analytical separation procedures consisted primarily of aromatic compounds and that these combined fractions represented 40.6 ($\pm$ 10)% of the hydrocarbons from oyster tissues but only 14.8 ($\pm$ 5.7)% of the hydrocarbons from water samples containing a range of hydrocarbon concentrations. The variation in the oyster samples could not be correlated with length of exposure or concentration of hydrocarbons in the animals.

It would therefore appear that a higher content of aromatic than other hydrocarbons was maintained in the tissues relative to the water supply and the authors point out that the ratio of resolved peaks to unresolved envelope continued to decrease for at least 2 weeks after transfer.

Since initial rates of uptake seemed to be related to wet weight rather than lipid content, the authors suggest that the oysters have a multi-compartment system for handling hydrocarbons, such that passage to, and subsequent accumulation in, the stable lipid compartment does not occur until other tissue compartments are "saturated" by exposure to high ambient concentrations. Loss of hydrocarbons from such a stable compartment would then be a very protracted process, possibly requiring even lower levels of ambient hydrocarbons than those experienced before exposure.

Ehrhardt and Heinemann (1974) analysed blue mussels (<u>Mytilus edulis</u>) for recent biogenic and petroleum-derived hydrocarbons. Mussels sampled in April 1973, after the spring phytoplankton bloom contained large quantities (4 894 mg/kg wet tissue weight) of hydrocarbons presumably mainly derived from the phytoplankton, and superimposed on a background of largely aromatic hydrocarbons (14.7 mg/kg) indicative of fossil origin. Mussels sampled in January 1974, before the spring bloom, contained very little recent biogenic hydrocarbons. The authors offer three possible explanations for their results: (1) The mussels exchange hydrocarbons with the water which contains a constant or rising concentration of fossil hydrocarbons in addition to seasonally varying concentrations of recent biogenic hydrocarbons; (2) The mussels are able to degrade recent biogenic hydrocarbons but are less efficient in degrading cyclic saturated and aromatic hydrocarbons from fossil fuels; (3) Before the start of the experiment the mussels were exposed to an oil pollution incident and subsequently exchanged saturated and olefinic hydrocarbons with the water more rapidly than cyclic and aromatic hydrocarbons. Blackman and Shelton (unpublished) maintained mussels (Mytilus edulis) for a year on a raft in an estuary suffering little oil pollution. The BaP levels of the drained shell contents changed from 2.6-28.0 µg/kg wet weight over this period, as the average flesh wet weight changed from 2.2-0.4 g. Stich (Blackman, pers.comm.) found BaP levels in Mytilus edulis to vary from 13-39 µg/kg wet weight and 68-133 µg/kg wet weight at two polluted sites over the period May to September.

Vaughan (1973) exposed oysters (Crassostrea gigas) in a flow-through system to 50 µg/ml metered inputs of a Kuwait and South Louisiana crude oils and a No. 2 fuel oil, either dispersed with 10% Corexit 7664 or in an undispersed condition and both previously mixed with a variable-speed agitator. Water and whole tissues were analysed for $nC_{12}-nC_{20}$ alkanes, 1- and 2-methyl naphthalenes and for 5 selected dimethyl naphthalenes. Oil concentrations within the water column were found to range from 3-15% of the metered input and chemical dispersion resulted in greater water concentrations except with the Kuwait oil where both dispersed and undispersed oil was present at only <1 µg/ml. Oysters exposed to the No. 2 fuel oil (40% aromatics) for 11 days showed a rapid uptake of the methyl substituted naphthalenes compared with n-alkanes. Uptake of both had levelled off or elimination had commenced between the 4th and 11th day of exposure, except perhaps for dimethyl naphthalenes from the undispersed oil. Exposure to undispersed Kuwait crude oil under the same conditions for 14 days followed by depuration in clean water for 14 days resulted in a rapid initial uptake of substituted naphthalenes followed by reduced levels for 2-4 days. This was probably due to disturbance of feeding and other activity, or to the spawning of the oysters on the second day. Then followed a further period of uptake but tissue concentrations had levelled off by the end of the 14 days. A relatively low and continuous rate of uptake from dispersed oil led to much higher tissue levels by day 14. In both cases tissue concentrations had returned almost to background levels after 14 days depuration. Exposure to South Louisiana crude oil under the same conditions for 15 days resulted in maximum tissue levels of substituted naphthalenes during uptake of roughly 10 x those from Kuwait crude oil, but much less of the latter was actually present in the water column. In dispersed oil, rapid initial uptake gave way to reduced tissue levels well before depuration in clean water, while in undispersed oil one batch showed the interrupted pattern of uptake (but without spawning) and maximum tissue levels were reached on or near the 15th day. The chemical dispersion increased the maximum tissue levels of substantial naphthalenes by 2 x and reduced the time to reach these levels. In undispersed oil, one day in clean water was sufficient to continue the process of depuration to near background levels. In dispersed oil, depuration took 5 days in clean water to reach near background levels. At the end of 14 days in the South Louisiana crude oil, the ratio of gonad to adductor muscle levels of n-alkanes was some 2-20, but for mono- and di-methyl naphthalenes some 6-100. Ratios in the controls were near unity and the ratio of gonad to remainder of (non-muscle) tissue levels were more uniformly 0.2-2.0 throughout. The ratio of n-alkanes:substituted naphthalenes decreased from the parent South Louisiana crude oil (18) to the undispersed oil-in-water (5.4) to dispersed oil-in-water (4.3-5.0) to whole tissues of oyster exposed to undispersed oil for 7 days (1.2) and for 15 days (0.1), but ratios for initial and final uptake varied greatly with the type of oil and between batches of oysters.

Crustacea

Working with crustacea, Cox and Anderson (1973) placed brown shrimp (Penaeus aztecus) of two sizes, greater than 2.0 g and less than 0.6 g wet weight in a 1.3 ppm water-soluble fraction of No. 2 fuel oil for 20 hours. For both size classes, maximum uptake of naphthalene, methyl naphthalenes and dimethyl naphthalenes occurred within the first hour and depuration began during the exposure period. Within one hour of exposure, larger shrimp had absorbed approximately 4 x the three diaromatics as did smaller shrimp, but eliminated these at a faster rate, having less body burden than smaller shrimps after 20 hours exposure. Complete depuration took 5-10 days for large and 25-33 days for small shrimp. Tatem and Anderson (1973) exposed the grass shrimp (Palaemonetes pugio) to a 2.61 ppm water soluble fraction of a No. 2 fuel oil containing 0.038 ppm dimethyl naphthalene (DMN). After 6 hours the body tissue concentration was 2.07 ppm DMN compared with 0.02 ppm in the water. After 24 hours the tissues contained only 0.32 ppm compared to 0.013 ppm in the water. After 435 hours replacement in clean filtered sea water the tissue concentration was 0.196 ppm. Lee (1975) exposed various groups of crustacean zooplankton to radioactively labelled PNAH's, octadecane and naphthalene. Groups of 6 individuals of copepod were exposed to 800 ml of sea water containing 80 μg 14C-naphthalene, 5 μg 14C-octadecane, 0.2 μg 3H-methyl cholanthrene, or 1 μg 3H-BaP, the latter being in 50 ppb of hydrocarbons from No. 2 fuel oil while groups of 3 amphipods were exposed to 800 ml sea water containing 15 μg of 14C-BaP. Rates of uptake were greatest over the first day, levelling off over 2-3 days and, where exposure was continued beyond 3 days, the body burdens showed no further increase or had already begun to fall. Individuals of large species (1-2.2 mg dry weight) took up 3 x the body burden of those from small species (0.3 mg dry weight) but temperature differences may have diminished this effect since Calanus plumchrus from temperate waters took up a maximum of 22 x 10^{-4} μg body burden of 3H-BaP whereas C. hyperboreus from Arctic waters took up only 11 x 10^{-4} μg. Body burdens of BaP, methyl-cholanthrene and octadecane reached 22 x 10^{-4}, 5 x 10^{-4} and 5 x 10^{-4} μg respectively in C. plumchrus. Studies of adsorption of 3H-BaP showed a linear increase over 12 hours but no further adsorption. After 24 hours 25% of the body burden of 4.4 x 10^{-4} μg in C. helgolandicus was adsorbed, but this fraction was completely lost after 8 days depuration in clean water. This presumably leads Lee to describe the uptake as ingestion of hydrocarbons despite an assumption from the literature that all hydrocarbons were present in true solution. All the crustacea studied, including copepods, amphipods, crab zoea and euphausiids could metabolize naphthalene, BaP, methyl cholanthrene and octadecane to hydroxylated and more polar derivatives. Octadecane was rapidly metabolized and excreted in contrast to slower metabolic losses of BaP and methyl cholanthrene. C. plumchrus metabolized and discharged methyl cholanthrene faster than BaP. The amphipod Parathemisto pacifica showed the most rapid degradation rates, over 50% of ingested naphthalene, octadecane, BaP or methyl cholanthrene being metabolized after 24 hours. The ctenophores and jellyfish studied when fed copepods containing labelled compounds, showed no metabolism of these compounds although a proportion was discharged. All crustacean species studied discharged ingested compounds when transferred to clean sea water, rates being gradual over the first three days, but increasing later such that less than 1% of that taken up was present after 8 days, but detectable amounts remained even after 17-28 days depuration. When copepods were allowed to feed on diatoms during depuration, the depuration rates were much faster than in starved animals and some preliminary work by Lee suggests that hydrocarbons are metabolized and discharged by copepods at a greater rate when adsorbed to ingested diatoms. The figures which Lee reports to support his conclusions above are somewhat complicated by the losses of labelled compounds due to other causes. Thus, 10-15% of labelled hydrocarbon was adsorbed to the walls of the containers and using gas-tight containers showed losses of up to 10% BaP and methyl cholanthrene to the vapour phase while after 4 days 50% of the naphthalene was similarly lost. Over 10 days approximately 10% of both BaP and methyl cholanthrene were oxidized in the containers of filtered sea water in the dark, but only 1% was lost by bacterial decomposition. These losses not only make Lee's results more difficult to assess, but indicate that the results of other studies using labelled aromatics and PNAH's must be treated with caution. However, the capacity of these crustacean zooplankton to metabolize and excrete PNAH's remains clear. This is a factor of some importance since zooplankton will provide the first link in many of the food chains leading to man as a consumer, and the ingestion of oil droplets by copepods has been demonstrated (Parker, Freegarde and Hatchard, 1971). Morris (1974), in the Mediterranean, found heavy (20-30%) contamination of mixed and crustacean zooplankton lipids by hydrocarbon typical of petroleum, of which roughly 50% resided in the unresolved envelope, but reported earlier (1973) that barnacles (Lepas fascicularis) growing on tar balls did not show gross pollution. Only 5% of the total lipid was composed of non-natural hydrocarbons, of which 43% resided in the "unresolved envelope", suggesting that pollutant aromatic hydrocarbons were being assimilated and discharged quite rapidly. Cox, Anderson and Parker (1975) exposed bottom cages of clams (Rangia cuneata), oysters (Crassostrea virginica) and shrimp (Penaeus setiferus) to a No. 2 fuel oil spread on the surface of a large pond, 1.2 m deep. UV analysis showed that total water concentrations of combined naphthalene, methyl- and dimethyl-naphthalenes reached a peak of some

0.3 ppm after 48 hours, falling gradually to background level (0.03 ppm) or below after 38-96 days. Sediment levels did not reach a peak until after 14 days. Tissue levels in the three species followed the water concentrations. Di-aromatics reached peaks of 25, 40 and 50 ppm from initial levels of 0.9, <0.001 and 30 ppm over 72, 96 and 24 hours in oysters, clams and shrimps, respectively. Shrimp levels then fell to 0.2 ppm, and clam levels to 0.5 ppm over 96 days, while oysters fell to 6 ppm over 2 days and remained at 4-6 ppm over the 96 days. Dimethyl-naphthalenes were found to be the most concentrated in both water and tissues over this period. Thirty eight days after oil was added, samples were removed to oil-free, charcoal-filtered laboratory water to depurate. Shrimp were fed an artificial diet but the molluscs were unfed. After falling to some 0.5 ppm from some 4 ppm in 48 hours, shrimp levels fell to background levels within 10 days. Oyster levels fell to 1-2 ppm from some 4 ppm in 48 hours and to background levels within 47-96 days while clams fell from 2-4.5 ppm to 1 ppm over 47 days. Only 1 clam and 1 oyster/sample were analysed.

Corner et al. (1976a) immersed Calanus in 14C-naphthalene solutions for 24 hours and found detectable quantities (3.6 picograms/animal) from concentrations as low as 0.1 µg/l. The uptake from labelled food (algae and barnacle nauplii) was more rapid and reached greater levels, than from water. Water concentrations of 100-700 x the food concentration were required to reach comparable uptakes from water, while depuration following sea water uptake reached less than 5% in 10 days, but depuration following uptake from food had reached only 30% after 7 days (c.f. Lee, 1975). Feeding with clean food after uptake from food produced only a slightly greater depuration rate, therefore only a small proportion of the loss was via the faeces, and of less than 10% released 24 hours after uptake from food, less than two thirds was present as unchanged naphthalene, while the 90% retained was all as unchanged naphthalene. Uptake from water followed a linear relation to concentration over the range 0.1-1000 µg/l, presented, (2-10)-100000 pg/animal taken up.

Corner, Kilvington and O'Hara (1973) force-fed spider-crabs (Maia squinado) of body weight 0.95 kg with 1.14 g naphthalene in lard over 5 days. Urine and faeces collected over 1-7 days from the start of the experiment showed 5.2% wet weight of apparently unchanged naphthalene in the faeces but only 73.2 µg/l in the urine. Only microgram quantities of naphthalene metabolites were present in 700 ml urine, even lower quantities being found in the hepatopancreas. Nevertheless, 4 hydroxy- and sulphate-metabolites of naphthalene were found which were common to mammalian metabolism of this compound, together with glucoside conjugates and an acetylcysteine derivative. The small quantities of administered naphthalene accounted for in the faeces and urine led Corner and his colleagues to suggest that the detected compounds may have been stored elsewhere and slowly released to the urine over more than 7 days, or that other compounds and pathways of release were present.

6.1.2.3 Summary of evidence for accumulation or discharge of PNAH's

Several points emerge from the evidence presented on uptake and discharge of hydrocarbons by marine fish, molluscs and crustacea. These may be summarized as:

(i) Fish and crustacea possess the requisite systems for the metabolism of PNAH's and their excretion as the more water-soluble hydroxylation products. Molluscs appear to lack these systems.

(ii) The general increase in metabolic activity to be expected when feeding, can promote discharge of PNAH's in crustacea but increased translocation and storage in fish.

(iii) There is a greater storage and persistence of aromatics and PNAH's in lipid-rich than in lipid-poor fish types; in lipid-rich than in lipid-poor oyster populations; in lipid-rich gonad of oysters and mussels than in their muscle.

(iv)   The greater rates of uptake of aromatic hydrocarbons than paraffins and the higher tissue levels reached may result in faster rates of elimination, when returned to clean conditions, or in a greater persistence of aromatics and PNAH's within tissues.

(v)   The greater part of the aromatics and PNAH's taken up are quickly discharged on return to clean conditions, but some 1-10% of the maximum uptake may persist for much longer periods, and is subject to greatly reduced rates of discharge.

(vi)   Reduction in uptake rates, maintenance of an equilibrium between uptake and discharge or the initiation of depuration may all take place under conditions of exposure to increased levels of oil or PNAH's before a return to clean external condition.

(vii) The size of organisms, form and presentation of the hydrocarbons and the dispersion or dissolution of the oil can all greatly affect the rates of uptake of aromatics and PNAH's. In crustacea, adsorption to the body surface can be an important route of uptake, but uptake from food may be more rapid, and more slowly eliminated, than from water.

(viii) Within a given experimental system, and within certain limits, the rate of uptake is proportional to the amount of hydrocarbons presented.

Many of these points support, or could be explained by, a hypothesis for the method of bio-magnification originally proposed by Ehrhardt. This proposes that there are limits to the capacity of organisms to excrete and metabolize PNAH's. Below these limits, the rates of excretion and/or metabolism are functions of the rates of intake such that increased rates of intake do not raise tissue concentrations above their established "background" levels. If the rates of intake exceed these threshold limits, accumulation of PNAH's in the tissues will occur. If the rates of intake again fall below the thresholds, depuration of accumulated PNAH's will take place, and if depuration continues for long enough, tissue concentrations will again reach background level. Thus, tissue concentrations during alternating periods of temporary acute exposure, chronic exposure or no exposure to increased rates of intake of PNAH's will reflect the integrated rates of uptake, metabolism and excretion over that period of exposure.

Stegeman and Teal's (1973) suggestion of at least a two-compartment model for the handling of hydrocarbons also finds support in the data and is theoretically described by Wilson (1975). The primary process in uptake and elimination can then be seen as akin to simple partitioning between body fluids and external medium. The secondary process, when levels in the body fluids rise beyond a certain point, is then the metabolism of compounds, in the case of PNAH's principally by hydroxylation in the liver or hepatopancreas. This process may involve temporary storage before excretion, but it would be only the quantities of hydrocarbons beyond the capacities of these metabolic processes which pass to the tissues and are stored intra-cellularly. Such stored hydrocarbons are perhaps only released for metabolism and excretion as these cell contents are turned over or used as energy sources or released as gametes. It must be admitted that both Ehrhardt's and Stegeman and Teal's suggestions assume that the hydrocarbons, or particularly the PNAH's, are recognized as foreign compounds and are not directly utilized or stored as an energy source "by mistake", nor pass to tissue lipids by a simple partition. The position of molluscs is anomalous if metabolic systems for PNAH hydrocarbons are indeed absent. This would imply, within the terms of the model, that above certain levels in the body fluids, any PNAH's are stored in the tissues and their release and elimination would be protracted and not necessarily accomplished when food reserves are utilized. Whereas fish and crustacea may need only short periods between pollution incidents to fully cleanse their tissues and prevent accumulation of PNAH's by a "load on top" process, molluscs may be able to prevent this only where incidents are very infrequent. In the situation of chronic oil discharge, it is to be presumed that appropriate "background" tissue levels would result for both molluscs and fish and crustacea. Such levels in the former would be much higher since there would be an absence of the response to increased PNAH levels of increased PNAH metabolism seen in mammals (e.g., Schlede et al., 1970) and presumed to occur in fish and crustacea from the initiation of depuration before transfer to clean conditions. It is perhaps because of this, rather than their position in the food chain, that such high PNAH levels have been recorded for filter feeding bivalve molluscs.

However, bivalve molluscs show a two-phase system of depuration in common with crustacea and active over a similar time-scale to reduce tissue levels of at least di- and tri-aromatics to background levels. Some bivalves have been shown to reduce tissue PNAH levels to background levels over 2-3 months. These facts suggest either that molluscs possess metabolic systems for aromatic hydrocarbons albeit at a slow rate of activity, or that crustacea (and perhaps fish) can also bring about their complete depuration without metabolism.

That the rate of PNAH metabolism below certain levels is a function of the rate of intake has been shown for the microsomal hydroxylation and biliary excretion of carcinogenic PNAH's in mammals (Schlede et al., 1970; Schlede, Kuntzman and Conney, 1970; Gelboin and Wiebel, 1971) and for increased levels of aryl hydrocarbon hydroxylase in man (Conney et al., 1971). The reduction in rate of uptake, or the onset of depuration before transfer to clean conditions may reflect this process in marine organisms, and the induction of such microsomal enzyme systems has been brought about in some fish and crustacea by exposure to oil hydrocarbons. However, Lee, Sauerheber and Dobbs (1972) thought that BaP might cause a diminution in general metabolic rate in marine fish such that the uptake rate falls.

6.1.3 Problem 2: Threshold dose and hazard to man of elevated PNAH contents

It has been shown that marine produce, and notably bivalve molluscs, can take up PNAH's following a pollution incident, and in the laboratory reach whole tissue levels of some 10^2-10^3 x background levels. The potential for bioaccumulation is severely limited by the ability of at least some finfish and crustacea to metabolize these PNAH's, and by all forms to increase the rate of PNAH excretion and to carry out rapid initial elimination of some 90-99% of the maximum uptake. Do the remaining elevated PNAH levels pose a significant increase in the hazard of cancer-induction in man as a consumer, over the longer period required for their elimination?

Data:

The tumorigenic potential of polluted marine produce has been evaluated by topical application of sub-cutaneous injection of tissue extracts, using mammals. This practice may be irrelevant to the problem of PNAH's in marine produce, since this will be ingested, usually after cooking, and may or may not form a regular proportion of the daily food intake. Gerarde (1960) points out that the PNAH's are poorly absorbed by the mammalian gastrointestinal tract when added to the diet. Whereas naphthalene is completely absorbed from even a single massive oral dose to the rat, 40-97% of the larger molecules of BaP, Chrysene, 1:2, 5:6-Dibenzanthracene (DBA) and 20-Methylcholanthrene (MCA) were excreted as faeces from single or repeated dosages. Even long-repeated ingestion, where stomach and rectal contents remain for long periods in contact with the gut wall, seems scarcely comparable with topical application or sub-cutaneous injection, but appropriate to continuous environmental exposure of man.

Although factors such as co-carcinogens and promotors may make a determination of zero hazard to man impossible in clinical trials of tissue extracts of contaminated produce, it is apparently still a matter for debate whether there is any dose-response relationship for cancer-induction in man, or a threshold dose below which carcinogens do not induce cancer (WHO, 1972). Gerarde (1960) reports a dose-response relationship for topical application to mice of BaP, 20-MCA and 1:2, 5:6 DBA, with an effective threshold for BaP. IARC monographs (1973) report several such studies which show that the threshold dose for BaP and BaAnth by topical application or sub-cutaneous injection varies greatly with the strain of mice and the solvent used. For any given experimental system, increased doses above the threshold increased not only the percentage of tumour-bearing animals but also shortened the latent period before such tumours appeared. Some evidence of a dose-response relationship was also found for topical application to rabbits and for sub-cutaneous injection in rats and hamsters. In contrast, oral administration of BaP to mice and hamsters has so far shown only a dose-response relationship between percentage of stomach tumour formation and the number of times that a given dose was repeated.

If there is an effective threshold for oral intake of PNAH's in man, the critical question becomes whether this threshold could be exceeded by the PNAH content in contaminated marine produce. We have found no reports of epidemiological studies which link gastrointestinal cancers in man with the ingestion of particular marine fish or shellfish produce except the possible link between a high incidence of stomach cancer and heavily smoked fish in Iceland and Slovenia, and its low incidence in the U.S.A. (Wynder et al., 1963). However, there was no correlation between the high incidence of gastric cancer in Japan and the consumption of smoked foods, despite the very high PNAH contents recorded there. The importance of increased PNAH levels in any one food item is rendered doubtful by the findings of Wynder et al. and of Hakama and Saxen (1967) that the incidence of gastric cancer correlates with a diet relatively high in starchy foods such as potatoes, rice and bread as a main calorie source, and relatively low in fresh vegetables and fruits, despite the greater PNAH content in the latter items. Further studies in this area may have been commissioned, but were not readily available for this report.

Medical opinion competent in the field of carcinogenesis is required (a) to evaluate the few reported field and laboratory tissue levels for marine produce in the light of such a threshold, (b) to assess the additional risk of increased levels above any threshold, and (c) to set any levels above which a fishery should be closed and which must be regained by depuration in the field or under transferred stock-holding conditions before such produce could again be exposed for consumption. Such a task must include a decision as to which "indicator" PNAH compound, or spectrum of compounds should be analysed for, and by which methods, in samples of produce suspected of contamination. This leads to the important consideration of the relationship of oil "taint" in produce to PNAH levels in tissues. Sullivan (1974), Blumer (1971a), Blumer et al. (1970), Blumer, Mullin and Guillard (1970) and others have argued that oil-derived hydrocarbons, including the carcinogenic PNAH's, may be present at elevated levels in produce with no associated taint, and that detection of any hydrocarbon compounds or assemblages characteristic of oil indicates a potential hazard to man and warrants the closure of a fishery. Paradis and Ackman (1975) have shown that commercially acceptable lobster meats contain small but identifiable quantities (2-4 ppm) of the same assemblage of extractable volatiles and background "envelope" characteristic of certain oils, and associated with a taint at higher concentrations (12-16 ppm). Although some lower aromatics have been unequivocally identified, and Mann (1969) has implicated some phenolics and naphthenics, the compounds responsible for oil taint are likely to include polar, sulphur-bearing molecules. Plaice detectably tainted with crude and high aromatic heating oils contained < 0.05 μg/kg BaP or BaAnth. Strongly tainted clams and cockles contained 16 and 11.7 μg/kg BaP while untainted equivalents contained 8.2 and 8.0 μg/kg respectively. Mussels slightly tainted experimentally contained 1.2-2.1 μg/kg BaP while their untainted equivalents contained 0.86 μg/kg (Shelton, unpublished).

Whatever the range of tainting compounds (see section 6.2) the content of the carcinogenic PNAH's is doubtful and it is apparent from the studies reported that the quantities of other compounds taken up and temporarily stored, need bear little or no relationship to the quantities of PNAH's taken up, and that elimination or retention of the former need not imply retention or elimination of the PNAH's. Similarly, the presence of hydrocarbon patterns indicative of an oil pollutant, or of increased levels of other contaminants associated with the pollutant cannot be used as an indication of significantly increased PNAH levels. Taint in itself may be a reason for rejection of produce, but its presence or absence is not necessarily an indication of the PNAH levels. In the absence of rapid methods for directly evaluating the tumourigenic potential of mammalian ingestion of polluted tissues, we can only conclude that the concentration of selected carcinogenic PNAH compounds or assemblages must be determined by direct analysis. These levels should form the basis for closure or reopening of a fishery or the exposure for consumption of polluted produce since transferred to cleaner waters, bearing in mind the former background levels and the likely ingestion rates of the produce. Evaluation of the potential of co-carcinogenic effects should not be ignored in these decisions.

Until more data are available on the levels of PNAH's in produce from polluted and unpolluted waters and until medical opinion on the significance of these levels and changes is available, we can only give a very crude example of the significance of ingestion of polluted produce, based on reported BaP levels. Although the range of reported levels varies from 2×10^{-2} - 28 μg/kg for relatively unpolluted sites, and from 6×10^{-2} - 1000 μg/kg for polluted waters, a reasonable background tissue level would be 1 μg/kg. If we assume that field conditions following a pollution incident allow a magnification figure from reported experiments of 300 x over many days of uptake, and that abatement of ambient levels or transfer to clean water brings about an initial rapid depuration to 10% of the maximum uptake, then the produce will be left with a gradually declining BaP load of 30 μg/kg. This is 30 x background level, compared with an expected maximum annual background variation of 10 x. For an average daily intake of fish and shellfish of 24 g in the U.K. (MAFF, 1971) to 84 g in Japan (National Food Survey Committee, 1971), consumption of this produce would mean an increased dose of 0.72-2.52 μg/day. If shellfish alone are considered, the average daily U.K. intake is only 1.5 g (MAFF, 1972) and the increased dose would be 0.045 μg/day. These increases should be compared with the average total daily food intake in the U.K. of 1.5 kg (MAFF, 1971), which from Table 6-2 of BaP contents in non-marine foods would normally contain less than 5 μg. Borneff and Fabian (1966) estimated an average (Western) food intake of total PNAH's of 10 mg/year, of which perhaps 2.5% (0.25-5%) is BaP, i.e. 250 μg, giving a daily intake of 0.68 μg.

This example should be considered as potentially useful only within the context of this present paper. We are aware of the unsatisfactory nature of an approach using average daily intake figures, which cannot be representative of the real distribution of intakes, and this model does not assess the worst possible case. The figures here arrived at should only be regarded as an indication of a possible first step in any risk analysis by governmental or international bodies to determine acceptable exposure levels.

6.1.4 Problem 3: Induction of carcinomas in marine produce

It has been alleged that oil pollution has induced carcinomas in marine produce and that ingestion of such cancerous tissue would present a significant health hazard to man.

Data:

Reports of the induction of carcinomas in marine organisms exposed to oil pollution have been greatly confused by unwarranted attribution to oil components and by misidentification of tumours or lesions as malignant. Russel and Kotin (1957) found oral papillomas in 10 out of 353 bottom-grubbing fish (Genyonemus lineatus) trawled 2 miles from a sewage outfall. These have been attributed to refinery wastes but no mention of this is made in the paper. Three of these fish were examined but no evidence of malignant neoplastic changes was found in the papillomas. No tumours were found in 1116 fish from non-polluted waters 50 miles away. Such papillomas may occur in response to mechanical, infectious or chemical irritation and have frequently been reported in fish spending all or part of their lives on the sea bed in waters subjected to terrestrial pollutants (Harshbarger, 1975; Stich, 1975; Deys, 1969). The higher incidence of papillomas in English sole (Parophyrys vetulus) in parts of San Francisco Bay is possibly linked to greater discharges of petrochemicals among other wastes, but again the papillomas are non-malignant and there is some evidence that the tumours later regress (Cooper and Keller, 1968). Young (1964) observed papillomas in several species of fish associated with sewer effluents, but again no direct links with oil residues were made. Finkelstein (1960) suggested that papillomatous tumours in Baltic eels might be attributable to carcinogenic substances including ship fuel oil. Thus, while carcinomas undoubtedly occur in bony and cartilaginous fish, showing all the signs of malignancy and appearing both analogous and homologous with mammalian tumours (Wellings, 1968), any links between true carcinomas and oil pollution in fish, while not directly proven at present, requires more data and evaluation.

Powell, Sayce and Tufts (1970) described hyperplasia of ovicells in bryozoan colonies growing in oil polluted waters, but once again pollution from timbers treated with creosote was present, and colonies experimentally grown in clean water on timber frames treated with creosote and asphalt showed the same hyperplasia. Straughan and Lawrence (1975) investigated the bryozoa exposed to chronic crude oil pollution from natural seepage at Coal Oil Point, but found no evidence of ovicell hyperplasia in the species occurring there.

In marine crustacea, several reports of tumours and lesions appear to be mis-identifications of malignancy, but carcinomas have been positively identified in crabs (Carcinas maenas) (Sparks, 1968). No suggestions of attribution to oil pollution have been seen, although sunken, particulate oil may be ingested by crustacea and remain in the foregut for long periods (Blackman, 1972).

Hueper (1963) found papillary tumours in about 20% of burrowing clams (Mya arenaria) from areas of Chesapeake Bay, but considered doubtful an attribution to the ship fuel oil known to be present. Harshbarger (1975) points out that there is an indistinct boundary between neoplastic disease and the process of inflammation repair and regeneration in molluscs. Pauley (1969) reviews the literature on neoplasia in molluscs and points out that, while many apparently neoplastic tumours and lesions in marine molluscs are actually inflammatory and hyperplastic and the definition of invertebrate neoplasms is imprecise, true neoplasms do occur. Examination of one of Hueper's specimens and of 500 further specimens from the same area failed to establish the presence of true neoplasms. Neoplasma due to an epizootic infection have been noted in Macoma baltica from Chesapeake Bay (Christensen, Farley and Kern, 1974). Several other studies of tumours in Mya revealed only a low incidence of true neoplasms, and these were benign, showing no sign of invasion or nitotic figures. Tumour prevalence in 6 species of marine bivalves from Sequim Bay, Washington, and its vicinity showed a maximum incidence of 0.55% in Saxidomus giganteus and Pauley comments that there is a disparity of tumour prevalence between fresh water pelecypods (maximum incidence 15.7%) and marine forms from the area.

Benign lesions and tumours have been found in the oysters Crassostrea gigas and Ostrea edulis and tumours have been reported in C. commercialis, C. virginica and C. gigas showing abnormally high numbers of mitotic figures and some invasion. Tumours in C. commercialis from estuaries receiving no industrial discharge were unquestionably invasive (Wolf, 1968). Benign tumours have also been reported in cuttlefish, and a sarcomacoid disease, probably neoplastic, in mussels (Mytilus edulis) with an incidence of 10% in Yaquina Bay, Oregon (Farley, 1969). Pauley reports that it has been found very difficult to induce cancer in marine molluscs using mammalian carcinogens, but that malignant tumours have been produced in cuttlefish using massive doses of 1,2-DBA. Metastases have been observed in a mollusc and it is therefore known that molluscan tissues are capable of undergoing malignant changes, but it is doubtful if a thorough search for metastases has been made (e.g., Wolf, 1968). Harshbarger (1975) states that of 1100 species represented in the collection of the Registry of Tumours in Lower Animals, 50% have neoplasms, of which two thirds occur in bony fish, and molluscs form the greatest fraction of the remainder. He comments that the scavengers, grazers and filter-feeders among the fish and molluscs, which show the highest incidences, presumably receive more exposure to environmental carcinogens from sediments. None of the reported incidences of true neoplasms above have been directly attributed to the presence of oil pollution.

However, Yevich (1975) reports that soft-shell clams (Mya arenaria) and mussels taken over 4 years from the area of a spill of No. 2 fuel oil mixed with JP-5 jet fuel show possible gonadal and connective tissue tumours. This spill came from a tank farm which drained into Long Cove, Searsport, Maine, over at least 3 months in 1971. Massive mortalities of Mya resulted: 40% in the first 6 months; and 1973 sampling showed the recent mortality rate to be 56%. Despite Yevich's description of possible tumours, samples of 100 surviving clams taken at intervals up to 3 years after the spill were said in 1974 to have shown incidences of neoplastic tumours, mostly in the gonads, of 16-26% compared with 0-2% in controls. Frequent and atypical mitoses were stated to be present and a few instances showed invasion. In 10% of tumour-bearing clams metastases occurred in other organs, including the gills, although a few primary tumours in gill tissues were also identified.

Yevich reported in 1972 to an investigative team that 5-10 months after the spill, samples of 400-600 clams from areas 1000 yards from the drainage pipe showed incidences of gonadal lesions of 4-5%, c.f. no incidence in controls. Mitotic figures were observed, and cell masses seen in other organs were thought to be metastases from the primary germinomas. Barry and Yevich (1972) published the results from surveys of 175 and 364 clams from Rose Island, Narragansett Bay, in 1969 and 1970, respectively, which showed incidences of 2.3 and 2.75% ovarian and testicular neoplasms. A previous survey at this site (Yevich and Barry, 1969), which has suffered no reported oil pollution, reported ovarian tumours in only 3 out of 1300 clams collected in 1968. Teal (Blackman, pers.comm.) commented that most of Narragansett Bay is presently exposed to chronic oil pollution intertidally.

An official report of the NIEHS meeting in 1975 by O.A. Bessey (Blackman, pers.comm.) describes Yevich's paper as showing possible tumours from both field and laboratory exposures to various fuel oils. A report of this meeting in "Marine Affairs" stated that Yevich had observed a high incidence of malignant tumours in soft shell clams and mussels resulting from oil pollutants and that laboratory experiments had confirmed these as the cause. Barry and Yevich (1975) reported: "The microscopic picture seen in the animals collected from the Long Cove oil spill site meets most of the criteria of a malignant tumour. These characteristics have not been seen in soft shell clams examined from other areas of the United States, such as Rhode Island, Maryland, Massachusetts, and California. The causative factors of these tumours is not known. Since soft shells have a planktonic larval stage which serves to distribute populations, it is unlikely that the gonadal tumours can be attributed to genetic factors. The development of malignant gonadal tumours could not be associated with any seasonal or cyclic change. It is more probable that the lesions are a result of some drastic and traumatic alteration in their environment. It may be that the combination of the jet fuel and its additives and the No. 2 fuel oil supplied the trauma necessary to bring about these tumours. The highest incidence of the tumours correlates with the major site of impact of the oil spill."

The identification of the tumours as carcinomas, and particularly the descriptions of malignancy have been questioned by several workers. Mr. Yevich has consulted with Dr. Clyde Dawe, M.D., a comparative oncologist for the National Cancer Institute, and Dr. John Harshbarger of the Registry of Tumours in Lower Animals, Smithsonian Institution. The microscopic slides of the neoplastic soft shell clams were also presented to 30 molluscan pathology experts for study and critical analysis. Everyone concerned and knowledgeable in molluscan pathology agreed with the analysis that the neoplasms which were found in animals from the oil spill site at Long Cove, Searsport, Maine, are malignant neoplasms of the reproductive tract with metastasis to various body organs. There is no doubt that these are true malignant neoplasms. However, at the present time we cannot say what carcinogenic agents induced the neoplasms (Thompson, pers.comm.).

The chronic pollution of the Long Cove site by weathered oils prior to this incident, the relatively large spill (14 t draining into a cove only 1000 yards wide), and the massive mortalities sustained should be noted. Compare the spill at West Falmouth (Blumer et al., 1970) where 4 t of No. 2 diesel fuel oil over 22 acres (88 000 m^2) of the Wild Harbour River "wiped out" the shellfish crop for 2 years.

Yevich (1975) reports that soft shell clams from the site of a Bunker C oil spill at Portland, Maine, and from the site of a JP-4 and No. 2 fuel oil spill at Brunswick, Maine, also showed possible connective tissue tumours. The total evaluation of Yevich's more recent reports is made more complicated by previous surveys of the occurrence of atypical hyperplasia in the gills and renal organs of M. arenaria (Barry, Yevich and Thayer, 1971). By reference of such features as abundant mitotic figures but absence of invasion to human carcinomas, he described these lesions as pre-cancerous, i.e., exhibiting cell activity unusual in benign conditions yet lacking certain characteristics of actual cancer. The distribution of the occurrence of such lesions is peculiar, since in their survey of Point Judith Pond, although the figures are not consistent, there is apparently a strong correlation between high incidence and waters "considered polluted by the Department of Natural Resources, State of Rhode Island". However, surveys elsewhere show only 2 out of 26 clams affected in the "polluted" waters off Red Wood City, San Francisco, while 12 out of 32 clams are affected in the "unpolluted" Tomales Bay. In view of the high incidence of hyperplastic and benign tumours in marine filter-feeding bivalve molluscs and the difficulty of inducing true neoplasms in them with mammalian carcinogens, it is of interest that "Mercenene", an anti-neoplastic agent against tumours in mice, hamsters and human amnion cell-cultures, has been isolated from marine clams M. mercenaria (Schmeer, 1968). The greatest activity was isolated from the crystalline style and hepatic organs, and extracts from other marine molluscs, M. compechiensis, Ostrea, Busycon, Loligo, showed similar but lower anti-neoplastic activity. However, gonadal carcinomas with invasion have been described in M. mercenaria (Yevich and Barry, 1969) but with no apparent link with water quality.

In Japan, where marine algae are cultivated on a massive scale for human consumption, there have been several instances of a "cancerous disease" in fields receiving industrial wastes. The causative agents in some outbreaks remain unclear, since the "tumours" may be associated with, or perhaps arise from, infestations of the chytrid parasite Olpidiopsis (Arasaki, Inouye and Kochi, 1960). The galls then formed are probably comparable to the "xenomas" formed in insects and vertebrates in response to microsporidian infections (Sprague, 1968) when only a few characters of true neoplasia are present. Ishio, Yako and Nakagana (1971, 1972) linked these algal cancers with wastes from the coal chemical industry at Ohmuta and isolated four compounds from muds in the culture area which produced the abnormal growth characteristic of the disease in Porphyra tenera including benzanthrone, dibenzanthrones and a new 7-ring compound. Boney and Corner (1962) and Boney (1974) found that treatment of sporelings of red algae with carcinogenic aromatic hydrocarbons caused considerable increases in growth with a dose-response relationship for solutions of strong carcinogens. Whereas low concentrations still stimulated cell production from the single atypical meristematic cell, similar concentrations of weak or inactive carcinogens had no effect or inhibited cell production. Davavin, Mironov and Tsimbal (1975) found that exposure of certain red algae to emulsions of crude oil in sea water led to a decrease in RNA and DNA content, apparently by inhibition of biosynthesis. As Boney points out, meristematic cells are not confined to the apices of Porphyra, but extend over the frond margin. Any auxin-like stimulation of cell proliferation will therefore produce the abnormal frond morphology of the "cancerous disease" but the "tumours" may not be true neoplasms resulting from alterations of cell nuclei. He reports that increased rates of cell production were accompanied by some reduction in cell sizes, and Katayama and Fujiyama (1957) found greatly increased DNA contents in fronds undergoing the early stages of "tumour" growth of the disease. Boney also points out that some experiments show considerable growth promotion following very brief periods of exposure to these aromatic compounds and that the meristematic cells must be particularly sensitive to this stimulation.

6.1.4.1 Summary of cancer induction in marine produce

As yet, no reports have been found directly linking true neoplasms in fish or crustacea with oil pollution alone. Low doses of compounds likely to occur in crude or refined oils or in petrochemical discharges produce malformations in edible algae that may be analogous to carcinomas in mammals. High incidences of neoplasms have been reported in filter- or deposit-feeding bivalve molluscs but these are not necessarily linked with terrestrial pollution. Despite the difficulty of deciding criteria that would make such neoplasms analogous or homologous with mammalian carcinomas a low proportion of these molluscan neoplasms appear to be malignant. There are at least three reports linking molluscan tumours of apparent or possible malignancy with spills of refined oils. However, even in the best documented case there is confusion of chronic discharge and acute spillage, the latter producing relatively massive doses and mortalities.

The conclusion must be that under certain conditions oil pollution could cause carcinomas in marin produce. In the absence of a consensus of qualified medical opinion, it is impossible to state whether ingestion of such cancerous tissue would present any hazard of cancer-induction in man. In the opinion of those we have consulted, such a hazard is in itself remote unless the hypothesis of viral causation of some cancers is proven and transmission of such viral agents is involved. The activated chemical carcinogens which induced the cancer are thought to be no longer available to the consumer. Of much greater importance, it appears, is the probability of simultaneously ingesting elevated doses of unmetabolized carcinogens which may still be present.

6.1.5 Mutagenicity and teratogenicity

The relationships between carcinogenicity, mutagenicity and teratogenicity are summarized in Health Hazards of the Human Environment (WHO, 1972), from which it is seen that carcinogenicity is but one form of mutagenicity and that ipso facto all carcinogens are mutagens. However, mutagenesis has been used more narrowly to mean transformation of the genetic material in germ cells to produce a mutant offspring. In this respect, data on PNAH's have only been found for BaP which has a relatively low Mutagenic Index for a median lethal dose of 750 mg/kg in mice (WHO, 1972). However, as with carcinogenicity, other nitrogen and sulphur-bearing compounds from crude and refined oils may be implicated. The closeness of the correlation between general mutagenicity and particularly carcinogenicity is shown by the fact that screening tests for mammalian carcinogens using mutagenic effect in certain microbial or tissue cell cultures are considered acceptable (e.g., Searle et al., 1975). However, the mode of presentation to the developing germ cells from oral intake means that even powerful carcinogens can be only weakly mutagenic, in this sense. No data relevant to oils has been found on teratogenesis. The list of chemicals (WHO, 1972) that were then indisputable teratogens was very small and included no compounds likely to occur in significant quantity in oils, although it is possible that some of the alkylating cytostatics might occur in petrochemical and refinery discharges. Because of their mode of action, carcinogens are not considered to be potential teratogens.

6.2 Loss of marine foods

6.2.1 Definitions

- Loss of marine foods is any reduction in the quantity or quality of marine produce available to man as a consumer, resulting from accidental or intentional discharge of oils or refined products, or natural seepage, to the marine environment.

- Tainting means the accumulation of oil or oil compounds to levels causing taste or odours in marine products consumed, or likely to be consumed, by man, rendering them less desirable as food.

6.2.2 Problems

(i) These losses may result from acute or chronic lethal effects of oils or oil clean-up agents to adults, juvenile or larval forms, or eggs. They may also result from such sub-lethal effects as reduced fecundity, feeding or growth rates, or interference with chemical communication systems. These effects are considered in Chapter 5.

(ii) Seafoods may be rendered apparently unavailable to the human consumer on a local scale by refusal to purchase or consume produce from areas known or believed to be affected by oil discharges, irrespective of demonstrable pollution or effects. This subject is outside the scope of the Working Group.

(iii) Marine produce may be rendered actually unavailable to the human consumer on a local scale by the refusal of collectors to obtain produce or risk contamination of gear in areas known or believed to be affected by oil discharges, or because of official closure of fisheries or banning of sales, on the ground of known or suspected health risks.

(iv) Oil discharges or clean-up methods may render marine produce unacceptable to human consumers due to visible external fouling with oil or the presence of an unacceptable "oily" smell or taste (oil taint), or visible signs of tissue damage or loss of "condition".

It is with subject (iv) and the official closure of fisheries or banning of sales that this Working Group is concerned.

6.2.3 Background

The extent of losses proven or alleged to be due to oil pollution has been impossible to evaluate fully, due to the inadequate documentation of incidents, claims, closures, or condemnations of produce. However, a few examples will serve to indicate the probable extent of the problem. The loss of 8 million gallons of heavy oil from the Mitsubishi Mizushima refinery in December 1974 resulted in the oil company being sued for 10^{10} Yen by fishermen, in compensation for damage to fishery products (Anon., 1975). Spooner (pers. comm.) reports that the spill of gas oil from a storage tank into Hong Kong harbour resulted in some £.Stg. 500 000 being paid as compensation to fish farmers for losses due to mortality, taint and unacceptable external appearance. Following the spill of Bunker C oil from the ARROW, the clam fishery in Chedabucto Bay has remained closed for 4 years (Anon., 1974). Whitman (1975) has edited reports of several recent spills, among which are the following. A spill of 1000 gal of Venezuelan crude oil in Easter Passage, Halifax harbour, in January 1973, caused the tainting of 1800 lb of lobsters in storage tanks. The UNIVERSE LEADER spilt 2 600 t of Kuwait crude oil into Bantry Bay on 22 October 1974. The Irish Department of Agriculture and Fisheries imposed a moratorium on the taking of scallops, mussels or sea urchins in less than 20 ft of water at low tide and the fishing for periwinkles along those shores affected by either oil or dispersant. This moratorium was to run until 9 January 1975 and, although no restrictions were laid on the fishery for herring, flatfish or prawns and subsequent surveys showed no damage to scallop beds, local fishermen reported daily losses up to £.Stg. 1 000 per boat. On 10 January 1975, the AFRAN ZODIAC spilled 391 t of Bunker C oil into Bantry Bay, and the moratorium on scallops, mussels and sea urchins was extended to September 1975. Local fishermen at the time of the second spill estimated their losses since October to reach some £.Stg. 300 000. Following the West Falmouth spill of No. 2 fuel oil, the scallop and oyster fishery from several areas, including one of 22 acres, was closed for at least 2 years on the grounds of health risk from identifiable oil components in tissues, and one of these areas was still closed 4 years after the incident.

Grant (1969) reports that from one area near Brisbane, Australia, alone, 78 short tons of sea mullet (_Mugil cephalus_) were lost between 1 May - 14 June 1968 due to condemnation on the grounds of Kerosene-like taint. Connell (1974) reports that the incidence of tainted fish extends over some 100 miles of coast and is complicated by spawning migrations mingling tainted and untainted fish, leading to the condemnation of the whole catch. Spooner (pers. comm.) reports that, after a spill of Arabian light crude oil into Tarut Bay, Saudi Arabia, in April 1970, oiled fish traps were unusable for at least 6 weeks. The enclosed nature of the affected water body resulted in serious tainting of fish and shrimp, but the fishery returned normal after 3 months. By contrast, the METULA spill of 51 500 t Arabian crude and

2 100 t Bunker C oil in the Magellan Straits in August 1974 caused relatively little economic damage to the local fishery (Whitman, 1975). Fishing areas for finfish and mussels had to be relocated, due to tainting problems. The local National Health Service determined that fish (Eleginops maclovinus) were unfit for human consumption soon after the spill, but were found to be palatable by early 1975.

6.2.4 Tainting

Tainting of marine foods has been of great concern to fishermen for many years, due to the fear of catches being refused on the market or at factories on delivery. This fear is increasing because of the increased exploration, production, transport and refining of oil.

Crustaceans, fish and molluscs exposed to oily conditions can acquire an objectionable oily taste.

The mechanisms of taste and smell of oil are described under the section of organoleptical investigation in an ICES report (Kerkhoff, 1974) as follows: "In an organoleptical test the impression obtained in the mouth is a total one, consisting of taste, tactile and olfactory sensations. The taste buds, located on the papillae of the tongue and the palate distinguish only four tastes - acid, bitter, salt and sweet. Most taste sensations are produced by a combination of taste and smell. The olfactory organ, situated in the upper portion of the nasal passage, forms the essential organ of smell and is of much more importance in perceiving the volatile flavour components in food than in the taste organs. For that reason the ability to taste oily products has to be associated with the appearance of volatile smell compounds in the oil."

Kerkhoff continues: "During the refining process of crude oil, the odour compounds are divided among the several distillate fractions. Light (gasoline) and heavy distillate fractions contain few, but the middle distillate fractions, like diesel oil, contain many of the odorous compounds present in crude oil. Gas oil is a middle distillate fraction with many volatile odorous compounds. Therefore, human senses are able to detect very low concentrations of gas oil."

The nose is a very sensitive detector and can detect surprisingly low concentrations of mineral oil. Martin (1970) gives a table of threshold odour concentrations for various oils in water, which shows that while diesel oil can be detected at 0.0005 ppm, fuel and crude oils, containing relatively fewer odour compounds, are only detected at 0.1-0.5 ppm.

6.2.5 Data

Kerosene taint in sea mullet (Mugil cephalus) was reported from Australian waters near Brisbane (Grant, 1969) where fishermen from 1 May to 14 June 1968 had losses through kerosene tainting of as much as 78 short tons in one area. This problem has been further examined by Vale et al. (1970); Shipton et al. (1970); Sidhu et al. (1972) and Connell (1974). The kerosene taste was shown by gaschromatography and mass-spectrometry to be very similar, qualitatively and quantitatively, to that of a commercial sample of kerosene. The components identified were n-tetradecane, naphthalene, 2-methylnaphthalene and 1-methylnaphthalene, and tentative identification of methylisopropylbenzene, 3-(2-methylphenyl) pentane, 2, 6-dimethyl-1, 2, 3, 4-tetrahydronaphthalene and 2, 3-dimethyl-1, 2, 3, 4-tetrahydronaphthalene. The source of the tainting was believed to be refinery effluents.

Nitta (1972) reports that fish kept in holding facilities in Osaka Harbour, Japan, became unmarketable due to oil discharges by ships. Studies on fish spoilage due to effluents from petrochemical plants at Yokkaichi showed all fish caught within 2 km of the harbour, half of the fish within 2.5-4 km and several species of animals, including sea eel, squilla and flatfish, within 4-15 km had offensive odours. Ogata and Miyake (1970) identified the tainting substance in the Okayama area as toluene, derived from petrochemical industrial wastes.

Blackman et al. (1973) report on a spill of 3 000 t of leaded petrol (gasoline) from which people had reported mackerel tasting of petrol, but this taint may have derived from refinery or ships' discharges in the area since no analysis was made (Blackman, pers.comm.).

The spill of 2 200 t of diesel oil occurred after grounding of a tanker near Finnsnes in the north of Norway in 1973. The oil was spread by surface currents and wind both northward and southward in the Gisund, a narrow strait. During the next two months local fishermen reported a smell and taste of oil in the fish caught. Common fish in the area are cod, saith, haddock, herring, flounder, sea trout and salmon. Samples of fish from the area taken two months after the incident were analysed organoleptically by a taste panel, gas-chromatographically after steam distillation and silicagel clean up and by GC/MS at the Institute of Marine Research in Bergen. Results from the taste panel and the chemical analyses showed oil contamination in good agreement with the reported tainting (Palmork and Wilhelmsen, 1974). They found oil hydrocarbons in the order of 150-200 µg/kg in cod liver. Samples of water and sediments from the area did not contain diesel oil components. It was later found that one 300 m embayed shore contaminated with a 20/80 oil-in-water emulsion had been successfully treated with a low toxicity dispersant (Whitman, pers.comm.).

The use of dispersants facilitates the uptake of oil and its components and facilitates the introduction of hydrocarbons to the lipid pool. It is therefore to be expected that the use of oil dispersants will increase likelihood of tainting. The solvent fraction of older dispersants contained tainting compounds of the same nature as those found in diesel and crude oils (Shipton et al., 1970).

Wilder (1970) describes laboratory experiments conducted to determine whether Bunker C oil alone or in combination with two dispersant products would taint lobster meat, under what conditions the taint occurs and how long it would persist. It was also the aim of the experiment to see if it was feasible to clean Bunker C oil from live lobsters. Results from these experiments were urgently needed because of the grounding of the tanker ARROW in the Chedabucto Bay, resulting in about 1 500 000 gallons of Bunker C oil leaking out in an area where the lobster fisheries employ about 700 fishermen.

External contamination of oil does not necessarily mean that tainting of flesh has occurred. Even ingestion of oil does not necessarily cause tainting of flesh (e.g., Wilder, 1970), but some species of crustacea and molluscs are consumed together with the gut content which may lead to rejection of the produce. Some of Wilder's results were that: Liberally smearing lobsters with oil produced no tainting of the meat (muscle) or tomalley (digestive gland, etc.) after 4 and 8 days return to clean running sea water. Lobsters readily ate bait smeared with Bunker C oil but no definite tainting could be detected in lobsters fed such bait after 4 and 8 days in clean running sea water. Meat and tomalley of lobsters immersed for 90 hours in sea water containing one part per 1 000 Bunker C oil or 1/1 000 Bunker C oil and 1/1 000 of one dispersant product acquire a very objectionable oily flavour. Tainting persists in meat for more than three weeks and in tomalley for more than a month. Tomalley was also more strongly flavoured than meat. The meat became tainted before boiling since lobsters boiled in water containing appreciable amounts of Bunker C oil did not become tainted. Meat or tomalley from lobsters immersed for 116 hours in sea water containing 1 part per 1000 of the other dispersant product or for short periods in 10% of that product did not become tainted. Wiping oil-smeared lobsters with the full strength of dispersant products effectively cleaned them externally and 24 hours after immediate replacement in clean, running sea water, meat and tomalley were not detectably tainted. However, since there were indications that this process facilitated passage of oil and dispersant to the gut and gills, tainting may subsequently have become detectable. Wilder's experiments are particularly interesting since, although external fouling with oil may itself be a ground for condemnation or rejection of produce, as he showed this does not necessarily mean that tainting of flesh has occurred. However, cooking of externally contaminated produce may lead to tainting of the flesh. As he also demonstrated, even ingestion of oil does not necessarily cause tainting of the flesh, but some species of crustacea and molluscs are eaten together with their gut contents which may lead to rejection, and boiling of whole shrimps (Crangon crangon) whose foreguts contained sunken Kuwait crude oil led to tainting of the meats despite the removal of the cephalothorax and contained oil before consumption (Blackman, pers.comm.).

After the TORREY CANYON disaster, tainting of mackerel and sea trout was reported (Simpson, 1968), although the tainting was believed to be due to the dispersants used rather than the Kuwait crude oil spilt, since these dispersants consisted of surfactants dissolved in Kerosene or light refined oils. Tainting of shellfish was also reported and experiments over six days were performed in areas where spraying of detergents was going on. The lobsters and crabs used in the experiment were boiled and the digestive gland ("cream"), tail meat and claw meat tasted for contamination. Results are shown in Table 6-4

TABLE 6-4

Tainting in the cream, tail meat and claw meat of lobster held
in pots off Porthleven (3-10 April) and Sennan (4-10 April)

	Position offshore (yards)	Whether tainted or not		
		Cream	Tail	Claw
Porthleven (1 lobster per pot)	200	no	no	no
	400	yes	yes	no
	800	no	no	no
Sennan (2 lobsters per pot)	200 (a)	yes	yes	no
	(b)	yes	yes	no
	400 (a)	yes	yes	yes
	(b)	yes	no	no
	800 (a)	yes	yes	no
	(b)	yes	no	no

From Simpson (1968)

The tainting gave a taste similar to paraffin or white spirit. It was further noticed that eggs of "berried" (egg-carrying) female lobster when eaten raw were found to be tainted, but these tastes did not appear until the eggs were actually bitten. Experiments were carried out to investigate more about the distribution of the tainting in lobster and to see how long it would last if kept in clean water. Table 6-5 shows the results of the experiments lasting seven days.

TABLE 6-5

Tainting of the uncooked eggs and the cooked meat of berried female lobsters
held in clean sea water from 12-19 May at a temperature of 49°F (9°C)

Sampling date	Lobster number	Uncooked eggs	Cooked meat			
			Cream	Coral	Tail	Claw
12 May	1	strong	strong	strong	no	no
	2	strong	strong	strong	no	no
	3	strong	strong	strong	slight	no
	4	strong	strong	strong	slight	slight
15 May	5	little	slight	slight	no	no
	6	–	slight	slight	no	no
16 May	7	little	slight	slight	no	no
	8	no	slight	slight	no	no
17 May	9-12	little	not cooked			
	13-17	no	not cooked			
	18	no	little	little	no	no
	19	no	little	little	no	no

From Simpson (1968)

Table 6-6 shows hydrocarbons identified in one of the dispersants used to clean oil spilt from the TORREY CANYON

TABLE 6-6

Hydrocarbons identified in a dispersant product

Name	Formula	Molecular weight	Boiling point °C
n-Nonane	C_9H_{20}	128	151
Ethylbenzene	C_8H_{10}	106	136
p-Xylene	C_8H_{10}	106	138
m-Xylene	C_8H_{10}	106	139
n-Decane	$C_{10}H_{22}$	142	174
o-Xylene	C_8H_{10}	106	144
iso-Propylbenzene	C_9H_{12}	120	152
n-Propylbenzene	C_9H_{12}	120	159
3-Ethyl-1-methylbenzene	C_9H_{12}	120	161
4-Ethyl-1-methylbenzene	C_9H_{12}	120	162
n-Undecane	$C_{11}H_{24}$	156	196
1, 3, 5-Trimethylbenzene	C_9H_{12}	120	164
2-Ethyl-1-methylbenzene	C_9H_{12}	120	165
1, 2, 4-Trimethylbenzene	C_9H_{12}	120	169
1, 2, 3-Trimethylbenzene	C_9H_{12}	120	176
1-Methyl-3-propylbenzene	$C_{10}H_{14}$	134	182
n-Dodecane	$C_{12}H_{26}$	170	216
1, 3-Dimethyl-5-ethylbenzene	$C_{10}H_{14}$	134	184
1, 4-Dimethyl-2-ethylbenzene	$C_{10}H_{14}$	134	187
1, 3-Dimethyl-4-ethylbenzene	$C_{10}H_{14}$	134	187
1, 3-Dimethyl-2-ethylbenzene	$C_{10}H_{14}$	134	190
2-Methyl-2-phenylbutane	$C_{11}H_{16}$	148	192
1, 2, 4, 5-Tetramethylbenzene	$C_{10}H_{14}$	134	195
2-Methyl-1-phenylbutane	$C_{11}H_{16}$	148	197
n-Tridecane	$C_{13}H_{28}$	184	235
1, 2, 3, 4-Tetramethylbenzene	$C_{10}H_{14}$	134	205
n-Tetradecane	$C_{14}H_{30}$	198	254
Naphthalene	$C_{10}H_8$	128	241
2-Methylnaphthalene	$C_{11}H_{10}$	142	241
1-Methylnaphthalene	$C_{11}H_{10}$	142	245
2, 6-Dimethylnaphthalene	$C_{12}H_{12}$	156	261
2, 3-Dimethylnaphthalene	$C_{12}H_{12}$	156	
1, 6-Dimethylnaphthalene	$C_{12}H_{12}$	156	

From Palmork and Vinsjansen (1972)

For comparison, Table 6-7 shows some of the hydrocarbons identified in a sea water extract of North Sea oil from the "Ekofisk" field.

TABLE 6-7

Hydrocarbons isolated from a seawater extract of "Ekofisk" crude

Name	Formula	Molecular weight	Boiling point °C
Toluene	C_7H_8	92	111
Ethylbenzene	C_8H_{10}	106	136
p-Xylene	C_8H_{10}	106	138
m-Xylene	C_8H_{10}	106	139
o-Xylene	C_8H_{10}	106	144
iso-Propylbenzene	C_9H_{12}	120	152
n-Propylbenzene	C_9H_{12}	120	159
1-Ethyl-3-methylbenzene	C_9H_{12}	120	161
1-Ethyl-2-methylbenzene	C_9H_{12}	120	165
1, 2, 4-Trimethylbenzene	C_9H_{12}	120	169
1, 2, 3-Trimethylbenzene	C_9H_{12}	120	176
1, 3-Diethylbenzene	$C_{10}H_{14}$	134	181
1, 2-Diethylbenzene	$C_{10}H_{14}$	134	183
1, 4-Diethylbenzene	$C_{10}H_{14}$	134	184
2, 4-Dimethyl-1-ethylbenzene	$C_{10}H_{14}$	134	188
1, 2, 4, 5-Tetramethylbenzene	$C_{10}H_{14}$	134	197
Naphthalene	$C_{10}H_8$	128	218
2-Methylnaphthalene	$C_{11}H_{10}$	142	241
1-Methylnaphthalene	$C_{11}H_{10}$	142	245
1, 2-Dimethylnaphthalene	$C_{12}H_{12}$	156	
1, 6-Dimethylnaphthalene	$C_{12}H_{12}$	156	
2, 6-Dimethylnaphthalene	$C_{12}H_{12}$	156	261
1, 2, 6-Trimethylnaphthalene	$C_{13}H_{14}$	170	146 (at 10 mm Hg)

From Palmork and Vinsjansen (1972)

The major tainting components of oil identified to this date are: phenols, dibenzothiophenes, naphthenic acids, mercaptans, tetradecans and the methylated naphthalenes. It must be remembered though that a minor component of the oil taint spectrum may end up as the major taint-producing compound in marine produce, from a particular oil exposure situation, due to differences in physical and chemical character altering their relative abundance as presented to be taken up or metabolized and eliminated by these organisms. Studies on the uptake and elimination by marine organisms of aromatic hydrocarbons in general are reported elsewhere (section 6.1). Detailed results are given here of studies involving identified oil-taint components.

Anderson and Neff (1974) worked on discharges of oil hydrocarbons from oysters which had been exposed to 400 ppm dispersed No. 2 fuel oil for eight hours. During 672 hours

(28 days) in clean sea water, samples of oysters were quantitatively analysed at intervals by means of gaschromatography/mass spectrometry. The results of this discharge experiment are summarized in Table 6-8. They also analysed unexposed oysters and found a total of 2.5 ppm of napthalene and alkylnaphthalenes. The most dominant components were the di- and tri-methylnapthalenes. At the end of the exposure (8 hours) the oysters had accumulated 312 ppm oil hydrocarbons in their tissues. When the oysters were returned to "oil-free" sea water more than 90 percent of the n-alkanes were discharged in 24 hours; the aromatic components, however, were released much more slowly. After 672 hours (28 days) of depuration only small amounts of mono-, di- and tri-methylnaphthalenes remained in the oyster tissues.

Anderson and Neff (1974) also performed similar experiments with three other oils, South Louisiana crude, Kuwait crude and Venezuelan Bunker C, and they report a similar uptake and discharge to that of the first experiments. In all cases, naphthalenes were accumulated to high concentrations and were retained longer than other aromatics and n-alkanes.

Groups of clams, shrimp and fish were exposed to the water soluble fraction of No. 2 fuel oil in synthetic sea water for 2 to 24 hours and then placed in oil-free sea water recirculated through activated charcoal. Samples of the organisms were analysed for total naphthalenes during and after exposure (Figure 6-1) (Anderson and Neff, 1974).

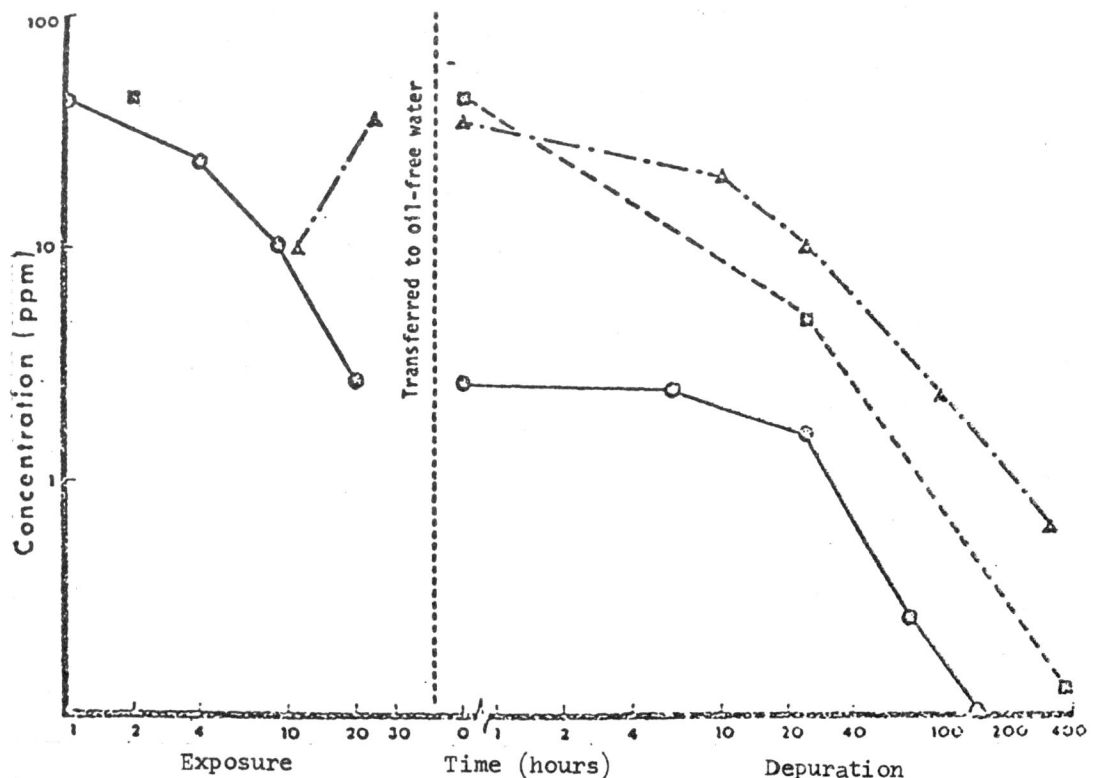

Figure 6-1. <u>The concentrations of total naphthalenes (TNs) in the tissues of clams, shrimp and fish at different times during exposure to water-soluble fractions of No. 2 fuel oil and following return to oil-free sea water</u>

△ - · -△   Clams, <u>Rangia cuneata</u>, exposure concentration 3.4 ppm TNs
O ――― O   Shrimp, <u>Penaeus aztecus</u>, exposure concentration 0.7 ppm TNs
□ ---- □   Fish, <u>Fundulus similus</u>, exposure concentration 1.9 ppm TNs

From Anderson and Neff (1974)

TABLE 6-8

Concentrations of different petroleum hydrocarbons in the tissues of
oysters Crassostrea virginica after 8 hours exposure to a No. 2 fuel oil-in-water dispersion
and at different times following return to oil-free sea water

Time (hours)	Petroleum Hydrocarbon Concentration (ppm µg/g wet weight)												
	n-P	N	1-MN	2-MN	DMN	TMN	B	MB	F	MF	DBT	P	MP	DMP	Total
Exposure															
0	-	0,2	0,1	0,3	1,0	0,8	-	-	-	-	-	-	-	-	2,4
8	235	14,7	8,7	15,0	21,8	9,1	0,3	0,5	1,0	1,2	0,3	1,9	1,9	0,3	312
Depuration															
3	156	12,0	8,4	12,0	22,7	10,8	0,3	0,4	0,7	0,7	0,3	1,3	1,3	0,2	228
6	68	7,3	5,1	7,3	13,2	5,7	0,1	0,2	0,4	0,2	0,1	0,6	0,6	0,1	109
24	18	6,5	5,7	7,5	14,8	9,5	0,2	0,2	0,5	0,7	0,2	1,2	1,3	0,3	67
120	10	8,2	4,7	6,8	13,4	4,9	0,1	0,1	0,2	0,1	0,1	0,4	0,4	0,2	54
672	-	-	-	0,1	0,5	0,9	-	-	-	-	-	-	-	-	1,5

n-P = n-paraffins; N = naphthalene; 1-MN, 2-MN = 1 methyl- and 2-methylnaphthalenes; DMN = dimethylnaphthalene;
TMN = trimethylnaphthalene; B = biphenyl; MB = methylbiphenyl; F = fluorene; MF = methylfluorene; DBT = dibenzo-
thiophene; P = phenanthrene; MP = methylphenanthrene; DMP = dimethylphenanthrene

From Anderson and Neff (1974)

The patterns of the uptake, retention and discharges of total naphthalenes (TN) were different for the three species of animals examined. The shrimp accumulated TNs extremely rapidly and reached a maximum after only 1 hour (43 ppm). They then began discharging the accumulated components back into the exposure water. The clams accumulated TN more slowly but continued to do so for the whole exposure time (24 hours) reaching 34 ppm. The fish accumulated TN very rapidly (2 hours) to the level of 43.2 ppm. In all three cases the maximum tissue concentration was one or more orders of magnitude higher than that of the exposure water. Fish and shrimp discharged the oil hydrocarbons more rapidly than did the clams and all three species used in the experiment were able to reduce tissue naphthalenes to low or undetectable levels within 14 days. In subsequent experiments Anderson and Neff (1974) showed that clams and fish were able to completely depurate accumulated hydrocarbons within three to four weeks. Lee, Sauerheber and Benson (1972) reported that after a 4 hour exposure to 32 μg 14C-naphthalene/l seawater, blue mussels (Mytilus edulis) showed a 60% decrease in tissue load of naphthalene taken up after 72 hours replacement in clean, static water. Lee, Sauerheber and Dobbs (1972) reported that after 1 hour exposure to 32 μg/l 14C-naphthalene, mudsuckers (Gillichthys mirabilis) and sculpins (Oligocottus maculosus) showed a 90% decrease in tissue load after 24 hours replacement in clean, static water. Table 6-10 gives some examples of tainting by petroleum products and oil dispersants in fish and shellfish.

The chemical verification of tainting can be difficult due to the natural hydrocarbons. Hydrocarbons are widespread in the environment due to their great chemical stability and many are produced by the living organisms themselves. Many of these biosynthesized hydrocarbons have been identified (Blumer et al., 1969, 1970; Blumer, Guillard and Chase, 1971) using modern techniques. Pristane and some n-alkanes and isoprenoid hydrocarbons are widespread in the food chain. The n-alkanes, with an odd number of carbon atoms, predominate. Blue mussel (Mytilus edulis) as an example contains around 1 mg natural hydrocarbons per mussel (Lee, Sauerheber and Benson, 1972). Squalene occurs widely in nature though usually in trace amounts, as a precursor to the synthesis of sterols (Lewis, R.W., 1971). In kerosene-tainted fish, mullet, the amount of squalene seems to increase considerably up to 100 times the normal concentration.

For many years the taste and/or odour were the main criteria for the refusal of contaminated or tainted fish and shellfish. In some incidents, oil pollution was blamed for the tainting of marine produce, where quite natural reactions or other pollutants were responsible for the tainting, for example the decomposition of dimethyl-β-propiothetin to dimethyl sulphide (Table 6-9).

TABLE 6-9

Other sources of tainting in fish and shellfish

Source	Off-flavour	References
Dimethyl-β- propiothetin	Like aromatic petroleum products	Motohiro, 1962 Ackman, Tocher and McLachlan, 1966 Ackman, Dale and Hingley, 1966 Ackman, Hingley and Ray, 1967
Dimethyl sulphide	Like aromatic petroleum products	Ackman and Hingley, 1972 Ronald and Thomson, 1964
Actinomycetes	Musty	Daugherty, Campbell and Morris, 1966
Chlorinated phenols	Strong phenolic taste	Mann, 1969

TABLE 6-10

Examples of tainting by petroleum products in fish and shellfish

Case	Species Affected	Oil Exposure	Taste Observations	Chemical Verification	Reference
1	*Scomber scombrus* Mackerel	Crude oil spill to coast (TORREY CANYON)	Tainting by detergent reported	None	Simpson, 1968
	Salmo trutta Sea trout	"	"	"	
	Hippoglossus hippoglossus Plaice	"	Oil taste reported in single specimen	"	
	Homarus vulgaris Lobster	"	Paraffin taste	"	
	Cancer pagurus Crab	"	Tainted	"	
2	*Aequipecton irradians* Scallop	No. 2 Diesel fuel oil spill to bay	Objectionable oily taste	Qualitative similarity between GLC peaks of scallop muscle extracts and fuel oil	Blumer, Souza and Sass, 1970
3	*Homarus americanus* Lobster	Bunker C oil spill in bay (ARROW)	Persistent oil flavour	None	Wilder, 1970
4	*Seriola quinqueradiata* Yellowtail	Lab: crude oil, 10-50 ppm, 5-13 days	Oily-smelling flesh	GLC peaks of fish flesh were almost identical to the crude oil	Deshimaru, 1971
5	*Cyprinus carpio* Carp	Lab: 4 oil dispersers 6-300 ppm, 48 hrs	Oily polluted smell	Close relationship between flesh GLC peaks and those of the dispersants	Deshimaru, 1971a
6	*Mugil cephalus* Mullet	Possibly oil refineries, industrial discharges, and/or boat spillage	Kerosene odour	Qualitative similarity between GLC peaks of mullet oil extracts and kerosene; also infra-red and NMR spectra.	Sidhu et al., 1972
7	Grey mullet	Oil effluents from petroleum refineries and other industries	Offensive oil odour	None	Nitta, 1972
	Sea eel	"	"	"	
	Squilla	"	"	"	
	Flatfish	"	"	"	

TABLE 6-10
(continued)

Case	Species Affected	Oil Exposure	Taste Observations	Chemical Verification	Reference
8	Scomber scombrus Mackerel	Petrol (gasolene) spill to coast	Petrol taste	None	Blackman et al., 1973
9	Mytilus edulis Blue mussel	Gas oil spill to estuary	Threshold: oily taste at hydrocarbon conc. in flesh $\geq$ 5 ppm	Qualitative similarity between GLC peaks of Mytilus flesh extracts and gas oil	Kerkhoff, 1974
10	Sebastes marinus	Diesel oil spill in fjord (BRITISH MALLARD)	Oily and kerosene flavour	—	Palmork and Wilhelmsen, 1974
	Salmo trutta	as above	as above	—	
	Clupea harengus Herring	as above	—	Identification by gaschromatography/mass spectrometry 100–200 mg/kg total oil	
	Liver of Gadus morhua Cod	as above	—	50–100 mg/kg total oil	
11	Lobster	Diesel oil spill fouling holding crates	Extractable volatiles and background envelope associated with taint at 12–16 ppm but not at 2–4 ppm	Temperature programmed gas chromatography	Paradis and Ackman, 1975

There have been too few studies on the tissue levels of oil components to establish tainting threshold levels. Those few that have been performed indicate thresholds of 5 ppm gas oil in spiked mussel tissue (Kerkhoff, 1974), 10-30 ppm in spiked tissue for a crude oil (Whittle and Mackie, 1975) and 4-12 ppm extractables from diesel oil in lobsters (Paradis and Ackman, 1975), while Whittle (Blackman, pers.comm.) has found that above 200-300 ppm of a crude oil in spiked tissue no further increases can be sensed by a trained taste panel. Other workers have reported that exposure to ambient water concentrations as low as 0.01 ppm of an ether extract of oil-industry wastes lead to tainting of the flesh of certain fish (Nitta et al., 1965).

6.2.6 Comments and conclusions

There are indications that the tissue lipid content and the amounts of free lipids increase the susceptibility to tainting (Whittle and Mackie, 1975; Palmork and Wilhelmsen, 1974). One should therefore expect that fatty fishes and lipid-rich organs, such as gonads, will become more strongly tainted and perhaps for a longer period of time (Wilder, 1970; Simpson, 1968). Therefore, the seasonal condition of the produce will also affect the susceptibility and the strength of taint, as lipid content and metabolic rates vary. At the low concentrations at which a taste panel would be employed to detect taint, there is unlikely to be any significant health hazard to panel members if the samples are not swallowed (reference is made to section 6.1 of this chapter). Particular types of fish and shellfish may be more at risk to tainting than others, as Sidhu et al. (1972) point out for mullet, being an estuarine fish and a bottom mud grazer and detritus feeder with relatively high body fat. This fish also shows no aversion to the presence of high ambient oil hydrocarbons and thus may remain in polluted waters for considerable periods. At present, the identity of the odorous components of oils has not been fully established. Therefore, for the present, samples of produce suspected of being tainted following an oil spill incident or from a site of chronic discharge will have to be tested by a taste panel at intervals, preferably correlated with tissue analysis, to identify major tainting components, (e.g., computerized GC/MS or sulphur specific detector in connexion with GC). Compilation of the results from analyses following such incidents should allow more accurate prediction of the time-scale for closing of the fishery and discharge to acceptable levels of oil components by organisms returned to clean conditions, and hence the time-scale to reach threshold levels of tainting by any identified components.

Part III: BIBLIOGRAPHY

1. REFERENCES TO PART II [1]

Abraham, G. and C. van Dam, On the predictability of waste concentrations. In Marine
1972   pollution and sea life, edited by M. Ruivo. West Byfleet, Surrey, Fishing
    News (Books) Ltd., p. 135-40
    1.2

Ackman, R.G. and J. Hingley, Dimethyl sulfide as an odor component in Nova Scotia fall
1972   mackerel. J.Fish.Res.Board Can., 29(7):1085-8
    6.2.5 (table)

Ackman, R.G., J. Dale and J. Hingley, Deposition of dimethyl-propiothetin in Atlantic cod
1966   during feeding experiments. J.Fish.Res.Board Can., 23:487-9
    6.2.5 (table)

Ackman, R.G., J. Hingley and A.W. Ray, Dimethyl-β- propiothetin and dimethyl sulfide in
1967   Labrador Cod. J.Fish.Res.Board Can., 24:457-61
    6.2.5 (table)

Ackman, R.G., C.S. Tocher and J. McLachlan, Occurrence of dimethyl-β-propiothetin in marine
1966   Phytoplankton. J.Fish.Res.Board Can., 23(3):357-64
    6.2.5 (table)

Ahmed, A.M. et al., Sampling errors in the quantitation of petroleum in Boston harbor water.
1974   Analyt.Chem., 46:1858-60
    3.3

Alexander, M., Microbial ecology. New York, John Wiley and Sons, Inc., p. 413-6
1971   2.1.9

Allen, H. Effects of petroleum fractions on the early development of a sea urchin. Mar.
1971   Pollut.Bull., 2:138-40
    5.1.6

Alyakrinskaya, I.O., Effect of oil on the survival and rate of growth of young (Black sea)
1966   mullets. Rybn.Khoz., 42:16-8 (in Russian)
    5.1.6

American Public Health Association, Standard methods for the examination of water and waste
1971   water. Washington, D.C., American Public Health Association, p. 254-6
    3.3

American Society for Testing Materials, Water: atmospheric analysis. Part 23. Method
1972   D-2778, 769 Philadelphia, ASTM
    3.3

Andelman, J.B. and M.J. Suess, Polynuclear aromatic hydrocarbons in the water environment.
1970   Bull.WHO, (43):479-508
    6.1.2.1

Anderson, G.E., The effects of oil on the gill filtration rate of Mya arenaria. Va.J.Sci.,
1972   23(2):45-7
    6.1.2.2

Anderson, J.W., Petroleum hydrocarbons and oyster resources of Galveston Bay, Texas. In
1975   Proceedings, Joint conference on prevention and control of oil spills. March
    25-27, 1975, San Francisco, California, p. 541-8

---

[1] The figures underlined at the end of each bibliographic entry refer to relevant subsections of part II.

Anderson, J.W. and J.M. Neff, Accumulation and release of petroleum hydrocarbons by edible
1974   marine animals. Paper presented to the International Symposium, Recent advances
       in the assessment of the Health Effects of Environmental Pollution, Paris,
       June 1974 (Information from: J.W. Anderson, Dept. of Biology, Texas A & M Universi-
       6.2.5; 6.2.5 (table)

Anderson, J.W. et al., Characteristics of dispersions and water-soluble extracts of crude and
1974   refined oils and their toxicity to estuarine crustaceans and fish. Mar. Biol.,
       27:75-88
       2.1.3; 6.1.2.1

Arasaki, S., A. Inouye and Y. Kochi, The disease of the cultured Porphyra with special
1960   reference to the cancer-disease and the Chytride-disease which occurred in the
       culture field in Tokyo-Bay during 1959-60. Bull.Jap.Soc.Sci.Fish., 26:1074-81
       6.1.4

Atema, J. and L.S. Stein, Sublethal effects of crude oil on lobster (Homarus americanus)
1972   behaviour. Tech.Rep.Woods Hole Oceanogr.Inst., (72-74):81 p.
       5.1.6; 5.3

Atlas, R.M. and R. Bartha, Degradation and mineralization of petroleum in seawater: limi-
1972   tation by nitrogen and phosphorus. Biotechnol.Bioeng., 14:309-18
       2.1.10

_____, Biodegradation of petroleum in seawater at low temperatures. Can.J.Microbiol.,
1972a 18:1851-5
       2.1.10

_____, Degradation and mineralization of petroleum by two bacteria isolated from
1972b coastal waters. Biotechnol.Bioeng., 14:297-308
       2.1.10

_____, Effects of some commercial oil herders, dispersants and bacterial inocula
1973   on biodegradation of oil in seawater. Publ.La.State Univ., (LSU-SG-73-01):283-90
       2.1.5

_____, Abundance, distribution and oil biodegradation potential of micro-organisms
1973a in Raritan Bay. Envir.Pollut., 4:291-300
       2.1.10

_____, Stimulated biodegradation of oil slicks using oleophilic fertilizers. Environ.
1973b Sci.Technol., 7:538-41
       2.1.10

Atlas, R.M., R. Bartha and A. van Leeuwenhoek, Inhibition by fatty acids of the biodegradation
1974   of petroleum. Biosci.Div.,Jet Propulsion Lab.,Pasadena,Calif., 39(2):257-71
       2.1.10

Aubert, M., Télémédiateur chimique et équilibre biologique océanique. 1. Théorie générale.
1971   Rev.Int.Océanogr.Méd., 21:5-16
       1.2

Aubert, M., R. Chaira and G. Malara, Etude de la toxicité de produits chimiques vis-à-vis de
1969   la chaine biologique marine. Rev.Int.Océanogr.CERBOM, 13/14:45-72
       5.4.1 (table)

Aubert, M. et al., Effets des pollutions chimiques vis-à-vis de télémédiateurs intervenant
1972   dans l'écologie microbiologique et planctonique en milieu marin. Rev.Int.Océanogr.
       Méd., 28:129-66
       1.2

Baker, J.M., The ecological effects of oil pollution on littoral communities. Proceedings
1971 of a symposium organized by the Institute of Petroleum, held at the Zoological
Society of London, edited by E.B. Cowell. London, Institute of Petroleum
5.1.9; 5.1.10; 5.4.1 (table)

_____, Ecological effects of refinery effluents. Rep.Oil Pollut.Res.Unit,Field Stud.
1973 Counc., 1972:7-13
5.1.10

_____, Recovery of salt marsh vegetation from successive oil spillages. Environ.Pollut.
1973a 4:223-30
5.1.10

_____, Biological effects of refinery effluents. In American Petroleum Institute,
1973b Proceedings of Joint conference on prevention and control of oil spills,
Washington, D.C., March 13-15, 1973. Washington, D.C., American Petroleum
Institute, p. 715-23
5.1.10

Barber, R.T. et al., Variations in phytoplankton growth associated with the source and
1971 conditioning of upwelling water. Inv.Reg., 35:171-93
1.2

Barger, W.R. and W.D. Garrett, Surface active organic material in the marine atmosphere.
1970 J.Geophys.Res., 75:4561-6
1.2

Barry, M.M. and P.P. Yevich, Incidence of gonadal cancer in the Quahog Mercenaria mercenaria.
1972 Oncology, 26:87-96
6.1.4

_____, The ecological, chemical and histopathological evaluation of an oil spill site.
1975 Part III. Histopathological studies. Mar.Pollut.Bull., 6(11):164
6.1.4

Barry, M.M., P.P. Yevich and N.H. Thayer, Atypical hyperplasia in the shell Clam Mya arenaria.
1971 J.Invert.Pathol., 17:17-27
6.1.4

Battelle Memorial Institute, Pacific Northwest Laboratories, Study of effects of oil discharges
1974 and domestic and industrial wastewaters on the fisheries of Lake Maracaibo,
Venezuela. Vol. 1. Ecological characterization and domestic and industrial
wastes. Richland, Washington, Battelle Pacific Northwest Laboratories, 184 p.
2.1.5; 5.1.4

_____, Study of effects of oil discharges and domestic and industrial wastewaters on
1974a the fisheries of Lake Maracaibo, Venezuela. Vol. 2. Fate and effects of oil.
Richland, Washington, Battelle Pacific Northwest Laboratories, 142 p.
2.1.5; 5.1.4

_____, Summary report on effects of oil discharges, domestic and industrial wastewaters
1974b on the fisheries of Lake Maracaibo, Venezuela. Richland, Washington, Battelle
Pacific Northwest Laboratories, 18 p.
5.1.4

Belterman, T., The AMER oil spill. Report 2-71. Annual report and review of events 1971.
1972 Cambridge, Mass., Smithsonian Institution Center for Short-Lived Phenomena,
June 1972
5.1.1

Bennett, M., H.J. Dee and N. Harkness, The determination of vegetable and mineral oils in
1973 the effluents and sewage sludges of the upper Tame basin. Water Res., 7:1849-59
6.1.1.4

Berridge, S.A., M.T. Thew and A.G. Loriston-Clarke, Formation and stability of emulsions of
1968 water in crude petroleum and similar stocks. J.Inst.Petrol., 54:333-57
2.1.4

Berridge, S.A. et al., The properties of persistent oils at sea. J.Inst.Petrol., 54:300-9
1968 2.1.1

Blackman, R.A.A., Effects of sunken crude oil on the feeding and survival of the brown shrimp
1972 Crangon crangon. International Council for the Exploration of the Sea
(CM 1972/K:13):8 p.
6.1.4

Blackman, R.A.A. et al., The DONA MARIKA oil spill. Mar.Poll.Bull., 4(12):181-2
1973 2.1.4; 6.2.5; 6.2.5 (table)

Blair, W.R., W.P. Iverson and F.E. Brinckman, Application of gas chromatographatonic
1974 absorption detection system to a survey of mercury transformations by Chesapeake
Bay microorganisms. Chemosphere, 3:167-74
4.3

Blaylock, J.W., R.M. Bean and R.E. Wildung, Determination of extractable organic material
1974 and analysis of hydrocarbon types in lake and coastal sediments. Spec.Publ.
U.S.Natl.Bur.Stand., (409):217-9
3.2 (table)

Bliss, C.I. and McK. Cattell, Biological assay. Ann.Rev.Physiol., 5:479-573
1943 5.4.1

Bloom, S.A., An oil dispersant's effect on the microflora of beach sand. J.Mar.Biol.Assoc.,
1970 U.K., 50:919-23
2.1.5

Blumer, M., Hydrocarbons in the digestive tract and liver of a basking shark. Science, Wash.,
1967 D.C., 156:390-1
1.2; 2.2; 6.1.2.1

—————, Oil pollution of the ocean. Oceanus, 15:2-7
1969 1.2

—————, Oil in the sea. Transcript of a radio broadcast presented by the American
1970 Chemical Society, October 2, 1970
5.1.6

—————, Verunreinigung der Gewässer durch Oel. Zum Problem der persistenten Chemika-
1971 lien in der Umwelt (Pollution of water by oil. On the problem of persistent
chemicals in the environment). Paper presented at Eidgenoessische Anstalt fuer
Wasserversorgung, Abwassereinigung und Gewaesserschutz, Zurich, Switzerland,
May 11, 1971. (Separatum No. 398):21 p.
2.1.9; 5.1.3; 5.1.6

—————, Scientific aspects of the oil spill problem. Environ.Affairs, 1:54-73
1971a 2.1.9; 5.1.3; 6.1.1.2; 6.1.3

—————, Submarine seeps: are they a major source of open ocean oil pollution? Science,
1972 Wash.,D.C., 176:1257-8
1.2

Blumer, M., Oil contamination and the living resources of the sea. In Marine pollution and
1972a sea life, edited by M. Ruivo. West Byfleet, Surrey, Fishing News (Books) Ltd.,
pp. 476-81
5.3

Blumer, M. and J. Sass, Indigenous and petroleum-derived hydrocarbons in a polluted sediment.
1972 Mar.Pollut.Bull., 3(6):92-3
2.1.6; 2.1.9

_____, The West Falmouth Oil Spill - data available in November, 1972. Massachusetts,
1972a Woods Hole Oceanographic Institution (WHOI-72-19)
2.1.9

_____, The West Falmouth oil spill - data available in November 1971. 2. Chemistry,
1972b Massachusetts, Woods Hole Oceanographic Institution, (WHOI-72-19:104)
2.1.9

_____, Oil pollution: persistence and degradation of spilled fuel oil. Science,
1972c Wash.,D.C., 176:1120-2
1.2; 2.1.6

Blumer, M. and D.W. Thomas, Pristane in the marine environment. Helgol.Wiss.Meeresunters.,
1964 10:187-201
1.2

_____, "Zamene", isomeric C_{19} monoolefins from marine zooplankton, fishes and mammals.
1965 Science,Wash.,D.C., 148:370-1
1.2; 6.1.2.1

Blumer, M., M. Ehrhardt and J.H. Jones, The environmental fate of stranded crude oil. Deep
1973 Sea Res., 20:239-59
1.2; 2.1.9

Blumer, M., R.R.L. Guillard and T. Chase, Hydrocarbons of marine phytoplankton. Mar.Biol.,
1971 8(3):183-9
6.2.5

Blumer, M., M.M. Mullin and R.R.L. Guillard, A polyunsaturated hydrocarbon (3,6,9,12,15,18-
1970 Heneicosahexaene) in the marine food web. Mar.Biol., 6(3):226-35
1.2; 6.1.2.1; 6.1.3

Blumer, M., M.M. Mullin and D.W. Thomas, Pristane in zooplankton. Science Wash.,D.C.,
1963 140:974
6.1.2.1

Blumer, M., G.Souza and J. Sass, Hydrocarbon pollution of edible shellfish by an oil spill.
1970 Mar.Biol., 5:195-202
2.2; 3.2; 5.1.2; 5.1.6; 6.1.2.1; 6.2.5 (table)

Blumer, M. et al., Phytol-derived C_{19} Di- and Triolefinic hydrocarbons in marine zooplankton
1969 and fishes. Biochemistry, 8:4067-74
1.2; 6.2.5

_____, The West Falmouth oil spill - persistence of the pollution eight months after
1970 the accident. Tech.Rep.Woods Hole Oceanogr.Inst., (70-44)
1.2; 6.1.2.1; 6.1.3; 6.1.4; 6.2.5

_____, A small oil spill. Environment, 13(2):2-11
1971 1.2; 2.1.9

Blumer, M. et al., Petroleum. In A guide to marine pollution, edited by E.D. Goldberg.
1972   New York, Gordon and Breach
         2.1.9; 3.2; 6.1.1.4; 6.1.2.1

_____, Interaction between marine organisms and oil pollution. Environ.Protect.Ag.
1973   Ecol.Res.Ser. (EPA-R3-73-042):97 p.
         1.2

Boehm, P.D. and J.G. Quinn, The solubility behaviour of No. 2 fuel oil in seawater.
1974   Mar.Pollut.Bull., 5:101-5
         5.1.6

Boney, A.D., Aromatic hydrocarbons and the growth of marine algae. Mar.Pollut.Bull.,
1974   5(12):185-6
         6.1.4

Boney, A.D. and E.D.S. Corner, On the effects of some carcinogenic hydrocarbons
1962   on the growth of sporelings of marine red algae. J.Mar.Biol.Assoc.U.K., 42:579-85
         6.1.4

Borneff, J. and B. Fabian, Kanzerogene Substanzen in Speisefett und Öl. Arch.Hyg.Bakteriol.,
1966   150:485-512
         6.1.3

Bourne, W.R.P., Atlantic puffin decline. Smithsonian Institution Center for Short-Lived
1971   Phenomena. Annual Report, Report 83-71
         5.2.1

Bourne, W.R.P. and L. Johnston, The threat of oil pollution to North Scottish seaboard
1971   colonies. Mar.Pollut.Bull., 2:117-9
         5.1.1; 5.2.1

Boylan, D.B. and B.W. Tripp, Determination of hydrocarbons in sea water extracts of crude
1971   oil and crude oil fractions. Nature,Lond., 230(5288):44-7
         5.1.6

Boyland, E. and J.B. Soloman, Metabolism of polycyclic compounds. 8. Acid labile precursors
1955   of naphthalene produced as metabolites of naphthalene. Biochem.J., 59:518-22
         2.2

Bridie, A.L. and J. Bos, Biological degradation of mineral oil in sea water. J.Inst.Petrol.,
1971   Lond., 57(557):270-7
         2.1.10

Brinckman, F.E. et al., Mercury distribution in Chesapeake Bay. Progress Wat.Technol.,
1975   7(Supp.):251-2
         4.3

Brocksen, R.W. and H.T. Bailey, Respiratory responses of juvenile chinook salmon and striped
1973   bass exposed to benzene, a water-soluble component of crude oil. In American
         Petroleum Institute, Proceedings of Joint conference on the prevention and
         control of oil spills, Washington, D.C., March 13-15, 1973. Washington, D.C.,
         American Petroleum Institute, p. 783-91
         5.1.3

Brown, R.A. et al., Distribution of heavy hydrocarbons in some Atlantic Ocean waters.
1973   In American Petroleum Institute, Proceedings of the Joint conference on prevention
         and control of oil spills, Washington, D.C., March 13-15, 1973. Washington, D.C.,
         American Petroleum Institute, p. 505-19

Brown, R.G., Sea birds and oil pollution. In Oil and the Canadian environment - Proceedings
1973   of the Conference, 16 May 1973, sponsored by the Insitute of Environmental
       Sciences and Engineering, edited by D. Mackay and W. Harrison, Toronto, Ont.,
       University of Toronto
       5.1.1

Brown, R.G. et al., Bird mortality from oil slicks off eastern Canada. February-April, 1970.
1973   Can.Field Nat., 87:225-34
       5.1.1; 5.2.1

Brownell, R.L., Jr., Whales, dolphins and oil pollution. In Biological and oceanographical
1971   survey of the Santa Barbara oil spill 1969-1970, compiled by D. Straughan.
       Los Angeles, University of Southern California, Allan Hancock Foundation,
       Vol.1:255-76
       5.1.2

Brownell, R.L., Jr. and B.J. LeBoeuf, California sea lion mortality: natural or artifact?
1971   In Biological and oceanographical survey of the Santa Barbara oil spill,
       1969-1970, compiled by D. Straughan, Los Angeles, University of Southern
       California, Allan Hancock Foundation, Vol.1:287-306
       5.1.2

Brunnock, J.F., D.F. Duckworth and G.G. Stephens, Analysis of beach pollutants. J.Inst.
1968   of Petrol., 54:310-25
       2.1.10

Burnett, F.L. and D.E. Snyder, Blue crab as a starvation food of oiled American eiders.
1954   Auk, 71:315-6
       5.1.1

Burns, K.A. and J.M. Teal, Hydrocarbons in the pelagic Sargassum community. Deep-Sea Res.,
1973   20:207-11
       1.2; 2.2; 6.1.2.2

Cabioch, L., The fight against pollution by oil on the coast of Brittany. In Water
1971   pollution by oil, edited by P. Hepple. London, The Institute of Petroleum,
       p. 245-9
       5.1.9

Cahnmann, H.J. and M. Kuratsune, Determination of polycyclic aromatic hydrocarbons in
1957   oysters collected in polluted water. Analyt.Chem., 29(9):1312-7
       1.2; 6.1.1.5

Cairns, J. and A. Scheier, The effects of temperature and water hardness upon the toxicity
1962   of naphthenic acids to the common bluegill sunfish Lepomis machrochirus and
       to the pond snail Physus heterostropha. Not.Nat.,Phila. (353):1-12
       5.4.1 (table)

Carruthers, W. et al., 1,2 Benzanthracene derivatives in a Kuwait mineral oil. Nature,
1967   Lond., 213(5077):691-2
       6.1.1.1

Catchpole, W.M., E. MacMillan and N. Powell, Specifications for cutting oils with special
1971   reference to carcinogenicity. J.Inst.Petrol.,Lond., 57(577):247-60
       6.1.1.2

Cerniglia, C.E. and J.J. Perry, Crude oil degradation by microorganisms isolated from the
1973   marine environment. Z.Allg.Mikrobiol., 13(4):299-306
       2.1.10

Chan, G.L., A study of the effects of the San Francisco oil spill on marine organisms. In
1973 American Petroleum Institute, Proceedings of Joint Conference on prevention and
control of oil spills. Washington, D.C., March 13-15, 1973. Washington, D.C.,
American Petroleum Institute, p. 741-81
5.1.6

Chia, F.S., Diesel oil spill at Anacortes. Mar.Pollut.Bull., 2:105-6
1971 5.1.6

Chipman, W.A. and P.S. Galtsoff, Effects of oil mixed with carbonised sand on aquatic animals.
1949 Washington. Spec.Sci.Rep.USFWS, (F1):52 p.
5.4.1 (tables)

Christensen, D.J., C.A. Farley and F.G. Kern, Epizootic neoplasms in the clam Macoma balthica
1974 (L) from Chesapeake Bay. J. Natl.Cancer Inst., 52(6):1739-49
6.1.4

Clark, R.B., Oil pollution and its biological consequences. A review of current scientific
1971 literature. Paper presented for Great Barrier Reef Petroleum Drilling Royal
Commissions. University of Newcastle-upon-Tyne, Department of Zoology, 111 p.
6.1.1.5

─────────, The biological consequences of oil pollution of the sea. In David Davies
1971a Memorial Institute of International Studies, Report of a conference on water
pollution as a world problem, Aberystwyth, 11-12 July 1970. London, Europa
Publications, p. 53-77
6.1.1.5

─────────, Impact of chronic and acute oil pollution on sea birds. In Background papers
1973 for a Workshop on Inputs, Fates and Effects of Petroleum in the Marine Environment.
Washington, D.C., p. 619-34
5.1.1; 5.2.1

Clark, R.B. and J.P. Croxall, Rescue operations on oiled seabirds. Mar.Pollut.Bull.,
1972 3:123-7
5.1.1

Clark, R.C., Jr. and J.S. Finley, Paraffin hydrocarbon patterns in petroleum-polluted
1973 mussels. Mar.Pollut.Bull., 4:172-6
5.1.6

─────────, Techniques for analysis of paraffin hydrocarbons and for interpretation of
1973a data to assess oil spill effects in aquatic organisms. In American Petroleum
Institute, Proceedings of Joint conference on prevention and control of oil spills,
Washington, D.C., March 13-15, 1973, American Petroleum Institute, p. 161-72
5.1.6

Clark, R.C., Jr., J.S. Finley and G.G. Gibson, Acute effects of outboard motor effluent
1974 on two marine shellfish. Environ.Sci.Technol., 8(2):1009-14
5.1.6

Clark, R.C. et al., Interagency investigations of a persistent oil spill on the Washington
1973 coast. Animal population studies, hydrocarbon uptake by marine organisms, and
algae response following the grounding of the troopship GEN. M.C. MEIGS. In
Proceedings of Joint conference on prevention and control of oil spills,
Washington, D.C., March 13-15, 1973, Washington, D.C. American Petroleum
Institute, p. 793-808
5.1.6; 6.1.2.1

Clendenning, K.A. and W.J. North, Effects of wastes on the giant kelp Macrocystis pyrifera.
1960 Proc.Intern.Conf.Waste Dispos.Mar.Environ., 1:82-91
5.4.1 (table)

Cobet, A.B., H.E. Guard and M.A. Chatigny, Considerations in application of microorganisms
1973   to the environment for degradation of petroleum products. Publ.La.State Univ.,
        (LSU-SG-73-01):81-7
        2.1.10

Colwell, R.R. and J.D. Nelson, Jr., Metabolism of mercury compounds in microorganisms.
1975   U.S.EPA Publ.No.600/3-15-007
        4.3

Colwell, R.R., J.D. Walker and J.D. Nelson, Jr., Microbial ecology and the problem of
1973   petroleum degradation in Chesapeake Bay. Publ.La.State Univ., (LSU-SG-73-01):
        184-98
        2.1.10

Colwell, R.R. et al., Microbial activities in the estuarine ecosystem. Paper presented to
1974   the First International Congress of International Association of Microbiological
        Societies, Tokyo, Japan. Publ.Sci.Counc.Jap.,Tokyo, 2:410-20
        2.1.10

_____, Microbiological studies of Atlantic Ocean water and sediment from potential
1976   off-shore drilling sites. Dev.Ind.Microbiol., 17:269-82
        2.1.10

Connell, D.W., Kerosene-like tainting in Australian mullet. Mar.Pollut.Bull., 2:188-90
1971   2.1.6

_____, A kerosene-like taint in the sea mullet Mugil cephalus (Linnaeus) 1. Composition
1974   and environmental occurrence of the tainting substance. Aust.J.Mar.Freshwat.Res.,
        25:7-24
        6.2.3; 6.2.5

Conney, A.H. et al., Effects of environmental chemicals on the metabolism of drugs, carcinogens
1971   and normal body constituents in man. Science,Wash.,D.C., 179:155-72
        6.1.2.3

Conover, R.J., Some relations between zooplankton and Bunker C oil in Chedabucto Bay follo-
1971   wing the wreck of the tanker ARROW. J.Fish.Res.Board Can., 28:1327-30
        2.2; 5.1.3

Conrad, B.F., J.D. Walker and R.R. Colwell, Utilization of crude oil and mixed hydrocarbon
1976   substrate by Atlantic ocean bacteria. Dev.Ind.Microbiol. (in press)
        2.1.10

Cook, W.L., J.K. Massey and D.G. Ahearn, The degradation of oil by yeasts and its effect on
1973   Lesbistes reticulatus. In D.G. Ahearn and S.P. Meyers (eds.), The microbial
        degradation of oil pollutants, Louisiana Univ.Pub.No.LSU SG 7301, p. 279-82
        2.1.10

Cooper, B.S., R.C. Harris and S. Thompson, Land-derived pollutant hydrocarbons. Mar.Pollut.
1974   Bull., 5(1):15-6
        6.1.1.4

Cooper, R.C. and G.A. Keller, Epizootiology of papillomas in English Sole, Parophys vetulus.
1968   Natl.Cancer Inst.Monogr., (31):173-86
        6.1.4

Corner, E.D.S., The fate of fossil fuel hydrocarbons in marine animals. Proc.R.Soc.Lond.,B,
1975   189:391-413
        6.1.2.1; 6.1.2.2

Corner, E.D.S., C.C. Kilvington and S.C.M. O'Hara, Qualitative studies on the metabolism
1973   of naphthalene in Maia squinado (Herbst). J.Mar.Biol.Assoc.U.K., 53:819-32
       6.1.2.2

Corner, E.D.S. et al., Hydrocarbons in zooplankton and fish. Proceedings Symposium on
1976   experimental studies on the biological effects of environmental pollutants.
       Society for Experimental Biology, SEB Seminar Series, Cambridge University
       Press (in press)
       6.1.2.1

_____, Petroleum compounds in the marine food web: Short-term experiments on the
1976a fate of naphthalene in Calanus. J.Mar.Biol.Ass.,U.K., 56(1):121-33
       6.1.2.2

Cox, B.A. and J.W. Anderson, Some effects of No. 2 fuel oil on the brown shrimp Penaeus
1973   aztecus. Am.Zool., 13(4):1308
       6.1 2.2

Cox, B.A., J.W. Anderson and J.C. Parker, An experimental oil spill: the distribution of
1975   aromatic hydrocarbons in the water, sediment, and animal tissue within a shrimp
       pond. In Proceedings, Joint conference on Prevention and Control of oil spills,
       March 25-27, San Francisco, California, p. 607-12
       6.1.2.2

Crapp, G.B., The biological effects of marine oil pollution and shore cleaning. Rep.Fld.
1969   Stud.Coun., 1969:27-42
       5.1.9

_____, The effects of oil pollution and emulsifier cleansing on littoral animals and
1971   plants. Rep.Oil Pollut.Res.Unit Field Stud.Counc., 1971:7-13
       5.1.11

Crocker, A.D., J. Cranshaw and W.M. Holmes, The effect of a crude oil on intestinal
1974   absorption in ducklings Anas platyrhynchos. Environ.Pollut., 7:165-77
       5.1.1

Croker, R.A. and A.J. Wilson, Kinetics and effects of DDT in a tidal marsh ditch.
1965   Trans.Amer.Fish.Soc., 94(2):152-9
       4.3

Crosby, E.S., W. Rudolfs and H. Heukelkian, Biological growths in petroleum refinery waste
1954   water. Ind.Eng.Chem., 46:296-300
       5.1.9

Daly, J.W., D.M. Jernis and B. Witkop, Arene oxides and the NIH shift. The metabolism,
1972   toxicity and carcinogenicity of aromatic compounds. Experientia, 28:1129-49
       2.2

Daugherty, J.D., R.D. Campbell and R.L. Morris, Actinomycete: isolation and identification
1966   of agent responsible for musty odors. Science,Wash.,D.C., 152(3727):1372-3
       6.2.5 (table)

Davavin, I.A., O.G. Mironov and I.M. Tsimbal, Influence of oil on nucleic acids of algae.
1975   Mar.Pollut.Bull., 6:13-4
       5.4.2; 6.1.4

Day, J.H. et al., The effect of oil pollution from the WAFRA on the marine fauna of the
1971   Cape Agulhas area. Zool.Afr., 6(2):209-19
       5.1.3

Deshimaru, O., Studies on the pollution of fish meat by mineral oils. 1. Deposition of
1971   crude oil in fish meat and the detection. <u>Bull.Jap.Soc.Sci.Fish.</u>, 37(4):297-301
   <u>6.2.5</u> (table)

_____, Studies on the pollution of fish meat by mineral oils. 2. Injury and pollution
1971a   brought forth on fish by oil dispersers. <u>Bull.Jap.Soc.Sci.Fish.</u>, 37(4):302-6
   <u>6.2.5</u> (table)

Deys, B.F., Papillomas in the Atlantic eel, <u>Anguilla vulgaris</u>. <u>Natl.Cancer Inst.Monogr.</u>,
1969   (31):187-94
   <u>6.1.4</u>

Diamond, L. and H.F. Clark, Comparative studies on the interaction of benzo (a) pyrene with
1970   cells derived from poikilothermic and homothermic vertebrates. 1. Metabolism of
   benzopyrene. <u>J.Natl.Cancer Inst.</u>, 45:1005-11
   <u>2.2</u>

Dickason, O.E., Smithsonian Institution Center for Short-Lived Phenomena, Event Report
1970   36:70/926. Washington, D.C.
   <u>5.1.2</u>

Dickman, M., Preliminary notes on changes in algal primary productivity following exposure
1971   to crude oil in the Canadian Arctic. <u>Can.Field Nat.</u>, 85:249-51
   <u>5.1.8</u>

Dietrich, G. and K. Kalle, Allgemeine Meereskunde. Gebr. Born-Araeger, Berlin
1965   <u>1.3</u>

DiSalvo, L.H., H.E. Guard and L. Hunter, Tissue hydrocarbon burden of muscle as a
1975   potential monitor of environmental hydrocarbon insult. <u>Environ.Sci.Technol.</u>,
   9(3):247-51
   <u>6.1.2.2</u>

Dooley, J.E. et al., Compound type separation and characterization studies for a 370° to
1973   535°C boiling distillate of Gach Saran, Iran, crude oil. Washington, D.C.,
   U.S. Dept. of the Interior, Bureau of Mines, Report of Investigations (7770)
   <u>6.1.1.2</u>

Downing, A.L. and G.A. Truesdale, Some factors affecting the rate of solution of oxygen in
1955   water. <u>J.Appl.Chem.,London</u>, 5:570-81
   <u>4.1</u>

Drapeau, G., Reconnaissance survey of oil pollution on south shore of Chedabucto Bay,
1970   March 24-25, 1970. Atlantic Oceanographic Laboratory, Bedford Institute,
   Dartmouth, Nova Scotia (Unpubl. Rep.)
   <u>5.4.3</u>

Drinkwater, B. et al., Santa Barbara's oiled birds in biological and oceanographic survey
1971   of the Santa Barbara oil spill. Allan Hancock Foundation, University of Southern
   California
   <u>5.1.1</u>

Duce, A., G. Quinn and L. Wade, Residence time of non-methane hydrocarbons in the atmos-
1974   phere. <u>Mar.Pollut.Bull.</u>, 5:59-61
   <u>6.1.1.4</u>

Duursma, E.K., The production of dissolved organic matter in the sea as related to the
1963   primary gross production of organic matter. Netherl.J.Sea Peas., 3:85-94
        1.2

Ehrhardt, M., Petroleum hydrocarbons in oysters from Galveston Bay. Environ.Pollut.,
1972   3(4):257-71
        1.2; 3.2; 6.1.2.1

Ehrhardt, M. and M. Blumer, The source identification of marine hydrocarbons by gas chroma-
1972   tography. Environ.Pollut., 3:170-94
        1.2

Ehrhardt, M. and J. Heinemann, Hydrocarbons in blue mussels from the Kiel Bight. Spec.Publ.
1974   U.S.Natl.Bur.Stand., (409):221-5
        6.1.2.2

Engel, R.H. and M.J. Neat, Toxicity of oil dispersing agents determined in a circulating
1971   aquarium system. In American Petroleum Institute/Environmental Protection
        Agency/U.S. Coast Guard, Proceedings of a Joint conference on prevention and
        control of oil spills. Washington, D.C., American Petroleum Institute, p. 297-303
        5.1.6

Environmental Protection Agency, Water Quality Office, Methods for chemical analysis of
1971   water and wastes. Analytical Quality Control Laboratory, Cincinnati, Ohio,
        No. 217
        3.3

ESSO Research and Engineering Co., Gasoline composition and vehicle exhaust gas polynuclear
1970   aromatic content. NAPCA Contract, Products Research Division, CPA-22-69-56
        6.1.1.2

Evans, D.R. and S.D. Rice, Effects of oil on marine ecosystems: a review for administrators
1974   and policy makers. Fish.Bull.NOAA/NMFS, 72:625-38
        5

Falk, H.L. et al., Intermediary metabolism of benzo (a) pyrene in the rat. J.Natl.Cancer
1962   Inst., 28:699-745
        2.2

Farley, C.A., Sarcomatoid proliferative disease in a wild population of blue mussels (Mytilus
1969   edulis). J.Natl.Cancer Inst., 43:509-16
        6.1.4

Farrington, J.W., Analytical techniques for the determination of petroleum contamination in
1973   marine organisms. Woods Hole, Massachusetts, Woods Hole Oceanographic Institution,
        (WHOI-73-57)
        3.4 (table)

Farrington, J.W. and G.C. Medeiros, Evaluation of some methods of analysis for petroleum
1975   hydrocarbons in marine organisms. In Proceedings of the 1975 Conference on
        Prevention and Control of Oil Pollution, San Francisco, California, 115 p.
        3.2; 3.2 (table)

Farrington, J.W. and J.G. Quinn, Petroleum hydrocarbons in Narragansett Bay. I. Survey of
1973   hydrocarbons in sediments and clams (Mercenaria mercenaria). Estuar.Coast.Mar.
        Sci., 1(1):71-9
        1.2; 6.1.1.4; 6.1.2.1

Farrington, J.W. and J.G. Quinn, Petroleum hydrocarbons and fatty acids in wastewater
1973a   effluents. J.Water Pollut.Control Fed., 45(4):704-12
         6.1.1.4

Fay, J.A., The spread of oil slicks on a calm sea. In Oil on the sea, edited by D.P. Hoult.
1969   Plenum Press, New York
         2.1.1

Fay, J.A. and D.P. Hoult, Physical processes in the spread of oil on a water surface. Final
1971   report, Joint API, EPA, USCG Conf. on prevention and control of oil spills,
         Wash., D.C., June 1971, p. 463-7
         2.1.1

Feldman, M.H., Petroleum weathering: some pathways, fate and disposition on marine waters.
1973   Environ.Protect.Ag.Ecol.Res.Ser., (EPA 660/3-73-013):22 p.
         2.1.8; 6.1.1.1

Finkelstein, E.A., Tumors of fish. Arkh.Patol.,Mosk., 22:56-61
1960   6.1.4

Floodgate, G.D., Biodegradation of hydrocarbons in the sea. In Water pollution microbiology,
1972   edited by R. Mitchell. New York, John Wiley and Sons, p. 153-71
         2.1.9

_____, Microbial degradation of oil. Mar.Pollut.Bull., 3:41-3
1972a   2.1.9; 2.1.10 (table)

Fogel, S., I. Chet and R. Mitchell, The ecological significance of bacterial chemotaxis.
1971   Bact.Proc., 28(G31)
         5.3

Forrester, W.D., Distribution of suspended oil particles following the grounding of the
1971   tanker ARROW. J.Mar.Res., 29(2):151-70
         5.1.3

Fort, E.R., B.O. Prescott and A. Walters, Mapping hydrocarbon seepages in water-covered
         regions. U.S. Patent 3.747.405
         3.2 (table)

Fossato, V.U. and E. Riviero, Oil pollution monitoring in the lagoon of Venice using the
1974   mussel Mytilus galloprovincialis. Mar.Biol., 25:1-6
         1.2

Foster, M., M. Neushul and R. Zingmark, The Santa Barbara oil spill. Part 2. Initial
1971   effects on intertidal and kelp bed organisms. Environ.Pollut., 2(2):115-34
         5.1.9; 5.1.10

Frank, V., Identification of petroleum oils by fluorescence spectroscopy. In Proceedings
1975   of the 1975 Conference on prevention and control of oil pollution. San Francisco,
         California, 87 p.
         3.3

Frankenfeld, J.W., Factors governing the fate of oil at sea: variations in the amounts and
1973   types of dissolved or dispersed materials during the weathering process. In
         American Petroleum Institute, Proceedings of Joint Conference on prevention and
         control of oil spills, Washington, D.C., March 13-15, 1973. Washington, D.C.,
         American Petroleum Institute, p. 485-95
         2.1.3

Garrett, W.D., Organic chemistry of natural sea surface films. Publ.Inst.Mar.Sci.Univ.
1970 Alaska, (1):469-77
1.2

_____, Impact of petroleum spills on the chemical and physical properties of the
1972 air/sea interface. Rep.Naval Res.Lab.,Wash.,D.C., (7372):15 p.
4.1

Gelboin, H.V. and F.J. Wiebel, Studies on the mechanism of aryl hydrocarbon hydroxylase
1971 induction and its role in cyto-toxicity and tumorigenicity. Am.N.Y.Acad.Sci.,
179:529-47
6.1.2.3

George, J.D., Sub-lethal effects on living organisms. Mar.Pollut.Bull., 1:109
1970 5.1.6

_____, The effects of pollution by oil and oil dispersants on the common intertidal
1971 polychaetes, Cirriformia tentaculata and Cirratulus cirratus. J.Appl.Ecol.,
8(2):411-20
5.1.6

Gerarde, H.W., Toxicology and biochemistry of aromatic hydrocarbons. London, Elsevier
1960 6.1.3

GESAMP, Report of the sixth session, 22-28 March 1974 - IMCO/FAO/UNESCO/WMO/WHO/IAEA/UN Joint
1974 Group of Experts on the Scientific Aspects of Marine Pollution. WHO, Geneva,
43 p.
5

Giger, W. and M. Blumer, Polycyclic aromatic hydrocarbons in the environment: isolation
1974 and characterization by chromatography visible, ultraviolet and mass spectrometry. Analyt.Chem., 46:1663
3.1 (table); 3.2 (table)

Gilfillan, E.S., Effects of seawater extracts of crude oil on carbon budgets in two species
1973 of mussels. In American Petroleum Institute, Proceedings of Joint conference on
prevention and control of oil spills, Washington, D.C., March 13-15, 1973.
Washington, D.C., American Petroleum Institute, p. 691-5
5.1.6

Goldschmidt, V.M., Geochemistry, edited by A. Muir, Oxford
1954 4.3

Gooding, R.M., Oil pollution on Wake Island from tanker R.C. STONER. NOAA/NMFS Spec.Sci.
1971 Rep.(Fish.), (636):12 p.
5.1.3

Gordon, D.C., Jr. and P.D. Keizer, Hydrocarbon concentrations detected by fluorescence spec-
1974 troscopy in seawater over the continental shelf of Atlantic Canada - background
levels and possible effects of oil exploration activity. Tech.Rep.Fish.Res.Board
Can., (448):24 p.
3.3

_____, Estimation of petroleum hydrocarbons in seawater by fluorescence spectroscopy:
1974a improved sampling and analytical methods. Tech.Rep.Fish.Res.Board Can., (481)
3.3

Gordon, D.C., Jr. and N.J. Prouse, The effects of three oils on marine phytoplankton photo-
1973 synthesis. Mar.Biol., 22:329-33
5.1.8

Gordon, D.C., Jr., P.D. Keizer and J. Dale, Estimates using fluorescence spectroscopy of the
1974   present state of petroleum hydrocarbon contamination in the water column of the
        northwest Atlantic ocean. Mar.Chem., 2:251-61
        3.2 (table)

Graf, W. and C. Winter,   3,4 Benzpyrene in Erdöl. Arch.Hyg.Bakt., 152:289-93
1968   6.1.1.1; 6.1.1.2; 6.1.1.3

Grant, E.M.,   "Kerosene" taint in sea mullet. Fish.Notes,Brisbane, 3(1):2-13
1969   6.2.3; 6.2.5

Greffard, J. and J. Meury, Note sur la pollution au Rade de Toulon par les hydrocarbures
1967   cancérigènes. Cah.Océanogr., 19(6):457-68
        6.1.1.5

Griffith, D. de G., Toxicity of crude oil and detergents to two species of edible molluscs
1972   under artificial tidal conditions. In Marine pollution and sea life, edited
        by M. Ruivo. West Byfleet, Surrey, Fishing News (Books) Ltd., p. 224-9
        5.1.6

Gusey, W.F. and Z.D. Maturgo, Petroleum production and fish and wildlife resources: the
1971   Gulf of Mexico, Louisiana. New York, Shell Oil Company, Environmental Conser-
        vation Department, 38 p.
        5.1.1; 5.1.4

Hakama, M. and E.A. Saxen,   Cereal consumption and gastric cancer. Int.J.Cancer, 2:265-8
1967   6.1.3

Hampson, G.R. and H.L. Sanders,   Local oil spill. Oceanus, 15:8-10
1969   5.1.3

Hansen, H.P., On photochemical degradation of surface films of petroleum hydrocarbons.
1975   Mar.Chem., 3(3):183-96
        1.2; 3

Hardy, E.,   Merseyside Naturalists' Association Bird Report 1958-59. Liverpool, Merseyside
1959   Naturalists' Association, p. 36-9
        5.1.1

Hardy, R. et al.,   Discrimination in the assimilation of n-alkanes in fish. Nature,Lond.,
1974   252:577-8
        6.1.2.2

Hargrave, B.T. and C.P. Newcombe,   Crawling and respiration as indices of sublethal effects
1973   of oil and a dispersant on intertidal snail Littorina littorea. J.Fish.Res.Board
        Can., 30:1789-92
        5.1.6

Harrison, R.M., R. Perry and R.A. Wellings,   Polynuclear aromatic hydrocarbons in raw,
1975   potable and waste waters. Water Research, 9:331-46
        6.1.1.4

Harshbarger, C.,   The registry of tumours in lower animals. Paper presented to NIEHS
1975   sponsored Marine Biomedical Research Meeting, February 13-14, 1975.
        Washington, D.C., Smithsonian Institution
        6.1.4

Hartung, R. and G.S. Hunt, Toxicity of some oils to waterfowl. J.Wildl.Mgmt., 30:564-70
1966   5.1.1

Hartung, R. and G.W. Klingler, Sedimentation of floating oils. Pap.Mich.Acad.Sci., 53:23-7
1968   2.1.9

Hawkes, A.L., A review of the nature and extent of damage caused by oil-pollution at sea.
1961   Trans.North Am.Wildl.Nat.Resourc.Conf., 26:343-55
    5.1.6

Hellmann, H., Das Verhalten von Rohölen auf Wasseroberflächen - Untersucht und dargestellt
1971   an den zeitlichen Veränderungen ihres Fliessverhaltens (The behaviour of crude
    oil on water surfaces as indicated by the temporal changes in flow characteristics).
    Erdöl Kohle Erdgas Petrochem., 24:417-22
    2.1.8

_____, Ausbreitung von Mineralölen auf Wasseroberflächen. Ihre praktische Bedeutung
1971a   für die Ölbekämpfung (Spreading of mineral oil on the water surface. Its practi-
    cal importance for combatting oil). Neue Deliwa Z., 5:189-92
    2.1.8

Hites, R.A. and K. Biemann, Water pollution: organic compounds in the Charles River, Boston.
1972   Science,Wash.,D.C., 178:158-60
    6.1.1.4

HMSO, Effects of polluting discharges on the Thames Estuary. Water Pollut.Res., Techn.
1964   Paper 11
    4.1

Holcomb, R.W., Oil in the ecosystem. Science,Wash.,D.C., 166:204-6
1969   2.1.9; 2.2

Hope-Jones, P. et al., Effect of HAMILTON TRADER oil on birds in the Irish Sea in May 1969.
1970   Brit.Birds, (63):97-110
    5.1.1

Horn, M.H., J.M. Teal and R.H. Backus, Petroleum lumps on the surface of the sea. Science,
1970   Wash.,D.C., 168(3928):245-6
    2.1.9; 2.2; 6.1.1.4

Hornig, A.W., Identification, estimation and monitoring of petroleum in marine waters by
1974   luminescence methods. Spec.Publ.U.S.Natl.Bur.Stand., (409):135-44
    3.3

Hueper, W.C., Environmental carcinogenesis in man and animals. Science,Wash.,D.C.,
1963   108:963-1038
    6.1.1.1; 6.1.4

Huey, C. et al., Transportation of persistant chemicals in aquatic ecosystems. Proceedings
1974   International Conference, National Research Council, Ottawa, Canada, 2:73-8
    4.3

Ianysheva, N.I. et al., A propos of the content of 3,4 benzopyrene in petroleum bitumen.
1963   Gig.Sanit., 28:71-3
    6.1.1.2

I.A.R.C., International Agency for Research on Cancer, Monographs on the evolution of
1973   carcinogenic risk of the chemical to man. Vol. 3. Certain polycyclic aromatic
    hydrocarbons and heterocyclic compounds.
    6.1.1.1; 6.1.1.4; 6.1.1.5; 6.1.3

Ibragimov, A.P., J.A. Davavin and S.A. Aripozhanov, The study of the sequence of pyrimidine
1971   oligonucleotides and physico-chemical properties of DNA cotton during performance
       and after irradiation of seeds. Radiobiologia, 11(1):28
       5.4.2

Iliffe, T.M. and J.A. Calder, Dissolved hydrocarbons in the eastern Gulf of Mexico loop
1974   current and the Caribbean sea. Deep-Sea Res., 21:481-8
       1.3

Ishio, S., T. Yako and H. Nakagana, Algal cancer and causal substances in wastes from the
1971   coal chemical industry. Adv.Wat.Pollut.Res., 5:1-8
       6.1.4

_____, Cancerous disease of Porphyra tenera and its causes. Proc.Int.Seaweed Symp.,
1972   7:373-6
       6.1.4

Iverson, W.P. and W.R. Blair, Approaches to the study of microbial transformation of metals.
1976   Proceedings Third International Biodegradation Symposium, J.M. Sharpley and
       A.M. Kaplan (Eds.), Applied Science Publications, London, p. 919-36
       4.3

Iverson, W.P. and F.E. Brinckman, Chemical and bacterial cycling of heavy metals in estuarine
1975   systems. T. Church (Ed.), ACS Symposium Series, No. 18, Marine chemistry of
       coastal environment, p. 319-42
       4.3

Jacobson, S.M. and D.B. Boylan, Effect of seawater soluble fraction of kerosene on chemotaxis
1973   in a marine snail, Nassarius obsoletus. Nature, Lond., 241:213-5
       5.3

Jannasch, H.W., Threshold concentrations of carbon sources limiting bacteria growth in sea
1970   water. Publ.Inst.Mar.Sci.Univ.Alaska, (1):321-8
       2.1.9

Jeffery, P.G., Oil in the marine environment. Stevenage, Warren Spring Laboratory
1972   (LR 156 (PC)):16 p.
       6.1.1.4

Jobson, A. et al., Effect of amendments on the microbial utilization of oil applied to soil.
1974   Appl.Microbiol., 27:166-71
       2.1.10

Joensen, A.H., North Sea oil spill and bird kill. Annu.Rep.Rev.Events Smithsonian Inst.
1973   Cent.Short-Lived Phenomena, 1973:36-8
       5.1.1

Johannes, R.E., Coral reefs and pollution. In Marine pollution and sea life, edited by
1972   M. Ruivo. West Byfleet, Surrey, Fishing News (Books) Ltd., p. 364-75
       5.1.6

Johannes, R.E., J. Maragos and S.L. Coles, Oil damages corals exposed to air. Mar.Pollut.
1972   Bull., 3:29-30
       5.1.6

Johnson, R., The decomposition of crude oil residues in sand columns. J.Mar.Biol.Assoc.U.K.,
1970   50:925-37
       2.1.10; 2.1.10 (table)

Kalugina et al., Effect of contamination on marine organisms of the Novorossiiskaya Bay.
   1967   Hydrobiol.J.,1
           5.2.2

Kaneko, T. and R.R. Colwell, Ecology of Vibrio parahaemmolyticus in Chesapeake Bay.
   1972   J.Bact., 113(1):24-32
           2.1.10

Kanter, R., D. Straughan and W.N. Jessee, Effects of exposure to oil on Mytilus californianus
   1971   from different localities. In Prevention and control of oil spills, American
          Petroleum Institute, 485-8
          5.1.6

Kasymov, A.G. and S.I. Granovskii, On the effect of oil on benthic animals of the Caspian Sea.
   1970   (In Russian). Gidrobiol.Zh., 6(5):96-9
          5.1.6 (table)

Katayama, T. and T. Fujiyama, Studies on the nucleic acid of algae with special reference to
   1957   the desoxyribonucleic acid contents of the crown-gall tissues developed on Porphyra
          tenera (KJELIM). Bull.Jap.Soc.Sci.Fish., 23:249-54
          6.1.4

Kator, H., Utilization of crude oil hydrocarbons by mixed cultures of marine bacteria. In
   1973   The microbial degradation of oil pollutants, edited by D.G. Ahearn and S.P. Meyers.
          Proceedings of a workshop, Atlanta, Georgia, 4-6 December 1972, Baton Rouge,
          Louisiana, Louisiana State University, Center for Wetland Resources, p. 47-65
          2.1.10

Kator, H., C.H. Oppenheimer and R.J. Miget, Microbial degradation of a Louisiana crude oil in
   1971   closed flasks and under simulated field conditions. In API/EPA/USCG Conference
          on prevention and control of oil spills. Washington, D.C., American Petroleum
          Institute, p. 287-96
          2.1.10

Katz, L.M., The effects of the water soluble fraction of crude oil on larvae of the decapod
   1973   crustacean Neopanope taxana (Sayi). Environ.Pollut., 5(3):199-204
          5.1.6

Kauss, P.R. et al., The toxicity of crude oil and its component to freshwater algae. In
   1973   American Petroleum Institute, Proceedings of a Joint conference on prevention and
          control of oil spills, Washington, D.C., March 13-15, 1973. Washington, D.C.,
          American Petroleum Institute, p. 703-14
          5.1.8

Kerhoff, M., Oil pollution of the shellfish areas in the Oosterschelde estuary: December
   1974   1973. Copenhagen, International Council for the Exploration of the Sea, Fisheries
          Improvement Committee, (CM 1974/E:13):11 p.
          6.2.4; 6.2.5; 6.2.5 (table)

Ketchum, B.H., Oil in the marine environment. In Background papers for a Workshop on
   1973   inputs, fates and effects of petroleum in the marine environment. Workshop held
          at Airlie, Va., May 21-25, 1974; National Academy of Sciences, Wash., D.C.,
          p. 709-25
          5.1.6

Kinney, P.J., D.K. Button and D.M. Schell, Kinetics of dissipation and biodegradation of
   1969   crude oil in Alaska's Cook Inlet. In Proceedings of the API/FWPCA Joint
          conference on prevention and control of oil spills, New York. Washington, D.C.,
          American Petroleum Institute, p. 333-40
          2.1.10

Kinsey, D.W., Small-scale experiments to determine the effects of crude oil films on gas
1973   exchange over the coral back-reef at Heron Island. Environ.Pollut., 4:167-82
          4.1

Kirby, M.A., Problems facing the oil industry in the Arctic. Mar.Pollut.Bull., 1:71-2
1970   5.1.2

Kittredge, J.S., Effects of the water-soluble component of oil pollution on chemoreception
1972   by crabs. U.S.Natl.Tech.Inf.Serv.Govern.Annu.Rep., 72(9):88
         5.3

_____, The effects of crude oil pollution on the behaviour of marine invertebrates.
1973   Durante, California, City of Hope National Medical Center, Division of Neuro-
          sciences, 14 p.
          5.3

Konarev, V.G. et al., Lable and metabolically active DNA. Dokl.Akad.Nauk SSSR, 166(2):480-2
1966   5.4.2

Koons, C.B. and P.H. Monaghan, Petroleum derived hydrocarbons in Gulf of Mexico waters.
1973   Trans.Gulf-Coast Assoc.Geol.Soc., 23:170-81
         2.1.11

Korringa, P., Biological consequences of marine pollution with special reference to the
1968   North Sea fisheries. Helgol.Wiss.Meeresunters., 17:126-40
         5.1.4

Krishtan, E.G., Some problems of biology of reproduction of Black Sea anchovy and other fish
1968   in the Novorossiiskaya Bay. In Biological study of the Black Sea and its
          resources, edited by M. Nauka.
          5.4.2

Kühnhold, W.W., Der Einfluss wasserlöslicher Bestandteile von Rohölen auf die Entwicklung
1969   von Heringsbrut. Ber.Dt.Wiss.Komm.Meeresf., 20:165-71
          5.4.1 (table)

_____, The influence of crude oils on fish fry. In Marine pollution and sea life,
1972   edited by M. Ruivo. West Byfleet, Surrey, U.K., Fishing News (Books) Ltd.,
          p. 315-7
          5.1.3

_____, Untersuchungen über die Toxizität von Rohölextrakten und Emulsionen auf Eier
1972a   und Larven von Dorsch und Hering (An examination of the toxicity of extracts of
          crude oil and crude oil emulsions to eggs and larvae of cod and herring).
          University of Kiel, Institute of Oceanography, Ph.D. Thesis, 174 p.
          5.1.3

Kuzin, A.M. et al., The change of DNA polymeric spectrum at radiation of its solutions.
1960   Rep.USSR Acad.Sci.,Nauka, Vol. 130
         5.4.2

Lacaze, J.C., Etude de la croissance d'une algue planctonique en présence d'un détergent
1967   utilisé pour la destruction des nappes de pétrole en mer. C.R.Hebd.Séances
         Acad.Sci,Paris,(D), 489
         5.1.8

Lalou, C., Concentration des benzo 3,4 pyrènes par les holothuries de la région de
1964   Villefranche et d'Antibes. Symp.Poll.Mar.Micro-org.Prod.Pétrol. 363-6
         6.1.1.4

LaRoche, B., R. Eisler and C.M. Tarzwell, Bioassay procedures for oil and oil dispersant
1973   toxicity evaluation. J.Water Pollut.Control Fed., 42:1981-9
   5.4.1; 5.4.1 (tables)

LaRock, P.A. and M. Severence, The bacterial treatment of oil spills: the facts considered.
1973   In Estuarine microbial ecology, edited by L.H. Stevenson and R.R. Colwell.
   Columbus, South Carolina, University of South Carolina Press
   2.1.10

LeBoeuf, B.J., Oil contamination and elephant seal mortality: a "negative" finding. In
1971   Biological and oceanographical survey of the Santa Barbara oil spill. 2.
   Los Angeles, Allan Hancock Foundation
   5.1.2

Lee, C.C., W.K. Craig and P.J. Smith, Water-soluble hydrocarbons from crude oil. Bull.
1974   Environ.Contam.Toxicol., 12(2):212-7
   6.1.2.1

Lee, R.F., Rate of petroleum hydrocarbons in marine zooplankton. In Proceedings of the
1975   Joint Conference on Prevention and control of oil spills, March 25-27, 1975,
   San Francisco, California, p. 549-54
   6.1.2.2

Lee, R.F. and A.A. Benson, Fates of petroleum in the sea: biological aspects. In
1973   Proceedings of a workshop on inputs, fates and effects of petroleum in the
   marine environment, 21-23 May 1973, Airlie, Virginia. Washington, D.C.,
   National Academy of Sciences, Vol. 2:541-51
   2.2

Lee, R.F., R. Sauerheber and A.A. Benson, Petroleum hydrocarbons: uptake and discharge by
1972   the marine mussel Mytilus edulis. Science,Wash.,D.C., 177, 344-6
   1.2; 2.2; 6.1.2.1; 6.1.2.2; 6.2.5

Lee, R.F., R. Sauerheber and G.H. Dobbs, Uptake, metabolism and discharge of polycyclic
1972   aromatic hydrocarbons by marine fish. Mar.Biol., 17:201-8
   2.2; 6.1.2.1; 6.1.2.2; 6.1.2.3; 6.2.5

Leenhardt, H., De l'action du mazout sur les coquillages. Rapp.Cons.Perm.Internat.Explor.
1925   Mer, 35:56-8
   5.4.1 (table)

Leppakoski, E., Effects of an oil spill in the northern Baltic. Mar.Pollut.Bull.,
1973   4(6):93-4
   5.1.10

Lewis, J.B., Effect of crude oil and oil-spill dispersant on reef corals. Mar.Pollut.Bull.,
1971   2:59-62
   5.1.6

Lewis, R.W., Squalene distribution in fish with normal and pathologically fatty livers.
1971   Int.J.Biochem., 2:609-14
   6.2.5

Lijinsky, W. and C.R. Raha, Polycyclic hydrocarbons in commercial solvents. Toxicol.Appl.
1961   Pharmacol., 3:469-73
   6.1.1.2

Lima-Zanghi, D. de, Bilan des acides gras du plancton marin et pollution par le benzo 3,4
1968   pyrène (Balance of fatty acids of marine plankton and pollution by benzo 3,4
        pyrène). Cah.Océanogr., 20:203-16
        6.1.1.4

Lindén, O., Acute effects of oil and oil/dispersant mixture on larvae of Baltic herring.
1975   AMBIO, 4(3):130-3
        5.1.3

――――――, The influence of crude oil and mixtures of crude oil/dispersants on the onto-
1976   genic development of Baltic herring. AMBIO,5(3):136-40
        5.1.3

――――――, Effects of oil on the amphipod Gammarus oceanicus. Environm.Pollut.,
1976a   10(4):239-50
        5.1.6

――――――, Effects of oil on the reproduction of the amphipod Gammarus oceanicus. AMBIO,
1976b   5(1):36-7
        5.1.6

Liu, D.L.S. and B.J. Dutka, Bacterial seeding techniques: novel approach to oil spill pro-
1972   blems. Ottawa, Canada Research and Development, July-August, 1972, p. 1-4
        2.1.10

Mackay, D. and R. Matsugu, Evaporation rates of liquid hydrocarbon spills on land and
1973   water. Can.J.Chem.Eng., 54:434-9
        1.2

Mackie, P.R., K.J. Whittle and R. Hardy, Hydrocarbons in the marine environment.
1974   1. n-alkanes in the Firth of Clyde. Estuar.Coast.Mar.Sci., 2:359-74
        3.1 (table);6.1.2.2

MacKin, J.G., A review of significant papers on effects of oil spills and oilfield brine
1973   discharges on marine biotic communities. College Station, Texas A and M Research
        Foundation, University of Texas, Project 737, 171 p.
        6.1.2.1

MAFF, Great Britain, Food Facts, 1971:25
1971   6.1.3

――――――, Food Facts, 1972:11
1972   6.1.3

Malacea, I., V. Cure and L. Weiner, Contributions to the knowledge of the noxious action
1964   of oil, naphthenic acids and phenols on certain fish and the crustacean Daphnia
        magna. Stud.Prot.Epur.Apelor., 5:353-97 (in Romanian, English summary)
        5.4.1 (table)

Mallet, L., Recherche des hydrocarbures polybenzéniques du type benzo 3,4 pyrène dans la
1961   faune des milieux marins (Manche, Atlantique et Méditerranée). C.R.Hebd.Séances
        Acad.Sci.,Paris, 253:168-70
        6.1.1.5

――――――, Pollution par les hydrocarbures en particulier du type benzo 3,4 pyrène des
1965   rivages Méditerranéens Français et plus spécialement de la Baie de Villefranche.
        Symp.Commn.Int.Explor.Scient.Mer Médit.,Monaco, 1964:325-330
        6.1.1.4

――――――, Présence des hydrocarbures du type benzo 3,4 pyrène dans les sédiments sous-
1965a   jacents au lit de la Seine en aval de Paris. Bull.Acad.Natl.Méd., (149):656-61
        6.1.1.4

Mallet, L. and M.L. Priou, On the retention of polybenzic hydrocarbons of the type benzo
1967   3,4 pyrene by the sediments, marine fauna and flora of the St. Malo Bay.
        C.R.Hebd.Séances Acad.Sci., Paris (D), 264:969-71
        6.1.1.5

Mann, H., Geschmacksbeeinflussungen bei Fischen. Fette-Seifen-Anstrichmitt., 12:1021-4
1969   6.1.3; 6.2.5 (table)

Martin, A.E., Water pollution by oil - some health considerations. In water pollution by
1970   Oil, edited by P. Hepple. London, Institute of Petroleum, p. 153-8
        6.2.4

Masek, V., 3,4 Benzopyrene and our crude petroleum fuel oil. Pracov.Lek., 13(7):363-6
1961   6.1.1.2

Mazmanidi, N.D., Diasamidze and Zambachidze, Oil effects on some species of molluscs and
1973   crustacea in the Black Sea. Materials of All Union Symposium on the Studies of
        the Black sand Mediterranean Seas, utilization and protection of their resources,
        Part 4. Kiev, Naukova Dumka
        5.1.6

McAuliffe, C.D., Solubility in water of paraffin, cycloparaffin, olefin, acetylene, cyclo-
1966   olefin, and aromatic hydrocarbons. J.Phys.Chem., 70:1267-75
        6.1.2.1

————————, Determination of C_1-C_{10} hydrocarbons in water. Spec.Publ.U.S.Natl.Bur.Stand.,
1974   (409):121-5
        3.2 (table)

McCarthy, R.D., Mammalian metabolism of straight chain saturated hydrocarbons. Biochem.
1964   Biophys.Acta, 84:74-9
        2.2

Medical Research Council, The carcinogenic action of mineral oils: a chemical and biological
1968   study. Her Majesty's Stationary Office, London, 251 p.
        6.1.1.1

Miget, R., Bacterial seeding to enhance biodegradation of oil slicks. Publ.La.State Univ.,
1973   (LSU-SG-73-01):291-309
        2.1.10

Miget, R.J. et al., Microbial degradation of normal paraffin hydrocarbons in crude oil.
1969   In Proceedings of the API/FWPCA Joint conference on prevention and control of
        oil spills, New York. Washington, D.C., American Petroleum Institute, p. 327-31
        2.1.10

Mincemin, J.M., Télémédiateurs chimiques et équilibre biologique océanique. 3. Etude in vitro
1971   de relations entre populations phytoplanctoniques. Rev.Int.Océanogr.Méd.
        22-23:165-96
        1.1

Mironov, O.G., The results of sanitary investigation of marine sediments along shores.
1961   Zdravookhr.Beloruss., 5

Mironov, O.G., Effect of low concentration of oil and oil products on the developing eggs of
1967   the Black Sea flatfish. Probl.Ichthyol., 7(3):577-80 (In Russian)
5.4.1.(table); 5.4.2

———, The effects of oil and oil products upon some molluscs in the littoral zone
1967a   of the Black Sea (In Russian; English summary). Zool.Zh., 46:134-6
5.1.6

———, Hydrocarbon pollution of the sea and its influence on marine organisms.
1968   Helgol.Wiss.Meeresunters., 17:335-9
5; 5.1.3; 5.1.7; 5.4.1 (table)

———, The development of some Black Sea fishes in sea water polluted by petroleu
1969   products. Probl.Ichthyol., 9(6):919-22
5.1.3; 5.4.2

———, Microorganisms growing on oil and oil products in western and central regions
1970   of the Mediterranean Sea. Rev.Int.Oceanogr.Med.CERBOM, 17:79-85
5.4.1;(table); 5.4.2

———, Oil-oxidising microorganisms in the sea. Kiev, Naukova Dumka, 234 p.
1971   (In Russian)
5.2.4

———, Biological resources of the sea and oil pollution. Moscow, Pishchevaya
1972   Promyshlennost, 105 p. (In Russian)
5.1.3; 5.1.6; 5.1.6 (table)

———, Effects of oil pollution on the flora and fauna of the Black Sea. In Marine
1972a   pollution and sea life, edited by M. Ruivo. London, Fishing News (Books) Ltd.,
    p. 222-4
5.1.7; 5.4.1 (table)

———, Oil pollution and life in the sea. Kiev, Naukova Dumka, 86 p. (In Russian)
1973   5.1.3

Mironov, O.G. and L.A. Lanskaya, Biology and distribution of plankton of the southern seas.
1967   Oceanographical Commission, Moscow (In Russian)
5.1.8; 5.4.1 (table)

Mironov, O.G. and I.M. Tsimbal, The development of macroscepii algae in oil polluted water.
1975   Biol.Sci., 5
5.1.9

Mironov, O.G. et al., Biological aspects of marine hydrocarbon pollution. In Marine
1975   pollution and marine waste disposal, edited by Parson and Frangipane. Oxford,
    Pergamon Press
5.1.8

Mitchell, C.T. et al., What oil does to ecology. J.Water Pollut.Contr.Fed., 42:812-8
1970   5

Mitchell, R., S. Fogel and I. Chet, Bacterial chemoreception: an important ecological
1972   phenomenon inhibited by hydrocarbons. Water Res., 6(10):1137-40
1.2; 5.3

Mize, C.E. et al., A major pathway for the mammalian oxidative degradation of phytanic acid.
1969   Biochem.Biophys.Acta, 176:720-39
2.2

Mommaerts-Billiet, F., Growth and toxicity tests on the marine nanoplanktonic alga
1973   Platymonastetrathele G.S. West in the presence of crude oil and emulsifiers.
    Environ.Pollut., 4:261-82
5.1.8

Morin et al., Metabolic heterogeneity of mouse liver DNA. Exp.Cell.Res., 13:204
1957   5.4.2

Morozova-Vodianitskaya, N.V., Observations of ecology of algae in the Novorossiiskaya Bay.
1927   Proc.Black Sea Res.Inst., Vol.52
          5.2.2

_____, Materials to sanitary biological analysis of seawater. Proc.Novorossiiskaya
1930   Biol.Station, 4:163-80
          5.2.2

Morris, R.J., Uptake and discharge of petroleum hydrocarbons by barnacles. Mar.Pollut.Bull.,
1973   4(7):197-9
          2.2; 6.1.2.1; 6.1.2.2

_____, Lipid composition of surface films and zooplankton from the Eastern Mediterra-
1974   nean. Mar.Pollut.Bull., 5:105-9
          2.2; 6.1.2.2

Morrow, J.E., Oil-induced mortalities in juvenile coho and sockeye salmon. J.Mar.Res.,
1973   31(3):135-43
          5.1.3

Motohiro, T., Studies on the petroleum odour in canned chum salmon. Mem.Fac.Fish.Hokkaido
1962   Univ., (1962):2-65
          6.2.5 (table)

Müller, D.G. and L. Jaenicke, Fucoserraten, the female sex attractant of Fucus serratus
1973   (Phaeophyta). FEBS Lett., 30(1):137
          1.2

Mulkins-Phillips, G.J. and J.E. Stewart, Effect of four dispersants on biodegradation and
1974   growth of bacteria on crude oil. Appl.Microbiol., 28:547-52
          2.1.5; 2.1.10

_____, Effects of environmental parameters on bacterial degradation of Bunker C oil,
1974a   crude oils and hydrocarbons. Appl.Microbiol., 28:915-22
          2.1.10

Nadeau, R.J. and T.H. Roush, A salt marsh microcosm: an experimental unit for marine
1973   pollution studies. In American Petroleum Institute, Proceedings of a Joint
          conference on prevention and control of oil spills, Washington, D.C.,
          March 13-15, 1973. Washington, D.C., American Petroleum Institute, p. 671-83
          5.4.1

National Academy of Sciences, Background papers for a Workshop on inputs, fates and effects
1973   of petroleum in the marine environment. Workshop held at Airlie, Va., May 21-25,
          1974; Wash.,D.C., National Academy of Sciences, 2 vol.:824 p.
          1.3

_____, Petroleum in the marine environment. NAS, Wash., D.C.
1975   1.3; 3

National Food Survey Committee, Annual report of the National Food Survey Committee.
1971   London, HMSO
          6.1.3

Nechaeva, Method of determination of nucleic acids in green plants. <u>Physiol.Rast.</u>, 13(5)
1966   <u>5.4.2</u>

Neff, J.M. and J.W. Anderson, Accumulation, release, and distribution of benzo (a)
1975   pyrene-C_{14} in the clam <u>Rangia cuneata</u>. <u>In</u> Proceedings, Joint conference on prevention and control of oil spills, March 25-27, 1975. San Francisco, California, p. 469-71
<u>6.1.2.2</u>

Nelson-Smith, A., The effects of oil pollution and emulsifier cleansing on shore life in
1968   southwest Britain. <u>J.Appl.Ecol.</u>, 5:97-107
<u>5.1.6</u>; <u>5.1.9</u>; <u>5.1.11</u>

_____, The effects of oil pollution on shore life. <u>In</u> Proceedings of the symposium
1970   on the effects of industry on the environment, Orielton, Pembroke, 1970. London, Field Studies Council, p. 36-42
<u>5.1.2</u>; <u>5.1.6</u>; <u>5.1.8</u>

_____, Effects of the oil industry on shore life in estuaries. <u>Proc.R.Soc.Lond.,(B)</u>,
1972   180:487-96
<u>5.1.1</u>; <u>5.2.1</u>

_____, Oil pollution and marine ecology. London, Paul Elek Scientific Books, 260 p.
1972a   <u>5.1.8</u>; <u>5.4.1</u> (table)

Nitta, T., Marine pollution in Japan. <u>In</u> Marine pollution and sea life, edited by
1972   M. Ruivo. London, Fishing News (Books) Ltd., p. 77-81
<u>6.2.5</u>; <u>6.2.5</u> (table)

Nitta, T. <u>et al</u>., Studies on the problems of offensive odors in fish caused by wastes from
1965   petroleum industries. <u>Bull.Tokai Reg.Fish.Res.Lab.</u>, 42:23-37
<u>6.2.5</u>

North, W.J., Tampico, a study of destruction and restoration. <u>Sea Frontiers</u>, 13:212-7
1967   <u>5.1.3</u>

_____, Position paper on effects of acute oil spills. <u>In</u> Background papers for a
1973   Workshop on inputs, fates and effects of petroleum on the marine environment, Workshop held at Airlie, Va., May 21-25, 1974; National Academy of Sciences, Wash., D.C., p. 745-65
<u>5.1.6</u>

North, W.J., M. Neushul and K.A. Clendenning, Successive biological changes observed in a
1964   marine cove exposed to a large spillage of mineral oil. <u>Symp.Pollut.Mar.Micro-org.Prod.Petrol.</u>, Monaco, p. 333-54
<u>5</u>; <u>5.1.3</u>; <u>5.1.6</u>; <u>5.1.9</u>; <u>5.4.1</u> (table); <u>5.4.3</u>

Notini, M. and A. Hagström, Effects of oils on Baltic littoral community, as studied in an
1974   outdoor model test system. <u>Spec.Publ.U.S.Natl.Bur.Stand.</u>, (409):251-4
<u>5.4.1</u>

Nuzzi, R., Effects of water soluble extracts of oil on phytoplankton. <u>In</u> American Petroleum
1973   Institute, Proceedings of Joint conference on prevention and control of oil spills, Washington, D.C., March 13-15, 1973. Washington, D.C., American Petroleum Institute, p. 809-13
<u>5.1.8</u>

Ogata, M. and Y. Miyake, Offensive-odor substance in fish in the sea along petrochemical
1970 industries. Jap.J.Public Health, 17:1125-30
6.2.5

Oppenheimer, C., Hydrocarbons in sea water and organisms. (First draft of the Final Report,
1974 Gulf Universities Research Consortium, Offshore Ecology Investigations)
Project OE,73 HJN. May 10, 1974
6.1.1.4

O'Sullivan, A.J., Some aspects of the HAMILTON TRADER oil spill. In Water pollution by
1971 oil, edited by P. Hepple, London, Institute of Petroleum, p. 307-16
5.1.3

O'Sullivan, A.J. and A.J. Richardson, The TORREY CANYON disaster and intertidal marine life.
1967 Nature,Lond., 214:448-542
5.1.3

Owens, E.H., The restoration of beaches contaminated by oil in Chedabucto Bay, Nova Scotia.
1971 Manuscr.Rep.Ser.Mar.Sci.Branch Dep.Energy Mines Resour.,Ottawa, (19)
5.4.3

Palmork, K.H. and A. Vinsjansen, Oljedispergeringsmidler og vannloselige oljekomponenter.
1972 En gasskromatografisk-massespektrometrisk undersokelse. Fisken og Havet(B),
(4):1-24
6.2.5; 6.2.5 (tables)

Palmork, K.H. and S. Wilhelmsen, Undersokelse av fisk fra oljeforurenset omrade av Gisundet.
1974 Fisken og Havet(B), (4):1-13
6.2.5; 6.2.5 (table); 6.2.6

Palmork, K.H., S. Wilhelmsen and T. Neppelberg, Report on the contribution of polycyclic
1973 aromatic hydrocarbons (PAH) to the marine environment from different industries.
International Council for the Exploration of the Sea, Fisheries Improvement
Committee, Copenhagen (CM 1973/E:33):21 p.
6.1.1.3

Pancirov, P.J. and R.A. Brown, Analytical methods for polynuclear aromatic hydrocarbons in
1975 crude oils, beating oils, and marine tissues. In Proceedings, Joint conference
on prevention and control of oil spills, March 25-27, 1975, San Francisco,
California, p. 103-14
6.1.1.2; 6.1.1.3

Paradis, M. and R.G. Ackman, Differentiation between natural hydrocarbons and low level
1975 diesel oil contamination in cooked lobster meat. J.Fish.Res.Board Can.,
32(2):316-20
6.1.3; 6.2.5; 6.2.5 (table)

Parker, C.A., The ultimate fate of crude oil at sea - uptake of oil by zooplankton.
1970 Rep.AML(Admir. Mater.Lab.)U.K., B198(M)
2.2

Parker, C.A., M. Freegarde and C.G. Hatchard, The effect of some chemical and biological
1971 factors on the degradation of crude oil at sea. In Water pollution by oil,
edited by P. Hepple. London, Institute of Petroleum, p. 237-44
6.1.2.2

Pauley, G.B., A critical review of neoplasia and tumour-like lesions in mollusks.
1969 Natl.Cancer Inst.Monogr., (31):509-40
6.1.4

Perdriau, J., Marine pollution by cancerogenous hydrocarbons of the type benzo-3,4 pyrene,
1964 biological casualties. Cah.Océanogr., 16:125-38
6.1.1.5

Pickering, Q.H. and C. Henderson, Acute toxicity of some important petrochemicals to fish.
1966 J.Water Pollut.Control Fed., 38:1419-29
5.4.1 ; 5.4.1 (table)

Pilpel, N., The natural fate of oil on the sea. Endeavour, 27(100):11-3
1968 2.1.10

Portmann, J.E., The toxicity of 120 substances to marine organisms. Shellfish Inf.Leafl.,
1970 MAFF.U.K., (19)
5.4.1

_____, Results of acute toxicity tests with marine organisms, using a standard
1972 method. In Marine pollution and sea life, edited by M. Ruivo. West Byfleet,
Surrey, Fishing News (Books) Ltd., p. 212-7
5.1.6

Powell, N.A., C.A. Sayce and D.F. Tufts, Hyperplasia in an estuarine bryozoan attributable
1970 to coal tar derivatives. J.Fish.Res.Board Can., 27:2095-6
6.1.4

Ramsdale, S.J. and R.E. Wilkinson, Identification of petroleum sources of beach pollution
1968 by gas - liquid chromatography. J.Inst.of Petrol., 54:326-32
2.1.10

Reisfeld, A., E. Rosenberg and D. Gutnick, Microbial degradation of crude oil: factors
1972 affecting the dispersion in sea water by mixed and pure cultures. Appl.Microbiol.,
24:363-8
2.1.10

Reish, D.J., Laboratory populations for long-term toxicity tests. Mar.Pollut.Bull.,
1973 4(3):46
5.4.1

Renzoni, A., Influence of crude oil derivatives and dispersants on larvae. Mar.Pollut.Bull.,
1973 4:9-13
5; 5.1.6

Rice, S.D., Toxicity and avoidance tests with Prudhoe Bay oil and pink salmon fry. In
1973 American Petroleum Institute, Proceedings of a Joint conference on the prevention
of oil spills, Washington, D.C., March 13-15, 1973. Washington, D.C., American
Petroleum Institute, p. 667-70
5.1.3; 5.3

Roberts, C.H., Oil pollution. J.Cons.perm.int.Expl.Mer, 1(3):245-75
1926 4.1

Robertson, B.R. et al., Hydrocarbon biodegradation. Occas.Publ.Inst.Mar.Sci.Univ.Alaska,
1973 (3):449-79
2.1.10

Robichaux, T.J. and H.N. Myrick, Chemical enhancement of the biodegradation of crude oil
1972   pollutants. J.Petrol.Technol., 24:16-20
2.1.10

Ronald, A.P. and W.A.B. Thompson, The volatile sulphur compounds of oysters. J.Fish.Res.
1964   Board Can., 21(6):357-64
6.2.5 (table)

Roubal, W.T. and T.K. Collier, Spin-labelling technique for studying mode of action of
1975   petroleum hydrocarbons on marine organisms. NOAA Fish.Bull., NOAA/NMFS 73(2):299-305
6.1.2.2

Russel, F.E. and P. Kotin, Squamous papilloma in the White Croaker. J.Nat.Cancer Inst.,
1957   18(6):857-61
6.1.4

Ryther, J.H., A comparison of the oxygen and C^{14} methods of measuring marine photosynthesis.
1954   J.Cons.CIEM, 20:25-37
5.4.1

Sanders, H.L., J.F. Grassle and G.R. Hampson, The West Falmouth oil spill. 1. Biology.
1972   Tech.Rep.Woods Hole Oceanogr.Inst., (72-20):48 p.
5.1.3; 5.1.6

Sayler, G.S. and R.R. Colwell, Partitioning of mercury and polychlorinated biphenyl by oil,
1976   water and sediment. Environ.Sci.Technol., (in press)
4.3

Scarratt, D.J. and V. Zitko, Bunker C oil in sediments and benthic animals from shallow depths
1972   in Chedabucto Bay, N.S. J.Fish.Res.Board Can., 29(9):1347-50
2.2; 6.1.2.2

Schlede, E., R. Kuntzman and A.H. Conney, Stimulatory effect of benzo(a) pyrene and pheno
1970   barbital pretreatment on the biliary excretion of benzo(a) pyrene metabolites in
the rat. Cancer Res., 30:2892-904
6.1.2.3

Schlede, E. et al., Effect of enzyme induction on the metabolism and tissue distribution
1970   of benzo(a) pyrene. Cancer Res., 30:2893-7
6.1.2.3

Schmeer, A.C., Morenene: an antineoplastic agent extracted from the marine clam Mercenaria
1968   mercenaria. Nat.Cancer Inst.Monogr., (31):581-92
6.1.4

Schwarz, J.R., J.D. Walker and R.R. Colwell, Hydrocarbon degradation at ambient and in situ
1974   pressure. Appl.Microbiol., 28:982-6
2.1.10

_____, Growth of deep-sea bacteria on hydrocarbons at ambient and in situ pressure.
1974a   Dev.Ind.Microbiol., 15:239-49
2.1.10

_____, Deep-sea bacteria: growth and utilization of n-hexadecane at in situ temperature
1975   ture and pressure. Can.J.Microbiol., 21:682-7
2.1.10

Searle, C.E. et al., Carcinogenicity and mutagenicity tests of some hair colourants and
1975   constituents. Nature, Lond., 255:506-7
       6.1.5

Seba, D.B. and E.F. Cocoran, Surface slicks as concentrators of pesticides in the marine
1969   environment. Pestic.Monit.J., 3:190-3
       4.3

Sebba, F., Control of oil slicks by sucking and sweeping. Chem.Ind., 1971:1157
1971   2.1.10

Shelton, R.G.J., Two recent problems in oil pollution research. Copenhagen, International
1971   Council for the Exploration of the Sea, Fisheries Improvement Committee,
       (CM 1971/E:12):6 p.
       6.1.2.2

Shelton, T.B. and J.V. Hunter, Anaerobic decomposition of oil in bottom sediments.
1975   J.Wat.Pollut.Contr.Feder., 47(9):2256-70
       2.1.6

Shiang-Chia, F., Diesel spill at Anacortes. Mar.Pollut.Bull., 2:105-6
1971   5.1.1

Shimkin, M.B., B.K. Koe and L. Zechmeister, An instance of the occurrence of carcinogenic
1951   substances in certain barnacles. Science,Wash.,D.C., 173:650-1
       6.1.1.5

Shipton, J. et al., Studies on Kerosene-like taint in mullet (Mugil cephalus). 2. Chemical
1970   nature of the volatile constituents. J.Sci.Food Agric., 21:433-6
       1.2; 6.2.5

Sidhu, G.S. et al., A kerosene-like taint in mullet (Mugil cephalus). In Marine pollution
1972   and sea life, edited by M. Ruivo. London, Fishing News (Books) Ltd., p. 546-50
       6.2.5; 6.2.5 (table); 6.2.6

Simpson, A.C., The TORREY CANYON disaster and fisheries. Lab.Leafl.MAFF.U.K., (18):1-43
1968   5.1.4; 6.2.5; 6.2.5 (table) 6.2.6

Sivadier, H.O. and P.G. Mikolaj, Measurement of evaporation rates from oil slicks on the
1973   open sea. In American Petroleum Institute, Proceedings of Joint conference on
       prevention and control of oil spills, Washington, D.C., American Petroleum
       Institute, p. 475-84
       2.1.3

Smith, J.E., TORREY CANYON, pollution and marine life. Cambridge Univ.Press. 196 p.
1970   5.1.3

Sparks, A.K., Review of tumours and tumour-like conditions in Protozoa, Coolenterata,
1968   Platyhelminthes, Annelida, Sipunculida, and Arthropoda, excluding insects.
       Natl.Cancer Institute Monogr., (31):671-82
       6.1.4

Spirin and Belosersky, Composition of nucleic acids at experimental variation of intestinal
1956   bacteria. Biochemistry, Nauka, 21:768
       5.4.2

Spooner, M.F., Some ecological effects of marine oil pollution. In Prevention and
1969   Control of Oil Spills. American Petroleum Institute, p. 313-6
       5.1.3; 5.4.3

_____, Oil spill in Tarut Bay, Saudi Arabia. Mar.Pollut.Bull., 1:166-7
1970   5.1.3; 5.1.9

Sprague, J.B., Measurement of pollutant toxicity to fish. 1. Bioassay methods for acute
1969   toxicity. Water Res., 3:793-829
       5.4.1

_____, Sublethal effects and 'safe' concentrations. Water Res., 5:245-66
1971   5.4.1

_____, Sublethal oil spill fumble. Mar.Pollut.Bull., 2:83-4
1971a 5.4.1

Sprague, J.B. and W.G. Carson, Toxicity tests with oil dispersants in connection with oil
1970   spill at Chedabucto Bay, N.S. Tech.Rep.Fish.Res.Board Can., (201)
       5.4.1

Sprague, V., Microsporida and tumours, with particular reference to the lesions associated
1968   with Iothyosperidium sp. Schwartz, 1963. Monogr.Natl.Cancer Inst., (31):237-50
       6.1.4

Stebbings, N.E., Recovery of salt marsh in Brittany 16 months after heavy pollution by oil.
1970   Environ.Pollut., 1:163-7
       5.1.9; 5.1.10

Stegeman, J.J. and J.T. Teal, Accumulation, release and retention of petroleum hydrocarbons
1973   by the oyster Crassostrea virginica. Mar.Biol., 22:37-44
       1.2; 2.2; 6.1.2.2; 6.1.2.3

Stich, H.F., Environmental pollution and fish papillomas. NIEHS sponsored Marine Biomedical
1975   Research Meeting, February 13-14, 1975. Washington, D.C., Smithsonian Institution
       6.1.1.4; 6.1.1.5; 6.1.4

Stichting Concawe, Effects of gasoline aromatics content on polynuclear aromatic exhaust
1974   emissions. The Hague, Holland, Report No. 6/74, September 1974
       6.1.1.2

Stock, A. and F. Cucuel, Die Verbreitung des Quecksilbers. Naturwiss., 22:319
1934   4.3

Strand, J.A. et al., Development of toxicity test procedures for marine phytoplankton.
1971   Prevention and control of oil spills. Washington, D.C., American Petroleum
       Institute, p. 279-86
       5.1.8; 5.4.1

Straughan, D., The Santa Barbara study. Proceedings API/FWPCA Joint conference on
1969   prevention and control of oil spills. New York, December 15-17, 1969.
       Washington, D.C., American Petroleum Institute, p. 309-11
       5

_____, Straughan, D., Oil pollution and wildlife and fisheries in the Santa Barbara
1971   channel. Trans.North Am.Wildl.Nat.Resourc.Conf., 36:219-29
       5.1.1; 5.1.3

_____, Biological and oceanographic survey of the Santa Barbara channel oil spill.
1971a Biology and Bacteriology, Vol. I. Allan Hancock Foundation, University of
       Southern California
       5.1.3; 5.1.9; 5.1.10

Straughan, D., Biological effects of oil pollution in the Santa Barbara channel. In
1972   Marine pollution and sea life, edited by M. Ruivo. West Byfleet, Surrey,
        Fishing News (Books) Ltd., p. 355-9
        5; 5.1.2

─────── , Factors causing environmental changes after an oil spill. J.Petrol.Technol.,
1972a   24:250-4
        5; 5.1.2

Straughan, D. and B.C. Abbott, The Santa Barbara oil spill: ecological changes and natural
1971   oil leaks. Water pollution by oil, edited by P. Hepple. London, Institute of
        Petroleum, p. 257-62
        5.1.6

Straughan, D. and D.M. Lawrence, Investigation of ovicell hyperplasia in bryozoans chroni-
1975   cally exposed to natural oil seepage. Water,Air and Soil Pollut., 5:39-45
        6.1.4

Strickland, J.D.H., Measuring the production of marine phytoplankton. Bull.Fish.Res.Board
1960   Can., (122):172 p.
        5.4.1

Strickland, J.D.H. and T.R. Parsons, A practical handbook of seawater analysis. Bull.Fish.
1965   Res.Board Can., (167):311 p.
        5.1.8; 5.4.1

Stroganov, N.S. and L.V. Kolosova, Influence of small concentrations of hexachlorbutadiene
1968   on water organisms. Proc.Moscow Soc.Nature Test, 30
        5.4.2

Sulimova, G.F. and A.G. Slyusarenko, Isolation of DNA from tissues of dry plants. DNA
1972   structure and the position of organisms in the system. Isd-vo MGU, M.
        5.4.2

Sullivan, J.B., Marine pollution by carcinogenic hydrocarbons. In U.S. National Bureau of
1974   Standards, Marine pollution monitoring (petroleum). Proceedings of a symposium
        and workshop, May 13-17, 1974. Washington, D.C., U.S. Government Printing Office,
        p. 261-3
        6.1.1.2; 6.1.1.5; 6.1.3

Sutton, C. and J.A. Calder, Solubility of higher-molecular-weight n-Paraffins in distilled
1974   water and seawater. Environ.Sci.Technol., 8(7):655-7
        6.1.2.1

Swedmark, M., A. Granmo and S. Kollberg, Effects of oil dispersants and oil emulsions on
1973   marine animals. Water Res., 7(11):1649-72
        5.1.3; 5.1.6; 5.4.1

Tagatz, M.E., Reduced oxygen tolerance and toxicity of petroleum products to juvenile
1961   American shad. Chesapeake Sci., 2:65-71
        5.1.3; 5.4.1 (table)

Talbot, J.W., The influence of tides, waves and other factors on diffusion rates in marine
1972   and coastal situations. In Marine pollution and sea life, edited by M. Ruivo.
        West Byfleet, Surrey, Fishing News (Books) Ltd., p. 122-30
        1.2

Tanis, J.J.C. and M.F. Mörzer Bruyns, The impact of oil pollution on sea birds in Europe.
1968   Proc.Int.Conf.Oil Pollut.Sea, Rome, p. 67-74
        5.1.1

Tatem, H.E. and J.W. Anderson, The toxicity of four oils to Palaemonetes pugio (Holthuis)
1973   in relation to uptake and retention of specific petroleum hydrocarbons.
        Am.Zool., 13(4):2614
        6.1.2.2

Teal, J.M. and J.W. Farrington, A comparison of hydrocarbons in animals and their benthic
        habitats. In Petroleum hydrocarbons in the marine environment, edited by
        A.D. McIntyre and K. Whittle. Rapp.P.-V.Reun.Cons.Int.Explor.Mer, 171
        6.1.2.2

Templeton, W.L., Ecological effects of oil pollution. J.Water Pollut.Control Fed.,
1971   43:1081-8
        2.1.10

Tendron, G., Contamination of marine flora and fauna by oil and biological consequences
1968   of the TORREY CANYON accident. In Report of proceedings of the International
        conference on oil pollution of the sea, 7-9 October, 1968. Winchester, U.K.,
        Wykeham Press, p. 114-21
        2.2; 5.1.3

Thomas, M.L.H., Effects of Bunker C oil on intertidal and lagoonal biota in Chedabucto Bay,
1973   Nova Scotia. J.Fish.Res.Board Can., 30:83-90
        5.1.9; 5.1.10; 5.4.3

Thompson, S. and G. Eglington, The presence of pollutant hydrocarbons in estuarine epipelic
1976   diatom populations. Estuar.Coast Mar.Sci., 4:417-25
        6.1.2.2

Todd, J.H., J. Atema and D.B. Boylan, Chemical communication in the sea. Mar.Technol.Soc.J.,
1972   6(4):54-6
        1.2; 5.3

Tokarskaya, Heterogeneity of natural DNA-RNA complexes in the nucleus of seeds. Rep.Acad.
1967   Sci.USSR
        5.4.2

Tuck, L.M., The Murres. Ottawa, Canadian Wildlife Service
1960   5.2.1

Turnbull, H., J.G. Demara and R.F. Weston, Toxicity of various refinery materials to fresh-
1954   water fish. Ind.Eng.Chem., 46:324-33
        5.4.1 (table)

Tye, R. et al., Carcinogens in a cracked petroleum residue. Arch.Environ.Health, 13(2):202-7
1966   6.1.1.2

United States, National Bureau of Standards, Marine pollution monitoring (petroleum).
1974   Proceedings of a symposium and workshop held at the National Bureau of Standards,
        Gaithersburg, Maryland, May 13-17, 1974. Washington, D.C., U.S. Government
        Printing Office, Spec.Publ.U.S.Natl.Bur.Stand., (409):316 p.
        3.1 (table)

United States, Public Health Service, Industrial waste guide: Oil refining. Ohio River
1939   Pollution Survey, Cincinnati
        5.4.1 (table)

Vale, G.L. et al., Studies on a kerosene-like taint in mullet (Mugil cephalus). I. General
1970   nature of the taint. J.Sci.Food Agric., 21:429-31
     6.2.5

Vaughan, B.E. (Ed.), Effects of oil and chemically dispersed oil on selected marine biota - A
1973   laboratory study. Publ.API (Am.Petrol.Inst.), (4191)
     6.1.2.1; 6.1.2.2

Walker, J.D. and R.R. Colwell, Microbial degradation of model petroleum at low temperatures.
1974   Microbiol.Ecol., 1:63-95
     2.1.10

_____, Microbial petroleum degradation: use of mixed hydrocarbon substrates.
1974a   Appl.Microbiol., 27(6):1053-60
     2.1.10

_____, Mercury-resistant bacteria and petroleum degradation. Appl.Microbiol.,
1974b   27(1):285-7
     4.3

_____, Petroleum hydrocarbons in Baltimore Harbor of Chesapeake Bay: distribution
1975   in sediment cores. Environ.Pollut., 9:231-8
     2.1.10; 4.3

_____, Degradation of hydrocarbons and mixed hydrocarbon substrate by microorganisms
1975a   from Chesapeake Bay. Progress Wat.Technol., 1:783-91
     2.1.10; 5.2.4

_____, Some effects of petroleum on estuarine and marine microorganisms. Can.J.
1975b   Microbiol., 21(3):305-13
     5.2.4

_____, Long-chain n-alkanes occurring during microbial degradation of petroleum.
1976   Can.J.Microbiol., 22(6):886-91
     2.1.10

_____, Oil, mercury and bacterial interactions. Environ.Sci.Technol.
1976a   2.1.10; 4.3

_____, Utilization of South Louisiana crude oil by estuarine microorganisms.
     Arch.Microbiol., (in press)
     2.1.10

Walker, J.D., H.F. Austin and R.R. Colwell, Utilization of mixed hydrocarbon substrate by
1974   petroleum-degrading microorganisms. J.Gen.Appl.Microbiol., 21:27039
     2.1.10

Walker, J.D., R.R. Colwell and L. Petrakis, Microbial petroleum degradation: Application
1975   of computerized mass spectronomy. Can.J.Microbiol., 21:1760-7
     2.1.9; 2.1.10; 3

_____, Degradation of petroleum by an alga, Prototheca zopfii. Appl.Microbiol.,
1975a   30:79-81
     2.1.10

_____, Biodegradation rates of components of petroleum. Can.J.Microbiol.,
1976   22:1209-13
     2.1.9; 2.1.10; 6.1.2.1

Walker, J.D., R.R. Colwell and L. Petrakis, Biodegradation of petroleum by Chesapeake
1976a Bay sediment bacteria. Can.J.Microbiol., 22:423-8
       2.1.10

Walker, J.D., P.A. Seesman and R.R. Colwell, Effects of petroleum on estuarine bacteria.
1975   Mar.Pollut.Bull., 5(12):186-8
       5.2.4

_____, Effect of South Louisiana crude oil and No. 2 fuel oil on growth of hetero-
1975a trophic microorganisms including proteolytic, lipolytic, chitinolytic and
       cellulolytic bacteria. Environ.Pollut., 9:13-33
       5.2.4

Walker, J.D. et al., Petroleum hydrocarbons: degradation and growth potential for marine
1975   sediment bacteria. Mar.Biol., 34:1-9
       2.1.10

_____, Extraction of petroleum hydrocarbons from oil-contaminated sediments.
1975a Bull.Environ.Contam.Toxicol., 13(2):245-8
       4.3

Wayne, B. and R. Mitchell, Chemotactic and growth responses of marine bacteria to algal
1972   extracellular products. Biol.Bull.Mar.Poll.Lab.Woods Hole, 143(2):265-77
       1.2

Weidemann, H. and H. Sendner, Dilution and dispersion of pollutants by physical processes.
1972   In Marine pollution and sea life, edited by M. Ruivo. West Byfleet, Surrey,
       Fishing News (Books) Ltd., p. 115-21
       1.2

Wellings, S.R., Neoplasia and primitive vertebrate phylogeny: Echinoderms, Prevertebrates
1968   and Fishes - a Review. Natl.Cancer Inst.Monogr., (31):59-128
       6.1.4

Wells, P.G., Influence of Venezuelan crude oil on lobster larvae. Mar.Pollut.Bull.,
1972   3(7):105-6
       5.1.6

Westlake, D.W.S. et al., Biodegradability and crude oil composition. Can.J.Microbiol.,
1974   20:915-28
       2.1.10

Whitman, J., Center for natural areas spill report; date 09/27/75. Center for Shortlived
1975   Phenomena, Smithsonian Institution, Cambridge, Massachusetts
       6.2.3

Whittle, K.J. and M. Blumer, Interactions between organisms and dissolved organic sub-
1970   stances in the sea; chemical alteration of the starfish Asterias vulgaris
       to oysters. Occ.Pub.Inst.Mar.Sci., Alaska I, 495-507
       1.2

Whittle, K.J. and P.R. Mackie, Hydrocarbons in zooplankton and fish. Part 2. Fish.
1975   Paper presented to the Symposium of the Society for Experimental Biology,
       Liverpool, April 1975
       6.2.5; 6.2.6

WHO, Health hazards of the human environment. Geneva, World Health Organization
1972   6.1.3; 6.1.5

Wilder, D.G., The tainting of lobster meat by Bunker C oil alone or in combination with
1970 the dispersant Corexit. Manuscr.Rep.Fish.Res.Board Can., (1087)
6.2.5; 6.2.5 (table) 6.2.6

Wilson, K.W., Toxicity of oil-spill dispersants to embryos and larvae of some marine fish.
1972 In Marine pollution and sea life, edited by M. Ruivo. West Byfleet, Surrey.
Fishing News (Books) Ltd., p. 9-18
5.1.3

Wilson, K.W., The laboratory estimation of the biological effects of organic pollutants.
1975 Proc.R.Soc.Lond., (B)., 189:459-77
6.1.2.3

Wilson, R.D. et al., Natural marine oil seepage. Science,Wash.,D.C., 184(4139):857-65
1974 1.2; 6.1.1.4

Wolf, P.H., Neoplastic growth in two Sydney Rock Oysters, Crassostrea commercialis.
1968 Natl.Cancer Inst.Monogr., (31):563-74
6.1.4

Wolff, P.M., W. Hansen and J. Joseph, Investigation and prediction of dispersion of pollutants
1972 in the sea with hydrodynamical numerical (HN) models. In Marine pollution and
sea life, edited by M. Ruivo. West Byfleet, Surrey, Fishing News (Books) Ltd.,
p. 146-50
1.2

Woodin, S.A., C.F. Nyblade and F.S. Chia, Effect of diesel oil spill on invertebrates.
1972 Mar.Pollut.Bull., 3:139-43
5.1.6

Wynder, E.L. et al., An epidemiological investigation of gastric cancer. Cancer, 16:1461-96
1963 6.1.3

Yevich, P.P., Histopathological effects of oil pollutants on marine life. Paper presented
1975 to the NIEHS sponsored Marine Biomedical Research Meeting, February 13-14, 1975.
Smithsonian Institution, Wash., D.C.,
6.1.4

Yevich, P.P. and M.M. Barry, Ovarian tumors in the Quahog Mercenaria mercenaria. J.Invert.
1969 Pathol., 14:266-7
6.1.4

Young, P.H., Some effects of sewer effluent on marine life. Calif.Fish Game, 50:33
1964 6.1.4

Youngblood, W.W. et al., Saturated and unsaturated hydrocarbons in marine benthic algae.
1972 Mar.Biol., 8:190-201
1.2

Zafiriou, O.C., K.J. Whittle and M. Blumer, Response of Asterias vulgaris to bivalves and
1972 bivalve tissue extracts. Mar.Biol., 13:137-45
1.2

Zahner, R., Über die Wirkung von Treibstoffen und Ölen auf Regenbogenforellen. Vom Wasser,
1962 29:142-77
5.4.1 (table)

Zats, V.J., Experimental studies of horizontal diffusion in the Black Sea coastal zone.
1972 In Marine pollution and sea life, edited by M. Ruivo. West Byfleet, Surrey,
Fishing News (Books) Ltd., p. 130-4
1.2

Zitko, V. and S.N. Tibbo, Fish kill caused by an intermediate oil from coke ovens.
1971 Bull.Environ.Contam.Toxicol., 6(1):24-5
5.1.3

ZoBell, C.E., Marine microbiology. Waltham, Mass., Chronica Botanica
1946 2.1.10

_____, Action of microorganisms on hydrocarbons. Bact.Rev., 10:1-48
1946a 5.3

_____, Microbial modification of crude oil in the sea. In Proceedings of the
1969 API/FWPCA Joint conference on prevention and control of oil spills, New York. Washington, D.C., American Institute of Petroleum, p. 317-26
2.1.5; 2.1.10

_____, Sources and biodegradation of carcinogenic hydrocarbons. In Proceedings of
1971 the Joint conference on prevention and control of oil spills. Washington, D.C., American Petroleum Institute, Environmental Protection Agency and the U.S. Coast Guard, p. 441-51
6.1.1.4; 6.1.1.5

_____, Bacterial degradation of mineral oils at low temperatures. Publ.La.State Univ.,
1973 (LSU-SG-73-01):153-61
2.1.9

Anon., Report of the Task Force - Operation Oil. Vol. 2. Ministry of Transport. Information
1970 Canada, Cas. No. T22-2470/2, Ottawa, 104 p.
5.1.1; 5.1.2; 5.1.3

Anon., Summary of physical, biological, socio-economic and other factors relevant to potential
1974 oil spills in the Passamaquoddy region of the Bay of Fundy. Fish.Res.Board Can., Techn. Rep. No. 428
6.2.3

Anon, Japanese oil leak. Mar.Pollut.Bull., 6(3):36
1975 6.2.3

2. ADDITIONAL RELATED LITERATURE

2.1 ANALYTICAL

2.1.1 Methods of Taking Samples

Bruce, H.E. and S.P. Cram, Sampling marine organisms and sediments for high precision gas
1974 chromatographic analysis of aromatic hydrocarbons. Spec.Publ.U.S.Natl.Bur.Stand.,
(409):181-2

Chang, W.J. and J.R. Jadamec, Evaluation of thin film oil samplers. Spec.Publ.U.S.Natl.
1974 Bur.Stand., (409):85-8

Chang, W.J. and W.A. Saner, Evaluation of boat deployable thin film oil samplers. Paper
1974 presented at the sixth annual Offshore Technology Conference, Houston, Texas,
6-8 May 1974, Pap.No. (OTC 1984):20 p.

Curtis, D.L. and A.D. Le Vantine, Development of a fixed site surface film oil sampler.
1974 Washington, D.C., U.S. Coast Guard, Office of Research and Development, Final
Rep., (CG-D-102-74):100 p.

Farrington, J.W., Some problems associated with the collection of marine samples and
1974 analysis of hydrocarbons. Tech.Rep.Woods Hole Oceanogr.Inst., (74-23):19 p.

Freegarde, M., A novel oil collection device. Effluent Water Treat.J., (10):203-6
1970

Jeffery, P.G., J. Nightingale and D.J.A. Woodley, Instrumentation for oil pollution
1973 measurement. Pollut.Monit., (13):31-3

Miget, R. et al., New sampling device for the recovery of petroleum hydrocarbons and
1974 fatty acids from aqueous surface films. Analyt.Chem., (46):1154-7

Schatzberg, P. and D.F. Jackson, Remote sampler for determining residual oil content of
1973 surface waters. In American Petroleum Institute, Proceedings of Joint
conference on prevention and control of oil spills, Washington, D.C.,
13-15 March 1973. Washington, D.C., American Petroleum Institute, p. 139-44

Straughan, D.M., Field sampling methods and techniques for marine organisms and sediments.
1974 Spec.Publ.U.S.Natl.Bur.Stand., (409):183-7

2.1.2 Detection

Ackman, R.G. and D. Noble, Steam distillation: a simple technique for recovery of petroleum
1973 hydrocarbons from tainted fish. J.Fish.Res.Board Can., 30(5):711-4

Aerojet Electrosystems Co., Development of a prototype airborne oil surveillance system.
1973 Vol. 1. System definition studies. Azusa, California, Aerojet Electrosystems Co.

_____, Development of a prototype airborne oil surveillance system. Vol. 2. Design
1973 report. Azusa, California, Aerojet Electrosystems Co.

_____, Development of a prototype airborne oil surveillance system. Vol. 3. Sub-
1973 system specifications. Azusa, California, Aerojet Electrosystems Co.

Ahmadjian, M. and C.W. Brown, Feasibility of remote detection of water pollutants and oil
1973   slicks by laser-excited raman spectroscopy. Environ.Sci.Technol., (7):452-3

Aukland, J.C., J.D. Sohn and L.E. Rasmussen, Multisensor detection and tracking of
1971   controlled oil spills. Hollywood, California, Spectran Inc., Microwave Sensor Systems Division, 138 p.

Axelsson, S., Registrering av oljeutsläpp med fjärranalysteknik (Oil slick mapping by
1972   remote sensing). Linköping, Sweden, Saab-Scania AB (STU-rapport 71-646/U538): 147 p.

_____, Registrering från flygplan av oljeutsläpp. Resultat från fältförsök (Airborne
1972   remote sensing of oil slicks. Results from a field experiment). Linköping, Sweden, Saab-Scania AB (STU-uppdrag 71-646/U538):45 p.

Axelsson, S. and E. Ohlsson, Remote sensing of oil slicks. Ambio, (2):70-6
1973

Beak, T.W., Marine oil analysis. Halifax, N.S., Fisheries Research Board
n.d.

Bittel, R. and G. Lacourly, Méthode d'approche pour l'évaluation des niveaux de pollution
1971   chimique des milieux marins et des chaînes alimentaires marines. Rev.Int. Océanogr.Méd.CERBOM, (22-3):129-42

Brown, C.W., P.F. Lynch and M. Ahmadjian, Novel method for sampling oil spills and for
1974   measuring infrared spectra of oil samples. Analyt.Chem., (46):183-4

Brown, D.E. and K. Pitt, Drop size distribution of stirred non-coalescing liquid-liquid
1972   system. Chem.Eng.Sci.,Lond., 27(3):577-83

Brown, R.A., Methods for polynuclear aromatic hydrocarbons. In Proceedings of a Workshop
1973   on inputs, fates and effects of petroleum in the marine environment, 21-25 May 1973, Airlie, Virginia. Washington, D.C., National Academy of Sciences, Vol. 1:152-7

_____, Measurement of hydrocarbons in water and sediment. In Proceedings of a
1973   Workshop on inputs, fates and effects of petroleum in the marine environment, 21-25 May 1973, Airlie, Virginia. Washington, D.C., National Academy of Sciences, Vol. 1:134-51

Cahnmann, H.J., Detection and quantitative determination of Benzo(a) pyrene in American
1955   (Colorado) shale oil. Analyt.Chem., 27(8):1235-40

Campbell, C.E., Remote sensing of oil pollution by ultraviolet fluorescence. Proc.Soc.
1972   Photo-optic.Instrum.Eng., (27):85-90

Catoe, C.E., Results of overflights of Chevron oil spill in Gulf of Mexico. Washington,
1970   D.C., U.S. Coast Guard, Office of Research and Development, 20 p.

_____, The applicability of remote sensing techniques for oil slick detection.
1972   Paper presented at the fourth annual Offshore Technology Conference, Houston, Texas, 1-3 May 1972, 16 p.

_____, Remote sensing techniques for detecting oil slicks. J.Petrol.Technol.,
1973   (25):267-78

Chandler, P.B., Oil pollution surveillance. Paper presented at the Joint conference on
1971   sensing of environmental pollutants, Palo Alto, California, 8-10 November 1971.
       Pap.Am.Inst.Aeronaut.Astronaut.,New York, (71-1073):8 p.

Edgerton, A.T. and P. Hinds, Microwave radiometric observations of controlled oil spills.
1971   In U.S. Coast Guard, Remote sensing of southern California oil experiment.
       Washington, D.C., U.S. Coast Guard, Office of Research and Development, p. 1-37

Edgerton, A.T., D.C. Meeks and D. Williams, Microwave emission characteristics of oil slicks.
1973   Paper presented at the Joint conference on sensing of environmental pollutants,
       Palo Alto, California, 8-10 November 1971. Pap.Am.Inst.Aeronaut.Astronaut.,
       New York, (71-1071):6 p.

Estes, J.E. and B. Golomb, Monitoring environmental pollution. J.Remote Sens., 1(2):8-13
1970

Estes, J.E. and L.W. Senger, The multispectral concept as applied to marine oil spills.
1972   Remote Sens.Environ., (2):141-63

Estes, J.E., L.W. Senger and P.R. Fortune, Potential applications of remote sensing
1972   techniques for the study of marine oil pollution. Geoforum, (9):69-81

Fantasia, J.F. and H.C. Ingrao, The development of an experimental airborne laser oil
1973   spill remote sensing system. In American Petroleum Institute, Proceedings of
       Joint conference on prevention and control of oil spills, Washington, D.C.,
       13-15 March 1973. Washington, D.C., American Petroleum Institute, p. 101-15

Fantasia, J.F., T.M. Hard and H.C. Ingrao, An investigation of oil fluorescence as a
1971   technique for the remote sensing of oil spills. Washington, D.C., U.S. Coast
       Guard, Final Rep., (DOT-TSC-USCG-71-7):118 p.

──────, The remote sensing of oil spills by laser-excited fluorescence. Proc.Amer.
1972   Tech.Meet.Inst.Environ.Sci., Mount Prospect, Ill., (18):342-56

Goolsby, A.D., Water pollution detection by reflectance measurements. Paper presented at
1971   the Joint conference on sensing of environmental pollutants, Palo Alto,
       California, 8-10 November 1971. Pap.Am.Inst.Aeronaut.Astronaut.,New York,
       (71-1069):6 p.

Gram, H.G., The remote detection and identification of surface oil spills. In Proceedings
1974   of second conference on Environmental Quality Sensors, Las Vegas, Nevada,
       10-11 October 1973. Section 6. Washington, D.C., Environmental Protection
       Agency, p. 21-33

Guinard, N.W., Remote sensing of ocean effects with radar. Paper presented at the
1971   Conference on propagation limitations in remote sensing, North Atlantic Treaty
       Organization, Advisory Group for Aerospace Research and Development (AGARD), 12 p.

──────, The remote sensing of oil slicks. In Proceedings seventh International
1971   symposium on remote sensing of the environment, Ann Arbor, Michigan,
       17-21 May 1971. Willow Run Laboratories, University of Michigan, p. 1005-26

──────, Radar detection of oil spills. Paper presented at the Joint conference on
1971   sensing of environmental pollutants, Palo Alto, California, 8-10 November 1971.
       Pap.Am.Inst.Aeronaut.Astronaut.,New York, (71-1072):8 p.

Gularte, R.C. and P.F. Poranski, In situ surface film detector. In Report of the
1970   oceanographic engineering summer project, by S.C. Daubin et al., Woods Hole,
       Massachusetts, Woods Hole Oceanographic Institution (Ref. No. 71-5):14-27

Hayre, H.S. and M.S. Sohel, Oil pollution and ray optics in an inhomogenous medium. Paper
1971   presented at the Institute of Electrical and Electronics Engineers '71
        Engineering in the ocean environment conference, San Diego, California,
        21-24 September 1971, 4 p.

Horvath, R., In situ detection of oil slicks utilizing differential evaporation. Phase 1.
1974   Feasibility study. Ann Arbor, Michigan, Environmental Research Institute of
        Michigan

Horvath, R. and S.R. Stewart, Analysis of multispectral data from the California oil
1971   experiment of October 1970. In U.S. Coast Guard, Remote sensing of the
        southern California oil pollution experiment. Washington, D.C., U.S. Coast
        Guard, Office of Research and Development, p. 91-107

Horvath, R., E.F. Lirette and D.M. Zuk, In situ detection of oil slicks utilizing
1974   differential evaporation. Phase 2. System design. Washington, D.C.,
        U.S. Coast Guard, Office of Research and Development, Final Rep., (CG-D-47-75):
        38 p.

Horvath, R., W.L. Morgan and R. Spellicy, Measurements program for oil-slick characteristics.
1970   Washington, D.C., U.S. Coast Guard, Office of Research and Development, 192 p.

Horvath, R., W.L. Morgan and S.R. Stewart, Optical remote sensing of oil slicks: signature
1971   analysis and systems evaluation. Washington, D.C., U.S. Coast Guard, Office of
        Research and Development, 117 p.

Jackson, P., Leak detection in underwater oil pipelines. Galveston, Texas, U.S. National
1973   Maritime Research Center, Cargo Handling and Terminals Program. (NMRC-272-
        23100-R2):38 p.

Johnson, G.L., L. Townsend and J. Stockham, Detection of oil contamination in sea water.
1971   Vol. 3. Engineering evaluation and improvement of the infrared oleometer.
        Washington, D.C., U.S. Maritime Administration, Rep., (Ma-Rd-930-72-05):73 p.

Keizer, P.D. and D.C. Gordon, Jr., Detection of trace amounts of oil in seawater by
1973   fluorescence spectroscopy. J.Fish.Res.Board Can., (30):1039-46

Ketchel, R.J. and A.T. Edgerton, Development of U.S. Coast Guard prototype airborne oil
1973   surveillance system. In American Petroleum Institute, Proceedings of Joint
        conference on prevention and control of oil spills, Washington, D.C.,
        13-15 March 1973. Washington, D.C., American Petroleum Institute, p. 127-37

Kim, H.H. and G.D. Hickman, An airborne fluorosensor for the detection of oil on water.
1973   Paper presented at the Hydrographic Lidar Conference, Wallops Island,
        Virginia, 12 September 1973

Klemas, V., Detecting oil on water: a comparison of known techniques. Paper presented at
1972   the Joint conference on sensing of environmental pollutants, Palo Alto,
        California, 8-10 November 1971. Pap.Am.Inst.Aeronaut.Astronaut., New York,
        (71-1068):6 p.

Kriebel, A.R., Development of a floating oil slick detector. Washington, D.C., U.S.
1973   Coast Guard, Office of Research and Development, Final Rep., (CG-D-42-74):131 p.

Krishen, K., Detection of oil spills using a 13.3-GHz radar scatterometer. J.Geophys.Res.,
1973   (78):1952-63

McCormack, K., G. Fournier and W. Knight, Oil-on-water sensor. Washington, D.C., U.S.
1974   Coast Guard, Office of Research and Development, Final Rep., (CG-D-87-74): 207 p.

Measures, R.M. and M. Bristow, The development of an airborne remote laser fluorosensor
1971   for use in oil pollution detection and hydrologic study. Toronto, University of Toronto, Institute for Aerospace Studies, UTIAS Report (175):32 p.

Meeks, D.C. et al., Microwave radiometric detection of oil slicks. Washington, D.C.,
1971   U.S. Coast Guard, Office of Research and Development, Rep., (1335-2):91 p.

Millard, J.P. and J.C. Arvesen, Effects of skylight polarization, cloudiness, and view
1971   angle on the detection of oil on water. Paper presented at the Joint conference on sensing of environmental pollutants, Palo Alto, California, 8-10 November 1971. Pap.Am.Inst.Aeronaut.Astronaut.,New York, (71-1075):7 p.

_____, Results of airborne measurements to detect oil spills by reflected sunlight.
1971   In U.S. Coast Guard, Remote sensing of the southern California oil pollution experiment. Washington, D.C., U.S. Coast Guard, Office of Research and Development, p. 38-62

_____, Airborne optical detection of oil on water. J.Appl.Opt., (11):102
1972

_____, Polarization: a key to an airborne optical system for the detection of oil
1973   on water. Science, Wash.,D.C., (180):1170-1

Mohr, D. et al., Oil spill surveillance system study. Environ.Protect.Ag.Technol.Ser.
1973   Wash.,D.C., (EPA-R2-73-215):215 p.

Olson, D.G. and G.P. Wright, An optimal prevention and detection model for pollution
1973   patrol. In American Petroleum Institute, Proceedings of Joint conference on prevention and control of oil spills, Washington, D.C., 13-15 March 1973. Washington, D.C., American Petroleum Institute, p.145-50

O'Neil, R.A. et al., Remote sensing laser fluorometer. Paper presented at the Hydro-
1973   graphic Lidar Conference, Wallops Island, Virginia, 12 September 1973

Oudin, J.L., Prévention et lutte contre la pollution au cours des opérations de forage et
1971   de production en mer. 5. Détection et identification des nappes d'hydrocarbures en mer. Rev.Inst.Fr.Pétrole, (26):811-6

Parker, P.L., Petroleum-stable isotope ratio variations. In Impingement of man on the
1971   oceans, D.W. Hood (Ed.). New York, Wiley-Interscience, p. 431-44

Pilon, R.O. and C.G. Purves, Radar detection and monitoring of oil slick. In U.S.
1971   Coast Guard, Remote sensing of the southern California oil pollution experiment. Washington, D.C., U.S. Coast Guard, Office of Research and Development, p. 63-90

_____, Radar imagery of oil slicks. IEEE Trans.Aerosp.Electron.Syst.,AES-9,
1973   (5):630-6

Rudder, C.L., A.G. Wallace and C.J. Reinheimer, Aerial detection of spill sources.
1973   Environ.Protect.Ag.Technol.Ser.,Wash.,D.C., (EPA-R2-73-289):23 p.

Schwemmer, G.K. and H.H. Kim, Mapping and identification of oil on water by the use of an
1974   airborne laser system. Spec.Publ.U.S.Natl.Bur.Stand., (409):95-6

Sindermann, C.J. and J. Wash, Some biological indicators of marine environmental degradation.
1972   Acad.Sci., 62(2):184-9

Smith, J.T., Jr., Oil slick remote sensing. Photogramm.Eng., (37):1243-8
1971

Stumpf, H.G. and A.E. Strong, ERTS-1 views an oil slick. Remote Sens.Environ., (3):87-90
1973

Terry, S.A., F.J. Buckmeier and J.A. Watson, Oil spill reconnaissance using remote-sensing
1971   techniques. Paper presented at the seventh annual meeting, Marine Technology
       Society, Washington, D.C., 16-18 August 1971, 16 p.

Thomson, K.P.B. and W.D. McColl, A remote sensing survey of the Chedabucto Bay oil spill.
1972   Sci.Ser.Can.Cent.Inland Water, Burlington, Ont., (26):15 p.

United States, Coast Guard, Remote sensing of southern California oil pollution experiment.
1971   Washington, D.C., U.S. Coast Guard, Office of Research and Development, 107 p.

Van Melle, M.J., H.H. Wang and W.F. Hall, Microwave radiometric observations of simulated
1973   sea surface conditions. J.Geophys.Res., (78):969-76

Vizy, K.N., Detecting and monitoring oil slicks with aerial photos. Photogramm.Eng.,
1974   (40):697-708

Welch, R.I., A.D. Marmelstein and P.M. Maughan, Aerial spill prevention surveillance during
1973   sub-optimum weather. Environ.Protect.Techno.Ag.Ser., (EPA-R2-73-243):62 p.

Wobber, F.J., Imaging techniques for oil pollution survey purposes. Photogr.Appl.Sci.
1971   Technol.Med., 6(4):16-23

Wright, D.E. and J.A. Wright, Evaluation of an infrared oil film monitor. Washington, D.C.,
1974   U.S. Coast Guard, Office of Research and Development, Final Rep., (CG-D-51-75):
       102 p.

_____, A new infrared instrument for monitoring oil films on water. Spec.Publ.U.S.
1974   Natl.Bur.Stand., (409):93-4

Zissiz, G.J. et al., Remote sensing techniques for oil slick measurements. In Proceedings
1971   of the International symposium on the identification and measurement of environ-
       mental pollutants, Ottawa, 14-17 June 1971, p. 265-70

Anon., Imbiber polymer beads that soak up oil spills. Manigeer Repr.
1973

2.1.3 Chromatography

Ackman, R.G. and D. Noble, Steam distillation: a simple technique for recovery of petroleum
1973   hydrocarbons from tainted fish. J.Fish.Res.Board Can., (30):711-4

Adlard, E.R., L.F. Creaser and P.H.D. Matthews, Identification of hydrocarbon pollutants
1972   on seas and beaches by gas chromatography. Analyt.Chem., (44):64-73

Blaylock, J.W. et al., Determination of n-alkane and methylnaphthalene compounds in shell-
1973   fish. In American Petroleum Institute, Proceedings of Joint conference on
        prevention and control of oil spills, Washington, D.C., 13-15 March 1973.
        Washington, D.C., American Petroleum Institute, p. 173-7

Cole, R.D., Recognition of crude oils by capillary gas chromatography. Nature, Lond.,
1971   (233):546-8

Done, J.N. and W.K. Reid, A rapid method of identification and assessment of total crude
1970   oils and crude oil fractions by gel permeation chromatography. Sep.Sci.,
        (5):825-42

Dubois, L., A. Corkery and J.L. Monkman, The chromatography of polycyclic hydrocarbons.
1960   Int.J.Air Pollut., (2):236-52

Hunter, L., H.E. Guard and L.H. DiSalvo, Determination of hydrocarbons in marine organisms
1974   and sediments by thin layer chromatography. Spec.Publ.Natl.Bur.Stand., (409):
        213-6

Jeltes, R., Verontreiniging van water door minerale olie; bepaling van sporen olie in water
1969   met gaschromatografie (Pollution of water by mineral oil: determination of trace
        oil in water with gas chromatography). H_2O, 2(17):403-6

Kawahara, F.K., Characterization and identification of certain petroleum products by means
1971   of gas chromatographic analyses of minor components. Cincinnati, Ohio, U.S.
        Environmental Protection Agency, Water Quality Office, 11 p.

_____, Gas chromatographic analysis of mercaptans, phenols, and organic acids in
1971   surface waters with use of pentafluorobenzyl derivatives. Environ.Sci.Technol.,
        (5):235-9

Levy, E.M., L.R. Weber and J.D. Moffatt, A method for the high temperature gas chromato-
1973   graphic analyses of petroleum residues. J.Chromat.Sci., (11):591-3

Mitchell, B.L., P.G. Simmonds and F.H. Shair, Oil spill identification with microencap-
1973   sulated compounds suitable for electron capture. Envir.Sci.Technol., (7):121-4

Pastorelli, L. and G. Chiavari, Determinazione gascromatografica de tracce di alcani e di
1971   ftalati in acqua di mare (Gas-chromatographic determination of alkanes and
        phthalates in seawater). Ann.Chim., (61):311-7

Zafiriou, O.C., Improved method for characterizing environmental hydrocarbons by gas
1973   chromatography. Analyt.Chem., (45):952-6

Zafiriou, O.C., M. Blumer and J. Myers, Correlation of oils and oil products by gas
1972   chromatography. Tech.Rep.,Woods Hole Oceanogr.Inst., (72-55):110 p.

Zafiriou, O.C., J. Myers and F. Freestone, Interference of oil spill emulsifiers with
1973   gas chromatography. Mar.Pollut.Bull., (4):87-8

Zafiriou, O.C. et al., Oil spill-source correlation by gas chromatography: an experimental
1973   evaluation of system performance. In American Petroleum Institute, Proceedings
        of Joint conference on prevention and control of oil spills, Washington, D.C.,
        13-15 March 1973. Washington, D.C., American Petroleum Institute, p. 153-9

Zsolnay, A., Determination of aromatic hydrocarbons in submicrogram quantities in aqueous
1973   systems by means of high performance liquid chromatography. Chemosphere,
        (2):253-60

Zsolnay, A., Determination of total hydrocarbons in seawater at the microgram level with
1974   a flow calorimeter. J.Chromatogr., (90):79-85

―――――, Determination of aromatic and total hydrocarbon content in submicrogram and
1974   microgram quantities in aqueous systems by means of high performance liquid
       chromatography. Spec.Publ.U.S.Natl.Bur.Stand., (409):119-20

2.1.4 Spectrophotometry

Brown, C.W., M. Ahmadjian and P.F. Lynch, Sampling of oil spills and fingerprinting by
1974   infrared spectroscopy. Spec.Publ.U.S.Natl.Bur.Stand., (409):91-2

Bryan, D.E. et al., Development of nuclear analytical techniques for oil slick identification
1970   (Phase 1). San Diego, California, Gulf General Atomic Inc., (GA 9889):134 p.

Carlberg, S.R. and C.B. Skarstedt, Determination of small amounts of non-polar hydrocarbons
1972   (oils) in sea water. J.Cons.CIEM, (34):506-15

Coakley, W.A., Comparative identification of oil spills by fluorescence spectroscopy
1973   fingerprinting. In American Petroleum Institute, Proceedings of Joint
       conference on prevention and control of oil spills, Washington, D.C.,
       13-15 March 1973. Washington, D.C., American Petroleum Institute, p. 215-22

Coomber, R.S., Measuring tankship oil effluent. Pollut.Monit., (4):21-5
1971/2

Cretney, W.J. and C.S. Wong, Fluorescence monitoring study at ocean weather station 'P'.
1974   Spec.Publ.U.S.Natl.Bur.Stand., (409):175-7

Feldman, M.H. and D.E. Cawlfield, Marine environmental monitoring: trace elements in
1974   persistent tar ball oil residues. Spec.Publ.U.S.Natl.Bur.Stand., (409):237-41

Filby, R.H. and K.R. Shah, Mode of occurrence of trace elements in petroleum and
1971   relationship to oil-spill identification methods. In Proceedings American
       Nuclear Society topical meeting, Nuclear methods in environmental research,
       Columbia, Missouri, 23-24 August 1971. Columbia, Missouri, University of
       Missouri, p. 86-96

Gordon, D.C., Jr., and P.D. Keizer, Hydrocarbon concentrations in seawater along the
1974   Halifax-Bermuda section: lessons learned regarding sampling and some results.
       Spec.Publ.U.S.Natl.Bur.Stand., (409):113-5

Gruenfeld, M., Extraction of dispersed oils from water for quantitative analysis by
1973   infrared spectrophotometry. Environ.Sci.Technol., (7):636-9

Guinn, V.P. and S.C. Bellanca, Neutron activation analysis identification of the source
1969   of oil pollution of waterways. Spec.Publ.U.S.Natl.Bur.Stand., (312):93-7

Guinn, V.P., D.E. Bryan and H.R. Lukens, The trace-element characterization of crude oils
1971   and fuel oils via instrumental neutron activation analysis. In International
       Atomic Energy Agency, Proceedings of a Symposium on nuclear techniques in
       environmental pollution, Salzburg, 26-30 October 1970. Vienna, IAEA, p. 347-59

Hellman, H. and H. Zehle, Quantitative Bestimmung von Kohlenwasserstoffen (Paraffinen) im
1973   Mikrogrammbereich durch IR-Spektroskopie (Quantitative determination of hydro-
       carbons (paraffins) in the microgram range by means of IR-spectroscopy).
       Z.Analyt.Chem., (265):245-9

Howard, H.W., Jr., The identification of naval fuels and natural fluorophors in seawater
1972   by fluorescence spectrometry. Monterey, California, U.S. Naval Postgraduate
       School, 81 p.

Hughes, D.R., A preliminary study of oil contamination in a marine environment. Melbourne,
1972   Australia, Department of Agriculture, Public Works Department, Ports and Harbors
       Division, Division of Agricultural Chemistry, 25 p.

Jadamec, J.R., Single wavelength fluorescence excitation for on-site oil spill identifi-
1974   cation. In Proceedings of second Conference on environmental quality sensors,
       Las Vegas, Nevada, 10-11 October 1973. Section 6. Washington, D.C.,
       Environmental Protection Agency, p. 1-20

Kawahara, F.K. and D.G. Ballinger, Characterization of oil slicks on surface waters.
1970   Ind.Eng.Chem.Prod.Res.Dev., (9):553-8

Levy, E.M., The identification of petroleum products in the marine environment by
1972   absorption spectrophotometry. Water Res., (6):57-69

Loof, S., A study of the distribution of mineral oil along the Swedish West-Coast compared
1974   to the Baltic by means of infrared spectrophotometry. Paper presented to the
       Conference of the Baltic Oceanographers, Paper no. (23)

Lukens, H.R., Instruction manual for oil slick identification by trace element patterns
1972   measured with neutron activation analysis. San Diego, California, Gulf
       Radiation Technology, (GULF-RT-A-10973):74 p.

Lukens, H.R. et al., Development of nuclear analytical techniques for oil-slick identifi-
1971   cation. Phase IIA, Final report. San Diego, California, Gulf Radiation
       Technology, (GULF-RT-A-10684):125 p.

Lynch, P.F. and C.W. Brown, Identifying sources of petroleum by infrared spectroscopy.
1973   Environ.Sci.Technol., (7):1123-7

Mark, H.B., Jr. et al., Infrared estimation of oil content in sediments in presence of
1972   biological matter. Environ.Sci.Technol., (6):833-4

Mattson, J.S., 'Fingerprinting' of oil by infrared spectrometry. Analyt.Chem., (43):1872-3
1971

Mattson, J.S. et al., Infrared spectroscopic analyses of hydrocarbon content of Santa
1970   Barbara channel sediments. San Diego, California, Gulf General Atomic Inc.,
       (GA.10270):13 p.

2.1.5 General

Carlberg, S.R. and C. Skarstedt, Determination of small amounts of non-polar hydrocarbon
1972   oil in sea water. J.Cons.CIEM, 34(3):506-15

Clark, R.C., Jr. and J.S. Finley, Analytical techniques for isolating and quantifying
1974 petroleum paraffin hydrocarbons in marine organisms. Spec.Publ.U.S.Natl.Bur.
Stand., (409):209-12

DiSalvo, L.H. et al., Hydrocarbons of suspected pollutant origin in aquatic organism of
1973 San Francisco Bay. Methods and preliminary results. Microbial degradation oil
pollutants, Workshop, p. 205-20

Farrington, J.W., Some problems associated with the collection of marine samples and
1974 analysis of hydrocarbons. Tech.Rep.Woods Hole Oceanogr.Inst., (74-23):19 p.

Farrington, J.W. and P.A. Meyer, Hydrocarbons in the marine environment. In Environmental
1975 chemistry, Senior reporter: G. Eglington, London, Chemical Society, p. 109-36

Farrington, J.W. et al., Intercalibration of analyses of recently biosynthesized hydro-
1973 carbons and petroleum hydrocarbons in marine lipids. Bull.Environ.Contam.
Toxicol., 10(3):129-36

_____, Analyses of hydrocarbons in marine organisms: results of IDOE intercali-
1974 bration exercises. Spec.Publ.U.S.Natl.Bur.Stand., (409):163-6

Gordon, D.C., Jr., P.D. Keizer and P.S. Chamut, Estimation of hydrocarbon concentrations
1974 in the water column of Come-By-Chance Bay; 1971-1973. Tech.Rep.Fish.Res.
Board Can., (442):15 p.

Grassle, J.F., Methods for studying the effects of marine oil spills. In Proceedings of
1973 a Workshop on inputs, fates and effects of petroleum in the marine environment,
21-25 May 1973, Airlie, Virginia. Washington, D.C., National Academy of
Sciences, Vol. 1:323-46

Gunkel, W., Experimental ecological investigations regarding the limiting factors of
1967 microbial oil degradation in the marine environment. Helgol.Wiss.Meeresunters.,
(15):210-25

Hellmann, H., Identifizierung von Ölverschmutzungen in Gewässern (Identification of oil
1971 pollution in water). Z.Binnenschiff.Wasserstrass., (12):459-66

Hertz, H.S. et al., Methods for trace organic analysis in sediments and marine organisms.
1974 Spec.Publ.U.S.Natl.Bur.Stand., (409):197-9

H.M.S.O., MAFF Working Party on monitoring of foodstuffs for mercury and other heavy metals.
1971 First Report. Survey of mercury in food. London, H.M.S.O.

Keizer, P.D. and D.C. Gordon, Jr., Detection of trace amounts of oil in sea water by
1973 fluorescence spectroscopy. J.Fish.Res.Board Can., (30):1039-46

Kikuchi, R., Determination of mineral oil in seaweed and sea mud. Bull.Tokai Reg.Fish.
1973 Res.Lab., (73):113-9 (In Japanese, with English summary)

Lacaze, J.-C., Use of a simple experimental device to study water pollution in situ:
1971 comparative effects of three anti-petroleum emulsive agents. Tethys, 3(4):705-16

LaRoche, G., Analytical approach in the evaluation of biological damage resulting from
1973 spilled oil. In Proceedings of a Workshop on inputs, fates and effects of
petroleum in the marine environment, 21-25 May 1973, Airlie, Virginia.
Washington, D.C., National Academy of Sciences, p. 346-74

Lichatowich, J.A. et al., Development of methodology and apparatus for the bioassay of oil.
1973   Prevention and control of oil spills. Washington, D.C., American Petroleum
        Institute, p. 659-70

Majori, L., F. Petronio and G. Nedoclan, L'inquinamento marino da idrocarburi nell'alto
1973   Adriatico. Nota 1. Su un metodo di campionamento e dosaggio degli idrocarburi
        marini superficiali (Seawater pollution from hydrocarbons in the high Adriatic
        sea. Note 1. A method for the sampling and dosage of surface seawater
        hydrocarbons). Ig.Mod., 66(2):23 p.

Makolaj, P.G., Environmental applications of the Weibull distribution function: oil
1972   pollution. Science, Wash.,D.C., (176):1019-21

Medeiros, G.C. and J.W. Farrington, IDOE-5 intercalibration sample: results of analysis
1974   after sixteen months storage. Spec.Publ.U.S.Natl.Bur.Stand., (409):167-9

Moore, S.F., Towards a model of the effects of oil on marine organisms. In Proceedings
1973   of a Workshop on inputs, fates and effects of petroleum in the marine environ-
        ment, 21-25 May 1973, Airlie, Virginia. Washington, D.C., National Academy
        of Sciences, Vol. 2:635-54

Oliver, J.D. and R.R. Colwell, Extractable lipids of gram-negative marine bacteria:
1973   phospholipid composition. J.Bact., 114(3):897-908

Perkins, E.J., Some problems of marine toxicity studies. Mar.Pollut.Bull., (3):13-4
1972

Roubal, W.T., In vivo and in vitro spin-labeling studies of pollutant-host interaction.
n.d.   In Mass spectrometry and NMR spectroscopy in pesticide chemistry, R. Haque
        and F.J. Biros (Eds.). New York, Plenum Publishing Corp., p. 305-24

Shelton, R.G.J., Dispersant toxicity test procedures. Prevention and control of oil spills.
1969   Washington, D.C., American Petroleum Institute, p. 181-91

Sprague, J.B., Sublethal oil spill fumble. Mar.Pollut.Bull., (2):83-4
1971

Sullivan, J.B., Is our approach to the oil pollution problem too crude? In Proceedings
1973   of a Workshop on inputs, fates and effects of petroleum in the marine environment,
        21-25 May 1973, Airlie, Virginia. Washington, D.C., National Academy of Sciences,
        Vol. 2:797-824

Walker, J.D. et al., Comparison of membrane filter counts and plate counts on heterotrophic
1977   and oil agar used to estimate populations of yeast, fungi and bacteria.
        Philadelphia, American Society for Testing and Materials (to be published by
        U.S.EPA, edited by Robert Bordner)

Warner, J.S., Quantitative determination of hydrocarbons in marine organisms. Spec.Publ.
1974   U.S.Natl.Bur.Stand., (409):195-6

Wasik, S.P., Determination of hydrocarbons in sea water using an electrolytic stripping
1974   cell. J.Chromatogr.Sci., (12):845

Wasik, S.P. and R.N. Boyd, Determination of aromatic hydrocarbons in sea water using an
1974   electrolytic stripping cell. Spec.Publ.U.S.Natl.Bur.Stand., (409):117-8

Wasik, S.P. and R.L. Brown, Determination of hydrocarbon solubility in sea water and the
1973   analysis of hydrocarbons in water-extracts. In American Petroleum Institute,
        Proceedings of Joint conference on prevention and control of oil spills,
        Washington, D.C., 13-15 March 1973. Washington, D.C., American Petroleum
        Institute, p. 223-7

Weiss, F.T., Identification of waterborne oils. In Proceedings of a Workshop on inputs,
1973   fates and effects of petroleum in the marine environment, 21-25 May 1973,
        Airlie, Virginia. Washington, D.C., National Academy of Sciences, Vol. 1:291-306

Zitko, V., Determination of residual fuel oil contamination of aquatic animals. Bull.
1970   Environ.Contam.and Toxicol., (5):559-64

Zitko, V. and W.V. Carson, The characterization of petroleum oils and their determination
1970   in the aquatic environment. Tech.Rep.Fish.Res.Board Can., (217):29 p.

2.2   EFFECTS

    2.2.1 Plants and Plankton

Baker, J.M., Oil pollution in salt marsh communities. Mar.Pollut.Bull., (1):27-8
1970

_____, The effects of oils on plants. Environ.Pollut., (1):27-44
1970

Bell, W. and R. Mitchell, Chemotactic and growth response of marine bacteria to algal
1972   extra-cellular products. Biol.Bull.Mar.Biol.Lab.,Woods Hole, (143):265-77

Brown, D.H., The effect of Kuwait crude oil and a solvent emulsifier on the metabolism
1972   of the marine lichen Lichina pygmaea. Mar.Biol., (12):309-15

DeCoursey, P.J. and W.B. Vernberg, The effect of dredging in a polluted estuary on the
1974   physiology of larval zooplankton. Belle W. Baruch Institute for Marine Biology
        and Coastal Research and Department of Biology. London, Pergamon Press

Dicks, B., Changes in salt-marsh vegetation around a refinery effluent outfall at Fawley.
1973   Rep.Oil Pollut.Res.Unit, Field Stud.Counc., (1972):14-20

Galtsoff, P.S. and V. Koehring, Effect of crude oil on diatoms. Bull.Bur.Fish.,Wash.,
1935   (48):193-204

Ganning, B. and U. Billing, Effects on community metabolism of oil and chemically
1974   dispersed oil on Baltic bladder wrack, Fucus vesiculosus. In Ecological
        aspects of toxicity testing of oils and dispersants, L.R. Beynon and E.B. Cowell
        (Eds.). Barking, Essex, Applied Science Publishers, p. 53-61

Gordon, D.C., Jr. and N.J. Prouse, The effects of some crude and refined oils on marine
1972   phytoplankton photosynthesis. Copenhagen, International Council for the
        Exploration of the Sea, Fisheries Improvement Committee, 1972 (CM 1972/E:33):10 p.

Kauss, P.R., T.C. Hutchinson and M. Griffiths, Field and laboratory studies of the effects
1972   of crude oil spills on phytoplankton. In Proceedings eighteenth annual
        Technical conference environmental progress in science and education, p. 22-6

Lacaze, J.-C., Effects of "Torrey Canyon" type pollution on the unicellular marine alga
1969   *Phaeodactylum tricornutum*. Rev.Int.Oceanogr.Med., (13-14):157-79

———, Effects of three emulsifying agents against crude oil on the primary
1973   productivity of an experimental community of benthic diatoms. Vie Milieu (B),
23(1):51-67

———, Ecotoxicology of crude oils and the use of experimental marine ecosystems.
1974   Mar.Pollut.Bull., (5):153-6

———, De la production primaire d'écosystèmes expérimentaux établis dans l'estuaire
1974   de la Rance. Effet d'un pétrole brut (On the primary productivity of experimental ecosystems set up in the Rance estuary. Effect of a crude petroleum).
C.R.Hebd.Séances Acad.Sci.,Paris (D), (278):2531-4

Mallet, L. and R. Lami, Recherche sur la pollution du plancton par les hydrocarbures
1964   polybenzéniques du type benzo 3,4 pyrène dans l'estuaire de la Rance. C.R.
Séances Soc.Biol., (158):2261-2

Mallet, L. and J. Sardou, Recherche de la présence de l'hydrocarbure polybenzénique
1964   benzo 3,4 pyrène dans le milieu planctonique de la région de la baie de
Villefranche (Alpes-Maritimes). C.R.Hebd.Séances Acad.Sci.,Paris (D),
(258):5264-7

Mironov, O.G. and L.A. Lanskaya, The influence of oil on the development of marine
1966   phytoplankton. Paper presented at the second International oceanographic
congress, Moscow, 5 p. (In Russian)

———, The survival of some plankton and benthoplankton algae in seawater,
1968   contaminated by oil products. Botanical Journal, Vol. 53, No. 5

———, Growth of marine microscopic algae in seawater contaminated with hydrocarbons.
1969   Biol.Morya, (17):31-8 (In Russian)

Nelson-Smith, A., Effects of oil on marine plants. In Water pollution by oil,
1971   P. Hepple (Ed.). London, The Institute of Petroleum, p. 273-80

Niaussat, P., L. Mallet and J. Ottenwalder, Appearance of benzo 3,4 pyrene in several
1969   stocks of marine phytoplankton cultivated *in vitro*. Eventual role of associated
bacteria. C.R.Hebd.Séances Acad.Sci.,Paris (D), (268):1109-12

Pincemin, J.M., Télémédiateurs chimiques, et équilibre giogique océanique. 3. Etude
1969   in vitro de relations entre populations phytoplanctoniques. Rev.Int.Océanogr.
Med.CERBOM, (22-23):165-96

Prouse, N.J. and D.C. Gordon, Jr., The effects of three oils on the growth of the dino-
1974   flagellate *Dunaliella tertiolucta* and the diatom *Fragilaria* sp. in axenic
batch cultures. Copenhagen, International Council for the Exploration of the
Sea, Fisheries Improvement Committee, 1974 (CM 1974/E:41):6 p.

Pulich, W.M., Jr., K. Winters and C. van Baalen, The effects of a No. 2 fuel oil and two
1974   crude oils on the growth and photosynthesis of microalgae. Mar.Biol., (28):87-94

Quigley, M.M. and R.R. Colwell, Properties of bacteria isolated from deep-sea sediments.
1968   J.Bact., (1968):211-20

Ravanko, O., The PALVA oil tanker disaster in the Finnish S.W. Archipelago. 5. The
1972   littoral and aquatic flora of the polluted area. Aqua Fenn., (1972):142-4

Schramm, W., Untersuchungen über den Einfluss von Ölverschmutzungen auf Meeresalgen. 1.
1972   Die Wirkung von Rohölfilmen auf den CO_2-Gaswechsel ausserhalb des Wassers
       (Investigations on the influence of oil pollution on marine algae. 1. The
       effect of crude-oil films on the CO_2 gas exchange outside the water). Mar.Biol.,
       (14):189-98

―――――, The effects of oil pollution on gas exchange in Porphyra umbilicalis when
1972   exposed to air. Proc.Int.Seaweed Symp., (7):309-15

Shiels, W.E., Effects of crude oil treated seawater on the metabolism of phytoplankton and
1973   seaweeds. Fairbanks, University of Alaska, MS Thesis

Shiels, W.E., J.J. Goering and D.W. Hood, Crude oil phytotoxicity studies. Occas.Publ.
1973   Inst.Mar.Sci.Univ.Alaska, (3):413-46

Thomas, J.P., Release of dissolved organic matter from natural populations of marine
1971   phytoplankton. Mar.Biol., 11(4):311-23

Tkachenko, V.N. and L.E. Aivazova, Influence of dissolved oil products on marine and fresh-
1974   water monocell algae. Tr.VNIRO, (100):68-73 (In Russian)

2.2.2 Shellfish

Atema, J. and L.S. Stein, Effects of crude oil on the feeding behaviour of the lobster
1974   Homarus americanus. Environ.Pollut., (6):77-86

Atema, J. et al., The importance of chemical signals in stimulating behaviour of marine
1973   organisms: effects of altered environmental chemistry on animal communication.
       In Bioassay techniques and environmental chemistry, G.E. Glass (Ed.).
       Ann Arbor, Michigan, Ann Arbor Science Publishers, p. 177-97

Avolizi, R.J. and M.A. Nuwayhid, Effects of crude oil and dispersants on bivalves. Mar.
1974   Pollut.Bull., (5):149-53

Brown, A.C., P. de B. Baissac and B. Leon, Observations on the effects of crude oil
1974   pollution on the sandy beach snail, Bullia (Gastropoda:Prosobranchiata).
       Trans.R.Soc.S.Afr., (41):19-24

Brunies, A., Mineralölgeschmack bei Miesmuscheln (Taint of mineral oil in mussels).
1971   Arch.Lebensmittelhyg., March issue:63-4 English translation available from
       British Library, Lending Division, Boston Spa, Yorkshire, Translation T 674

Connor, P.M., Further investigations into the toxicity of oil and dispersants. Copenhagen,
1972   International Council for the Exploration of the Sea, Fisheries Improvement
       Committee, 1972 (CM 1972/E:14):6 p.

Corner, E.D.S., C.C. Kilvington and S.C.M. O'Hara, Qualitative studies on the metabolism
1973   of naphthalene in Maia squinado (Herbst). J.Mar.Biol.Assoc.U.K., (53):819-32

Davavin, I.A., DNA of the Black Sea mussels in clean water and in oil polluted water.
1973   Hydrobiol.J., 6

Davenport, J., A comparison of the effects of oil, BP 1100 and oleophilic fluff upon the
1973   porcelain crab, Porcellana platycheles. Chemosphere, (1):3-6

Dicks, B., Some effects of Kuwait crude oil on the limpet Patella vulgata. Environ.Pollut.,
1973   (5):219-29

Eisler, R., Latent effects of Iranian crude oil and a chemical oil dispersant on Red Sea
1973   molluscs. Isr.J.Zool., (22):97-105

Fossato, V.U., Elimination of hydrocarbons by mussels. Mar.Pollut.Bull., (6):7-10
1975

Galtsoff, P.S., Experimental studies of the effect of oil on oysters. Review of the
1935   literature. Bull.Bur.Fish.,Wash., 48(18):158-9

Galtsoff, P.S. and R.O. Smith, The effect of oil on feeding of oysters. Bull.Bur.Fish.,
1935   Wash., 48(18):167-93

Galtsoff, P.S. et al., Effects of crude oil pollution on oysters in Louisiana waters.
1935   Bull.Bur.Fish.,Wash., (18):143-210

Gilfillan, E.S., Decrease of net carbon flux in two species of mussels caused by extracts
1975   of crude oil. Mar.Biol., (29):53-7

Harger, J.R.E. and D. Straughan, Biology of sea mussels Mytilus californianus (Conrad)
1972   and M. edulis (Linn.) before and after the Santa Barbara oil spill (1969).
        Water Air Soil Pollut., (1):381-8

Howe, C. and S.M. Ottway, Some effects of crude oil, oil fractions and products on the
1971   prawn Leander squilla. Rep.Oil Pollut.Res.Unit Field Stud.Counc., (1971):22-8

Jeffries, H.P., A stress syndrome in the hard clam, Mercenaria mercenaria. J.Invert.
1972   Pathol., (20):242-51

Kanter, R., Susceptibility to crude oil with respect to size, season and geographic
1974   location in Mytilus californianus (Bivalvia). Los Angeles, California,
        University of Southern California, Sea Grant Program (USC-SG-4-74):43 p.

Karinen, J.F. and S.D. Rice, Effects of Prudhoe Bay crude oil on molting Tanner crabs,
1974   Chionoecetes bairdi. Mar.Fish.Rev., 36(7):31-7

Kittredge, J.S., Effects of the water-soluble component of oil pollution on chemoreception
1971   by crabs. Duarte, California, City of Hope National Medical Center, 5 p.

Leenhardt, H., De l'action du mazout sur les coquillages. Rapp.Cons.Perm.Internat.Explor.
1925   Mer, (35):56-8

Mazmanidi, N.D., Diasamidze and Zambachidze, Oil effects on some species of molluscs and
1973   crustacea in the Black Sea. Materials of All Union Symposium on the Studies
        of the Black and Mediterranean Seas, utilization and protection of their
        resources. Part 4. Kiev, Naukova Dumka

Mills, E.R. and D.D. Culley, Jr., Toxicity of various off-shore crude oils and dispersants
1972   to marine and estuarine shrimp. Proc.Southeast.Game Fish.Comm., 25(1971): 642-50

Mitchell, P.H., The effect of water-gas tar on oysters. Bull.Bur.Fish.,Wash., (32):201-6
1912

Nuwayhid, M.A., The effect of Arabian light crude oil and Corexit on the respiration of
1973   the bivalved molluscs *Brachidontes variabilis* and *Donax trunculus*. Beirut,
American University of Beirut, 1973. M.Sc. Thesis, 175 p.

Ottway, S.M., The toxicity of oil to animals subjected to salinity stress. Rep.Oil Pollut.
1973   Res.Unit, Field Stud.Counc., (1972):24-30

Parker, C.A., The ultimate fate of crude oil at sea. Interim report. 5. Uptake of oil
1969   by zooplankton. Rep.AML (Admir.Mater.Lab.)U.K., (B/198(M)):15 p.

Parsons, R., Note on the recovery of *Littorina saxatilis* from pollution with refinery
1973   effluent when under osmotic stress. Rep.Oil Pollut.Res.Unit, Field Stud.Counc.,
(1972):31

_____, Some sublethal effects of refinery effluent upon the winkle *Littorina*
1973   *saxatilis*. Rep.Oil Pollut.Res.Unit, Field Stud.Counc., (1972):21-3

Prytherch, H.F., Preliminary field investigations, 1933. Bull.Bur.Fish.,Wash., 48(18):146-9
1935

_____, Survival of oysters in oil-polluted water. Bull.Bur.Fish.,Wash., 48(18):159-67
1935

Smith, R.O., Survey of oyster bottoms in areas affected by oil well pollution 1934. Bull.
1935   Bur.Fish.,Wash., 48(18):150-7

St. Amant, L.S., Investigations of oily taste in oysters caused by oil drilling operations.
1957   Bienn.Rep.La.Wildl.Fish.Comm., (7):4 p.

Takahashi, F.T. and J.S. Kittredge, Sublethal effects of the water soluble component of
1973   oil: chemical communication in the marine environment. Publ.La.State Univ.,
(LSU-SG-73-01):259-64

Teal, J.T. and J.J. Stegeman, Accumulation, release and retention of petroleum hydrocarbons
1973   by the oyster *Crassostrea virginica*. In Proceedings of a Workshop on inputs,
fates and effects of petroleum in the marine environment, 21-25 May 1973, Airlie,
Virginia. Washington, D.C., National Academy of Sciences, Vol. 2:571-602

2.2.3 Fish

Blackman, R.A.A., Effects of sunken oil on the feeding of plaice on brown shrimps and other
1974   benthos. Copenhagen, International Council for the Exploration of the Sea
(CM 1974/E:24):7 p.

Blackman, R.A.A. and P.R. Mackie, Preliminary results of an experiment to measure the
1973   uptake of n-alkane hydrocarbons by fish. Copenhagen, International Council
for the Exploration of the Sea (CM 1973/E:23):9 p.

Blanton, W.G. and M.C. Robinson, Some acute effects of low boiling petroleum fractions on
1973   the cellular structures of fish gills under field conditions. Publ.La.State
Univ., (LSU-SG-73-01):265-73

Cardwell, R.D., Acute toxicity of No. 2 diesel oil to selected species of marine inverte-
1973   brates, marine sculpins, and juvenile salmon. Ann Arbor, Michigan, University
Microfilms, Order No. (73-22:557)

Dicks, B., Some effects of Kuwait crude oil on the limpet Patella vulgata. Environ.Pollut.,
1973   (5):219-29

Ebeling, A.W. et al., Santa Barbara oil spill: fishes. In Proceedings of the Santa Barbara
1972   oil symposium: offshore petroleum production, an environmental inquiry,
        Santa Barbara, California, 16-18 December 1970, R.E. Holmes and F.A. DeWitt, Jr.,
        (Eds.). Washington, D.C., National Science Foundation, p. 138-64

Eckardt, R.E., Industrial carcinogens. New York, Grune
1959

Falk, M.R. and M.J. Lawrence, Acute toxicity of petrochemical drilling fluid components
1973   and wastes to fish. Tech.Rep.Ser.Can.Resour.Manage.Branch, (CEN T-73-1):112

Findlay, E.R. and P. Kotin, Squamous papilloma in the white croaker. J.Natl.Cancer Inst.,
1957   18(6):857-60

Häkkilä, K. and A. Niemi, Effects of oil and emulsifiers on eggs and larvae of northern
1973   pike (Esox lucius) in brackish water. Aqua Fenn., (1973):44-59

Kloth, T.C. and D.E. Wohlschlag, Size-related metabolic responses of the pinfish,
1972   Lagodon rhomboides, to salinity variations and sublethal petrochemical
        pollution. Contrib.Mar.Sci.Univ.Tex., (16):125-37

Krishnaswami, S.K. and E.E. Kupchanko, Relationship between odor of petroleum refinery
1969   wastewater and occurrence of "oily" tasteflavor in rainbow trout Salmo gairdnerii.
        J.Water Pollut.Control Fed., 41(5)Part 2:169-89

_____, Relationship between odor of petroleum refinery wastewater and occurrence of
1973   "oily" tasteflavor in rainbow trout, Salmo gairdnerii. J. Water Pollut.Control
        Fed., J.WPCF., 41(5)Part 2:189-99

Kristoffersson, R., S. Broberg and A. Oikari, Physiological effects of a sublethal concen-
1973   tration of phenol in the pike (Esox lucius L.) in pure brackish water. Ann.
        Zool.Fenn., (10):392-7

Kühnhold, W.W., Investigations on the toxicity of seawater-extracts of three crude oils on
1974   eggs of cod (Gadus morhua L.). Ber.Dtsch.Wiss.Komm.Meeresforsch., (23):165-80

Lichatowich, J.A., J.A. Strand and W.L. Templeton, Development of toxicity test procedures
1972   for marine zooplankton. Am.Inst.Chem.Eng.Symp.Ser., (68):372-8

Mackie, P.R., A.S. McGill and R. Hardy, Diesel oil contamination of brown trout (Salmo
1972   trutta L.). Environ.Pollut., 3(1):9-16

Mankki, J. and J. Vauras, Littoral fish populations after an oil tanker disaster in the
1974   Finnish southwest archipelago. Ann.Zool.Fenn., (11):120-6

Matthews, J.E. and L.H. Myers, Static bioassay test of petroleum refinery wastewaters
1973   using redear sunfish (Lepomis microlophys). Paper presented at Workshop on
        pathological effects of chemicals on aquatic organisms, Pensacola, Florida,
        8-9 May 1973

Meehan, W.R., L.A. Norris and H.S. Sears, Toxicity of various formulations of 2,4-D to
1974   salmonids in southeast Alaska. J.Fish.Res.Board Can., 31(4):480-5

Meschin, F.L., Histological modifications of organs and tissues of Lebistes reticulatus (P)
1973   under phenol acute intoxication. Phenol influence on hydrobionts. Leningrad,
        Publishing House "Nauka"

Mitchell, D.M. and H.J. Bennett, The susceptibility of bluegill sunfish Lepomis macrochirus,
1972   and channel catfish Ictalurus punctatus to emulsifiers and crude oil. Proc.La.
Acad.Sci., (35):20-6

Miyake, Y., Studies on abnormal odor fish by oil pollution in Mizushima industrial area.
1967   Part 1. Distribution of abnormal odor fish. Okayama-Igakkai-Zasshi, 81(3-4):
193-7

Moore, H.J., Summary report on the effects of oil discharges and domestic and industrial
1975   waste waters on the fisheries of Lake Maracaibo, Venezuela. Richland,
Washington, Battelle Pacific Northwest Laboratories, Research Contract No.
212B00899

Morrow, J.E., Effects of crude oil and some of its components on young coho and sockeye
1974   salmon. Environ.Protect.Ag.Ecol.Res.Ser.,Wash.,D.C., (EPA-660-3-73-018):43 p.

Motohiro, T. and N. Inoue, n-paraffins in polluted fish by crude oil from "J" wreck.
1973   Bull.Fac.Fish.Hokkaido Univ., 23(4):204-8

Quinn, J.G. and T.L. Wade, Hydrocarbon analyses of IDOE intercalibration samples of cod
1974   liver oil and tuna meal. Mar.Mem.Ser.Univ.R.I., (33)

Saward, D., Some physiological and behavioural aspects of the effects of phenol upon plaice.
1972   Copenhagen, International Council for the Exploration of the Sea, Fisheries
Improvement Committee (CM 1972/E:23):5 p.

Shelton, R.G.J., Oil pollution and commercial fisheries in Britain. In Water pollution by
1971   oil, P. Hepple (Ed.). London, Institute of Petroleum, p. 377-9

Sipos, J.C. and R.G. Ackman, Association of dimethyl sulfide with the "blackberry" problem
1964   in cod from the Labrador area. J.Fish.Res.Board Can., (21):423-5

Sprague, J.B. and D.E. Drury, Avoidance reactions of salmonia fish to representative
1969   pollutants. Adv.Water Pollut.Res., (4):169-79

Sturdevant, D.C., The influence of crude oil on the serum proteins of the sockeye salmon.
1972   Fairbanks, University of Alaska, M.S. Thesis, 43 p.

Yoshida, K. and N. Uezumi, On the problem of offensive-odor fish in petrochemical
1961   industrial area. 1. Hygiene in Life, Jap., 5(4):11-7 (In Japanese)

Yoshida, K., N. Uezumi and K. Kosama, On the problem of offensive-odor fish in petro-
1967   chemical industrial area. 2. Hygiene in Life, Jap., 11(2):8-13 (In Japanese)

2.2.4 Other Organisms

Andrews, A.R. and G.D. Floodgate, Some observations on the interactions of marine protozoa
1974   and crude oil residues. Mar.Biol., (25):7-12

Barnett, C.J. and J.E. Kontogiannis, The effect of crude oil fractions on the survival of
1975   a tidal pool copepod, Tigriopus californicus. Environ.Pollut., (8):45-54

Bean, R.M., J.R. Vanderhorst and P. Wilkinson, Inter-disciplinary study of the toxicity of
1974   petroleum to marine organisms. Richland, Washington, Battelle Pacific Northwest Laboratories, 48 p.

Bender, J.E., J.L. Hyland and T.K. Duncan, Effect of an oil spill on benthic animals in
1974 the lower York River, Virginia. Spec.Publ.U.S.Natl.Bur.Stand., (409):257-9

Bourne, W.R.P., Effects of oil pollution on bird populations. Field Stud., (2):99
1968

Bugbee, S.L. and C.M. Walter, The response of macro-invertebrates to gasoline pollution in
1973 a mountain stream. In Prevention and control of oil spills, American Petroleum
Institute, p. 720-31

Bury, R.B., The effects of diesel fuel on a stream fauna. Calif.Fish.Game, 58(4):291-5
1972

Cardwell, R.D., Acute toxicity of No. 2 diesel oil to selected species of marine inverte-
1973 brates, marine sculpins and juvenile salmon. Ann Arbor, Michigan, University
Microfilms, Order No. (73-22:557)

Carthy, J.D. and D.R. Arthur, The biological effects of oil pollution on littoral
1968 communities. London, Institute of Petroleum, Field Studies Council

Carvacho, A.B., Efecto de la contaminación del mar con petróleo en poblaciones de Crustáceos
1971 Decápodas litorales (Effect of pollution of the sea by petroleum on populations
of littoral decapod crustacea). Not.Mens., 15(180):7-12

Chet, I., S. Fogel and R. Mitchell, Chemical detection of microbial prey by bacterial
1971 predators. J.Bact., (106):863-7

Chia, F.S., Killing of marine larvae by diesel oil. Mar.Pollut.Bull., (4):29-30
1973

Cohen, Y., Effects of crude oil on the Red Sea Alcyonarian Heteroxenia fuscescens.
1973 Jerusalem, Hebrew University of Jerusalem, Department of Zoology, M.S. Thesis

Cowell, E.B., J.M. Baker and G.B. Crapp, The biological effects of oil pollution and oil
1972 cleaning materials on littoral communities, including salt marshes. In Marine
pollution and sea life, M. Ruivo (Ed.). Fishing News (Books) Ltd., p. 359-64

Cubit, J., The effects of the 1969 Santa Barbara oil spill on marine intertidal inverte-
1972 brates. In Proceedings of the Santa Barbara oil symposium, offshore petroleum
production, an environmental inquiry, Santa Barbara, California,
16-18 December 1970, R.E. Holmes and F.A. DeWitt, Jr. (Eds.). Washington, D.C.,
National Science Foundation, p. 131-6

Davenport, J., A comparison of the effects of oil, B.P. 1100 and oleophilic fluff upon the
1973 porcelain crab Porcellana platycheles. Chemosphere, (1):3-6

Eisler, R., G.W. Kissill and Y. Cohen, Recent studies on biological effects of crude oils
1974 and oil-dispersant mixtures to Red Sea macrofauna. Environ.Protect.Ag.Environ.
Monit.Ser.,Wash.,D.C., (EPA-600/4-74-004):156-79

Evans, W.G., Thalassotrechus barbarae (Horn) and the Santa Barbara oil spill (Coleoptera:
1970 Carabidae). Pan-Pacif.Enterp., (46):233-7

Fauchald, K., A survey of the benthos off Santa Barbara following the January 1969 oil
1972 spill. Rep.CCOFI, (16):125-9

Fogel, S., I. Chet and R. Mitchell, Chemotactic responses of marine bacteria. Bact.Proc.
1971 G 31

Gibson, D.T. et al., Oxidation of the carcinogens benzo (a) pyrene and benzo (a) anthracene
1975 to dihydrodiols by a bacterium. Science, Wash.,D.C., (189):295-7

Gray, J.S. and R.J. Ventilla, Pollution effects on micro- and meiofauna of sand. Mar.
1971 Pollut.Bull., (2):39-43

Great Britain, Medical Research Council, The carcinogenic action of mineral oils: a
1968 chemical and biological study. London, H.M.S.O.

Howe, C. and S.M. Ottway, Some effects of crude oil, oil fractions and products on the prawn
1971 Leander squilla. Rep.Oil Pollut.Res.Unit, Field Stud.Counc., (1971):22-8

Jacobson, S.M. and D.B. Boylan, Effect of seawater soluble fraction of kerosene on chemo-
1973 taxis in a marine snail, Nassarius obsoletus. Nature, Lond., (241):213-5

Karinen, J.F. and D.R. Stanley, Effects of Prudhoe Bay crude oil on molting tanner crabs,
Chionoecetes bairdi. Mar.Fish.Rev., (In Press)

Kasymov, A.G. and A.D. Aliev, Experimental study of the effect of oil on some representatives
1973 of benthos in the Caspian Sea. Water Air Soil Pollut., (2):235-45

Katayama, T. and T. Fujiyama, Studies on the nucleic acid of algae with special reference
1957 to the Desoxyribonucleic acid contents of the Crown-gall tissues developed on
Porphyra tenera (KJUNM). Bull.Jap.Soc.Sci.Fish., (23):249-54

Katz, M., The effects of pollution upon aquatic life. In Water and Water Pollution
1971 Handbook, L.L. Ciaccio (Ed.). (1):297-326

Kontogiannis, J.E. and C.J. Barnett, The effect of oil pollution on survival of the tidal
1973 pool copepod Tigriopus californicus. Environ.Pollut., (4):69-79

Levell, D., Effects of oil pollution and cleaning on sand and mud fauna - interim report,
1973 January 1973. Rep.Oil Pollut.Res.Unit, Field Stud.Counc., (1972):32-4

Mironov, O.G. and S.U. Andreeva, The effect of oil pollution on the development of some
1973 Black Sea infusoria. Biol.Nauki, (5):19-21 (In Russian)

Moore, H.J., Summary report on the effects of oil discharges and domestic and industrial
1975 waste waters on the fisheries of Lake Maracaibo, Venezuela. Richland,
Washington, Battelle Pacific Northwest Laboratories, Research Contract No.
212B00899

Ottway, S.M., Zoological studies on shore communities. The comparative toxicities of
1971 crude oils. In The ecological effects of oil pollution on littoral communities,
E.B. Cowell (Ed.). New York, Elsevier Publishing Co., p. 172-80

Percy, J.A., Oil and Arctic marine invertebrates. In Oil and the Canadian environment.
1973 Proceedings of the Conference, sponsored by the Institute of Environmental
Sciences and Engineering, University of Toronto, Toronto, Ont., 16 May 1973,
D. MacKay and W. Harrison (Eds.). Toronto, University of Toronto, p. 71-4

Richards, T.L., The influence of petroleum fractions in the polychaetons annelid.
1970 Western Society of Naturalists Annual Meeting 51st Abstracts, 14-5

Rigdon, R.M. and J. Neal, Absorption and excretion of Benzpyrene in the cockroach
1963 (Periplaneta americana). Experientia, (19):474-7

Simpson, J.G. and W.G. Gilmartin, Elephant seal and sealion mortality on San Miguel Island.
1971 In U.S. Naval Undersea Research and Development Center, Proceedings of the
      Symposium on environmental preservation, San Diego, California, 20-21 May 1970.
      San Diego, U.S. Naval Research and Development Center, p. 105-10

Spooner, M.F. and C.J. Corkett, A method for testing the toxicity of suspended oil droplets
1974 on planktonic copepods used at Plymouth. In Ecological aspects of toxicity
      testing of oils and dispersants. Proceedings of a Workshop, L.R. Beynon and
      E.B. Cowell (Eds.). Barking, Essex, Applied Science Publishers, p. 69-74

St. Amant, L.S., The petroleum industry as it affects marine and estuarine ecology.
1972 J.Petrol.Technol., (24):385-92

Tsuruga, H., The effects of oil pollution on the marine organism. Paper presented at the
n.d. Meeting of the Joint IMCO/FAO/Unesco/WMO/WHO/IAEA Group of Experts on the
      Scientific Aspects of Marine Pollution, February 1970 (GESAMP/27):7 p.

University of California, Santa Barbara, Santa Barbara oil spill: short-term analysis of
1971 macroplankton and fish. Water Pollut.Control Res.Ser.,Wash.,D.C., (15080 EAL
      02/71):68 p.

Webber, H.H., Survey of mortality at two areas of South Beach of Guemes Island, 28 April to
1971 15 May 1971. In Biological assessment of diesel spill in the vicinity of
      Anacortes, Washington, May 1971. Final report. Appendix A, by J.A. Watson et al.
      Dallas, Texas, Texas Instruments Inc., 12 p.

2.2.5 Carcinogens

Andelman, J.B. and M.J. Suess, Polynuclear aromatic hydrocarbons in the water environment.
1970 Bull.Wld.Hlth.Org., (43):479-508

Badger, G.M., The carcinogenic hydrocarbons: chemical constitutions and carcinogenic
1948 activity. Brit.J.Cancer, (2):309-50

Berenblum, I. and R. Schoental, Carcinogenic constituents of shale oil. Brit.J.Exp.Pathol.,
1943 (24):232-9

Bingham, E., W. Horton and R. Tye, The carcinogenic potency of certain oils. Arch.Environ.
1965 Health, (10):449-51

Borneff, J., Kanzerogene Substanzen in Wasser. München. Med.Wschr., (105):1237-42
1963

Borneff, J. and R. Fisher, Kanzerogene (sic) Substanzen in Wasser und Boden. 9. Unter-
1963 suchungen von Filterschlamm eines Seewasserwerkes auf polyzyklische aromatische
      Kohlenwasserstoffe. Arch.Hyg.Bakt., (146):183-97

_____, Kanzerogene Substanzen in Wasser und Boden. 12. Polyzyklische aromatische
1963 Kohlenwasserstoffe in Oberflächenwasser. Arch.Hyg.Bakt., (146):572-85

Cavelieri, E. and M. Calvin, Molecular characteristics of some carcinogenic hydrocarbons.
1971 Science, Wash.,D.C., 68(6):1251-3

Cicatelli, M.S., Il benzo 3,4 pyrene, idrocarburo cancerogeno, nell'ambiente marino.
1966 (Benzo 3,4 pyrene, cancerogenous hydrocarbon in the marine environment). Arch.
      Zool.Ital., (51):747-74

Clayson, D.B., Chemical carcinogenesis. Boston, Little, Brown and Company, 467 p.
1962

Conney, A.H., E.C. Miller and J.A. Miller, Substrate-induced synthesis and other properties
1958   of Benzpyrene hydroxylase in rat liver. J.Biol.Chem., (228):753-66

Cromwell, N.H., Chemical carcinogens, carcinogenesis and carcinostasis. Amer.Scientist,
1965   (53):213-36

Daudel, P. and R. Daudel, Chemical carcinogenesis and molecular biology. New York,
1966   Interscience Publisher, 158 p.

Eckardt, R.E., Industrial carcinogens. New York, Grune
1959

Ehrhardt, J.P., Natural biosynthesis of carcinogenic hydrocarbons in a (sic) isolated
1972   lagoon: possible pathogenic properties. L'Evolut.Méd., 16(4):269-72

Eizen, E.G. and I.K.H. Arro, Cancerogenic compounds in some Estonian shale tars. Vopr.
1959   Onkol., 5(2):160-3

Fabian, B., Kanzerogene Substanzen in Speisefett und Öl. 5. Untersuchungen an verschieden
1968   zubereiteten Bratwürsten. Arch.Hyg.Bakt., (152):251-4

Falk, H.D. and P. Kotin, Chemistry, host entry and metabolic fates of carcinogens. Clin.
1963   Pharmacol.Therap., (4):88-103

Gelboin, H.V., A microsome dependent binding of benzo (a) pyrene to DNA. Cancer Res.,
1969   (29):1272-6

Gelboin, H.V. and F.J. Wiebel, Studies on the mechanism of aryl hydrocarbon hydroxylase
1971   induction and its role in cytotoxicity and tumorogenicity. Science, Wash.,D.C.,
        (179):529-47

Giaccio, M., Idrocarburi policiclici aromatici in alcune specie ittiche (Polycyclic
1971   aromatic hydrocarbons in some species of fish). Quad.Merceol., (10):21-6
        (In Italian with English summary)

Graf, W. and H. Diehl, Über den naturbedingten Normalpegel kanzerogener, polyzyclischer
1966   Aromate und seine Ursache. Arch.Hyg.Bakt., (150):49-59

Great Britain, Medical Research Council, The carcinogenic action of mineral oils: a
1968   chemical and biological study. London, H.M.S.O.

Grover, P.L. et al., In vitro transformation of rodent cells by k-region derivatives of
1971   polycyclic hydrocarbons. Proc.Natl.Acad.Sci., 68(6):1098-101

Haddow, A., Chemical carcinogens and their mode of action. Brit.Med.Bull., (14):79-92
1958

Hartwell, J.L., Survey of compounds which have been tested for carcinogenic activity.
1951   Washington, D.C., Public Health Service Publication 149, 2nd ed., 583 p.

Heidelberger, C., Studies on the molecular mechanism of hydrocarbon carcinogenesis.
1964   J.Cell.Comp.Physiol.(Suppl.), (64):129-48

Horton, A.W. et al., Composition versus carcinogenicity of distillate oils. Amer.Chem.Soc.
1963   Div.Petrol.Chem.Preprints, 8(4):C59-C65

Hueper, W.C., Carcinogens in the human environment. Arch.Pathol.Am.Med.Assoc., (71):
1961   237-67, 355-80

―――――, Environmental and occupational cancer hazards. Clin.Pharmacol.Therap.,
1962   (3):776-813

Hueper, W.C. and W.D. Conway, Chemical carcinogenesis and cancer. Springfield, Illinois,
1964   Charles C. Thomas, 744 p.

Lange, E., Carcinoid-like tumors in the pseudobranch of Gadus morhua L. Comp.Biochem.
1973   Physiol., 45A:477-81

Mallet, L., Pollution des milieux marins par les hydrocarbures cancérigènes (Pollution of
1960   marine environments by cancerogenous hydrocarbons). Bull.Acad.Méd.,Paris,
        (144):271-4

―――――, Pollution des milieux vitaux par les hydrocarbures polybenzéniques du type
1964   benzo 3,4 pyrène (Pollution of important environments by polybenzic hydro-
        carbons of the type benzo 3,4 pyrene). Gaz.Hôp.,Paris, Numéro Spécial
        10 Juin, (136):803-8

―――――, Pollutions marines par les hydrocarbures polycondensés du type benzo 3,4
1967   pyrène des côtes nord et ouest de France. Leur incidence sur le milieu
        biologique, et en particulier le plancton. Cah.Océanogr., (19):237-43

Mallet, L. and J. Sardou, Recherche de la présence de l'hydrocarbure polybenzénique
1964   benzo 3,4 pyrène dans le milieu planctonique de la région de la baie de
        Villefrance (Alpes-Maritimes). C.R.Hebd.Séances Acad.Sci.,Paris (D), (258):
        5264-7

Niaussat, P., Pollution by biosynthesis in situ of carcinogenic hydrocarbons of a lagoon
1970   biogenesis: in vitro reproduction of this phenomenon. Rev.Int.Océanogr.Med.,
        CERBOM, (17):87-98

Niaussat, P. and J. Ottenwalder, Apparition de benzo 3,4 pyrène dans des cultures in vitro
1969   de phyto-plancton marin. Importance des souillures bactériennes associées
        (Apparition of benzo 3,4 pyrene in in vitro cultures of marine phytoplankton.
        Incidence of associate bacteria). Rev.Hyg.Méd.Soc., (17):487-96

Niaussat, P., C. Auger and L. Mallet, Apparition relative de quantités d'hydrocarbures
1970   cancérigènes dans des cultures pures de Bacillus badius, en fonction de la
        présence, dans le milieu de certains composés chimiques. C.R.Hebd.Séances
        Acad.Sci.,Paris (D), (270):1042-5

Nigrelli, R.F., Spontaneous neoplasms in fishes. VI. Thyroid tumors in marine fishes.
1952   Cancer Res., (12):286

Piccinetti, C., Diffusione dell'idrocarburo cancerogeno benzo 3,4 pyrene nell'Alto e
1967   Medio Adriatico (Diffusion of the carcinogenic hydrocarbon benzo 3,4 pyrene
        in the north and middle Adriatic Sea). Arch.Océanogr.Limnol., 15(Suppl):
        169-84

Roe, J.C., Cancer inducing agents. Science J., (2):38-42
1966

Scaccini, A. and M. Scaccini-Cicatelli, Données préliminaires sur la diffusion de l'hydro-
1969   carbure cancérigène benzo 3,4 pyrène dans les matériaux du fond et dans certains
        organismes des mers italiennes (Preliminary data on the diffusion of the
        cancerogenous hydrocarbons benzo 3,4 pyrene in bottom materials and in certain
        organisms of the Italian seas). Rapp.P.-V.Réun.CIESM, (19):761-3

Scaccini, A. et al., Dosaggi spettrofotofluorimetrici ed osservazioni mediante microscopia
1970   a fluorescenza del benzo 3,4 pirene su organi e tessuti di Carassius auratus
        (Spectrophotofluorometric dosage and fluorescence microscope observations on
        benzo 3,4 pyrene in organs and tissues of Carassius auratus). Note Lab.Biol.
        Mar.Fano, (3):105-44

Scaccini-Cicatelli, M., Study of the phenomena of accumulation of benzo 3,4 pyrene; in the
1965   organism Tubiflex. Boll.Pesca.Piscic.Idrobiol., (20):245-50

_____, Sui fenomeni di accumulo del benzo 3,4 pirene nell'organismo di Tubifex
1966   (On the phenomena of accumulation of benzo 3,4 pyrene in the organism Tubifex).
        Boll.Soc.Ital.Biol.Sper., (42):957-9

_____, Benzo 3,4 pyrene, a cancerogenous hydrocarbon in the marine environment.
1966   Archo.Zool.Ital., (51):747-74

Scharrer, B. and M.S. Lockhead, Tumours in the invertebrates: a review. Cancer Res.,
1950   (10):403-19

Selkirk, J.K., E. Huberman and G. Heidelberger, An epoxide as an intermediate in the micro-
1971   somal metabolism of the chemical carcinogen, Dibenz (a,h.) anthracene. Biochem.
        Biophys.Res.Commun., 43(5)

Shabad, L.M., Studies in the U.S.S.R. on the distribution and circulation of carcinogenic
1967   hydrocarbons in the human environment and the role of their deposition in
        tissues in carcinogenesis. Cancer Res., 37(1):132

_____, On the distribution and fate of the carcinogenic hydrocarbon benz(a) pyrene
1968   (3,4 benzpyrene) in the soil. Z.Krebsforsch., (70):204-10

Shabad, L.M. and P.R. Dikun, On the distribution, circulation and fate of carcinogenic
1966   hydrocarbons in man's environment. IX. Tokyo, International Cancer Congress
        Abstracts, p. 1252

Shubik, P. and K.L. Hartwell, Survey of compounds which have been tested for carcinogenic
1969   activity. Publ.U.S.Natl.Cancer Inst.Wash.,D.C., Suppl. 2, No. 149, 655 p.

Sisler, F.D. and C.E. ZoBell, Microbial utilization of carcinogenic hydrocarbons. Science,
1947   Wash.,D.C., (106):521-2

Suess, M.J., Presence of polynuclear aromatic hydrocarbons and the possible health
1970   consequences. Rev.Intern.Océanogr.Méd., 18-9, 181-90

Vasserot, J., Pollution of edible marine animals by carcinogen hydrocarbons. Penn.Bed,
1962   (3):183-6

Zarazil, J. and F. Picha, The occurrence of the carcinogenic compounds 3,4 benzypyrene and
1966   arsenic in the soil. Neoplasma, Brit., (13):49-55

Zechmeister, L. and B.K. Koe, The isolation of carcinogenic and other polycyclic aromatic
1952   hydrocarbons from barnacles. Arch.Biochem.Biophys., (35):1-11

2.2.6 Case Histories

Abbott, B.C. and D. Straughan, Biological and oceanographic effects of oil spillage in the
1969   Santa Barbara Channel following the 1969 blowout. Mar.Pollut.Bull., (13):4-9

Adlard, E.R., European experience in the identification of water-borne oil. In Proceedings
1973   of a Workshop on inputs, fates and effects of petroleum in the marine environment,
       21-25 May 1973, Airlie, Virginia. Washington, D.C., National Academy of Sciences,
       Vol. 1:111-3

Ages, A.B., Oil reconnaissance in the Magdalen Islands, 1970. Dartmouth, Nova Scotia,
1971   Atlantic Oceanographic Laboratory, Bedford Institute, AOL Report (1971-8):22 p.

─────────, The VANLENE accident, March 1972. Pacif.Mar.Sci.Rep.B.C., Report
1972   No. (72-4):17 p.

Alpine Geophysical Associates Inc., Oil pollution incident Platform Charlie, Main Pass,
1971   Block 41 Field, Louisiana. Water Pollut.Control Res.Ser., (15080 FTU 05/71):
       142 p.

Anderson, E.K. et al., Preliminary report on ecological effects of the Santa Barbara oil
1969   spill. Western Oil and Gas Association, 11 p. (Unpubl.Rep.)

Ashby, E., Pollution in some British estuarine and coastal waters. In Report of the
1972   Royal Commission on environmental pollution, (3):128 p.

Baldwin, M.F., The Santa Barbara oil spill. Univ.Col.Law Rev., (42):33-76
1970

Bellamy, D.L. et al., Some effects of pollution from the TORREY CANYON disaster on
1967   littoral and sublittoral ecosystems dominated by attached macrophytes.
       Nature, Lond., (216):1170-3

Berry, W.L., Pollution control aspects of the Bay Marchand fire. J.Petrol.Technol.,
1972   (24):241-9

Binet, D. and E. Marchal, Hydrocarbon residues in Ivory Coast waters. Océanogr.Biol.
1974   O.R.S.T.O.M.,Abidjan, 8 p.

Blackman, R.A.A. et al., The DONA MARIKA oil spill. Mar.Pollut.Bull., 4(12):181-2
1973

Bourcart, J. and L. Mallet, Pollution marine des rives de la région centrale de la mer
1965   Tyrrhénienne (baie de Naples) par les hydrocarbures polybenzéniques du type
       benzo 3,4 pyrène (Marine pollution on the shores of the central region of the
       Tyrrhenian Sea (Bay of Naples) by polybenzin hydrocarbons of the type benzo 3,4
       pyrene). C.R.Hebd.Séances Acad.Sci.,Paris, (260):3729-34

Bourne, W.R.P., Dewdale oil spill. Mar.Pollut.Bull., 3(5):66-7
1972

Boyce, F., The Santa Barbara Channel oil spill: report of a visit to Santa Barbara,
1969   California on 11, 12 and 13 February 1969. Collect.Repr.Can.Cent.Inland Waters,
       2(2-1):32 p.

Brown, D.H., Field and laboratory studies on detergent damage to lichens at the Lizard,
1974   Cornwall. Cornish Stud., (2):33-40

Brundall, L., Platform "A": the oil spill that spread around the world. Expl.Econ.Petrol.
1972   Ind., (10):77

Burns, K.A. and J.M. Teal, Hydrocarbon incorporation into the salt marsh ecosystem from
1971   the West Falmouth oil spill. Tech.Rep.Woods Hole Oceanogr.Inst., (71-69):24 p.

Campbell, W.J. and S. Martin, Oil and ice in the Arctic Ocean: possible large-scale
1973   interaction. Science, Wash.,D.C., (181):56-9

Carlberg, S.R., On the Baltic oil pollution problem. Paper presented to the second
1973   Swedish-Soviet Symposium on the control of the Baltic Sea pollution

Clark, R.B., Postmortem on the ARROW. Mar.Pollut.Bull., (2):68
1971

Climberg, R., S. Mann and D. Straughan, A reinvestigation of Southern California rocky
1973   intertidal beaches $3\frac{1}{2}$ years after the 1969 Santa Barbara oil spill: a
   preliminary report. Washington, D.C., American Petroleum Institute, p. 687-701

Colwell, R.R., T. Kaneko and T. Staley, Vibrio rarahaemolyticus - an estuarine bacterium
1972   resident in Chesapeake Bay. Mar.Technol.Soc.J., (6):87-94

Connell, J.H., Biological and oceanographic survey of the Santa Barbara Channel oil spill.
1973   Submitted to the California State Lands Commission

Copeland, B.J. and T.J. Bechtel, Species diversity and water quality in Galveston Bay,
1971   Texas. Water Air Soil Pollut., 1(1):89-105

Cowell, E.B., Effects of oil pollution on salt marsh communities in Pembrokeshire and
1969   Cornwall. J.Appl.Ecol., (6):133-42

Cox, G.V. and W.G. Cox, Organics isolated from upper Long Island Sound following an oil
1972   spill in New Haven Harbor. Paper presented at the Long Island Sound conference,
   Adelphi University, 22 January 1972, 19 p.

Craigie, J.S. and J. McLachlan, Observations on the littoral algae of Chedabucto Bay
1970   following the ARROW oil spill. ARL-8, NRC #114-89

Croker, R.A., Post oil-spill intertidal survey, Great Bay, N.H., 3-7 June 1969. A report
1969   to the executive committee of the Jackson Estuarine Laboratory. Durham,
   New Hampshire, University of New Hampshire, 19 p.

Cundell, A.M. and R.W. Traxler, The isolation and characterization of hydrocarbon-
1973   utilizing bacteria from Chedabucto Bay, Nova Scotia. In Proceedings of the
   Joint conference on prevention and control of oil spills, Washington, D.C.,
   13-15 March 1973. Washington, D.C., American Petroleum Institute, p. 421-6

Day, J.H., Oil pollution in South African seas. S.Afr.J.Sci., 68(5):130-1
1971

Dickman, G.H., Report on major oil spill in San Francisco Bay on 18 January 1971,
1971   caused by a collision between the M/T ARIZONA STANDARD and the M/T OREGON STANDARD
   Washington, D.C., U.S. Coast Guard, 10 p. (Unpubl.Rep.)

Dorrings, P., The importance of oil pollution in the North Sea. In Proceedings of a
1973   Workshop in inputs, fates and effects of petroleum in the marine environment,
   21-25 May 1973, Airlie, Virginia. Washington, D.C., National Academy of
   Sciences, Vol. 2:726-30

Duce, R.A. et al., Enrichment of heavy metals and organic compounds in the surface micro-
1972   layer of Narragansett Bay, Rhode Island. Science, Wash.,D.C., (176):161-3

Feldman, M.H., The 50-mile ballast-oil dumping prohibited zone off Alaska reconsidered in
1970   the light of available data gleaned from significant incidents. Corvallis,
       Oregon, U.S. Department of the Interior, Working Paper No. 77

Foster, M., A.C. Charters and M. Neushul, The Santa Barbara oil spill. Part 1. Initial
1971   quantities and distribution of pollutant crude oil. Environ.Pollut., (2):97-113

Frost, L.C., TORREY CANYON disaster: the persistent toxic effects of detergents on cliff-
1974   edge vegetation at the Lizard Peninsula, Cornwall. Cornish Stud., (2):5-14

Furon, R., La 'marée noire' de 1967 et la pollution permanente des mers (The 'black sea'
1968   of 1967 and permanent pollution of the sea). Rev.Gén.Sci.Pur.Appl., (75):157-60

Gaines, T.H., Oil pollution control efforts - Santa Barbara, California. J.Petrol.Technol.,
1970   (22):1511-4

Gordon, D.C., Jr. and P.A. Michalik, Concentration of Bunker C fuel oil in the waters of
1971   Chedabucto Bay, April 1971. J.Fish.Res.Board Can., 28(12):1912-4

Gordon, D.C., Jr., P.D. Keizer and P.S. Chamut, Estimations of hydrocarbon concentrations
1974   in the water column of Come-by-Chance Bay; 1971-73. Tech.Rep.Fish.Res.Board
       Can., (442):15 p.

Green, D.R. et al., The Alert Bay oil spill: a one-year study of the recovery of a
1974   contaminated bay. Pacif.Mar.Sci.Rep.B.C., (74-9)

Greenham, M.S., Operation oil: Chedabucto Bay 1970. Operation Scour: report on the
1973   re-pump of the sunken tanker ARROW. In Task Force ARROW Oil Spill, Report
       of the Task Force - Operation oil (Clean-up of the ARROW oil spill in
       Chedabucto Bay) to the Minister of Transport. Ottawa, Ministry of Transport,
       Information Canada, Vol. 4:115-38

Gunkel, W., Distribution and abundance of oil-oxidizing bacteria in the North Sea.
1973   Publ.La.State Univ., (LSU-SG-73-01):127-39

Haila, Y., M.T. Palvan öljyonnettomuus (The PALVA oil tanker catastrophe). Suomen Riista,
1970   (22):7-13

Hann, R.W., Jr., VLCC METULA oil spill. Washington, D.C., U.S. Coast Guard, Office of
1974   Research and Development, Final Rep., (CG-D-54-75):65 p.

Heino, A., The PALVA oil tanker disaster in the Finnish southwestern Archipelago. 1.
1972   The extent of oil pollution in the sea area between Utö and Kökar in 1969 and
       1970. Aqua Fenn., (1972):116-21

Herlinveaux, R.H., Preliminary report on the oil spill from the grounded freighter
1972   VANLENE, March 1972. Pacif.Mar.Science Rep.B.C., (72-11):21 p.

H.M.S.O., Effects of polluting discharges on the Thames Estuary. Water Pollut.Res.,
1964   Techn. Paper 11

Hogle, R.D., Note on the Santa Barbara oil disaster. Adv.Water Pollut.Res., 5(2)
1971

Holmes, R.W. and F.A. DeWitt, Jr., Santa Barbara oil symposium, Santa Barbara, University
1970 of California, 377 p.

Irons, D.E., Case studies of the USN Supervisor of salvage: salvage related oil pollution
1973 incidents. In American Petroleum Institute, Proceedings of the Joint
conference on prevention and control of oil spills, Washington, D.C.,
13-15 March 1973. Washington, D.C., American Petroleum Institute, p. 597-600

Jardas, I. and I. Munjko, Preliminary observations of oil and phenol distribution in the
1972 central Adriatic. Bilj.Inst.Oceanogr.Ribarst., (29):6 p.

Kinney, P.J. et al., Quantitative assessment of oil pollution problems in Alaska's
1970 Cook Inlet. Alaska, Rep.Univ.Alaska, (R-69-16)

Lalou, C., Contribution à l'étude de la pollution des sédiments de la Méditerranée
1963 occidentale par les benzo-pyrènes (Contribution to the study of the pollution
of sediments in the western Mediterranean by benzo-pyrenes). Rapp.P.-V.Réun.
CIESM, (17):711-8

Lalou, C., L. Mallet and M. Heros, Sur la répartition en profondeur de benzo 3,4 pyrène
1962 dans une carrotte de la Baie de Villefranche-sur-Mer. C.R.Hebd.Séances Acad.
Sci.,Paris, (255):145-7

Levy, E.M., The presence of petroleum residues off the east coast of Nova Scotia, in the
1971 Gulf of St. Lawrence and the St. Lawrence River. Water Res., 5(9):723-33

_____, Evidence for the recovery of the waters off the east coast of Nova Scotia
1972 from the effects of a major oil spill. Water Air Soil Pollut., 1(2):144-8

Levy, E.M. and A. Walton, Dispersed and particulate petroleum residues in the Gulf of
1973 St. Lawrence. J.Fish.Res.Board Can., 30(2):261-7

Long, E.R., Studies of marine fouling and boring off Kodiak Island, Alaska. Mar.Biol.,
1972 14(1):52-7

Loucks, R.H. and D.J. Lawrence, Reconnaissance of an oil spill. Mar.Pollut.Bull.,
1971 (2):92-4

Mallet, L., Pollution par les hydrocarbures des alluvions déposées dans l'estuaire de
1962 la Seine à partir de Rouen. Bull.Acad.Natl.Méd., (146):569-75

_____, Study of pollution of the North Sea and Arctic Ocean by polyaromatic hydro-
1963 carbons of the 3,4 benzypyrene type. Bull.Acad.Natl.Méd., (147):320-5

_____, Pollutions marines par les hydrocarbures polycondensés du type benzo 3,4
1967 pyrène des côtes nord et ouest de France. Leur incidence sur le milieu
biologique et en particulier sur le plancton. Cah.Océanogr., (19):237-43

Mallet, L. and M.T. Le Theule, Recherche du benzo 3,4 pyrène dans les sables vaseux marins
1961 des régions côtières de la Manche et de l'Atlantique. C.R.Hebd.Séances Acad.Sci.,
Paris, (252):565-7

Mallet, L., L.V. Perdriau and J. Perdriau, Pollution by polybenzic hydrocarbons of the type
1963 benzo 3,4 pyrene of the western region of the glacial Arctic Ocean. C.R.Hebd.
Séances Acad.Sci.,Paris, (256):3487-9

Mallet, L., L.V. Perdriau and S. Perdriau, Study of pollution by polybenzic hydrocarbons
1964 of the type benzo 3,4 pyrene in the North Sea and in the frozen Arctic Ocean.
Bull.Acad.Natl.Méd., (147):320-5

Mark, H.B., Jr., Santa Barbara Channel oil spill study: analysis of bottom sediment. Paper
1971 presented at the 161st American Chemical Society meeting, Los Angeles, California,
62 p.

Markham, W.E., Project Oil. Report on ice conditions in Lennox Passage. Dartmouth,
1970 Nova Scotia, Atlantic Oceanographic Laboratory, 4 p. (Unpubl.Rep.)

Meek, R., G. Pearson and A.W. Ebeling, Quality testing of selected bottom fishes trawled
1969 after Santa Barbara oil leakage. Manuscript report to Santa Barbara Fisheries
Association, 2 p.

Mersey Estuary Oil Pollution Scientific Study Group, Oil pollution in the Liverpool
1969 pilotage area following the HAMILTON TRADER incident 30 April 1969. Mersey
Estuary Oil Pollution Scientific Study Group, 22 p.

Michalik, P.A. and D.C. Gordon, Jr., Concentration and distribution of oil pollutants in
1971 Halifax Harbour - 10 June to 20 August 1971. Tech.Rep.Fish.Res.Board Can.,
(284):26 p.

Moore, H.J., Study of the effects of oil discharges and domestic and industrial wastewaters
1974 of the fisheries of Lake Maracaibo, Venezuela. Richland, Washington, Battelle
Pacific Northwest Laboratories Research Report, Vols. 1 and 2 (#212B00899)

Morris, B.F., Petroleum: tar quantities floating in the northwestern Atlantic taken with
1971 a new quantitative neuston net. Science, Wash.,D.C., (173):430-2

Mulkins-Phillips, G.J. and J.E. Stewart, Distribution of hydrocarbon-utilizing bacteria
1974 in northwestern Atlantic waters and coastal sediments. Can.J.Microbiol.,
(20):955-62

Mustonen, M. and P. Tulkki, The PALVA oil tanker disaster in the Finnish SW Archipelago.
1972 4. Bottom fauna in oil polluted area. Aqua Fenn., (1972):137-41

Nelson, J.D., Jr. and R.R. Colwell, Metabolism of mercury compounds by bacteria in
1973 Chesapeake Bay

Nelson, R.F., The Bay Marchand fire. J.Petrol.Technol., (24):225-33
1972

Niaussat, P., J.P. Ehrhardt and J. Ottenwalder, Présence de benzo 3,4 pyrène dans les eaux
1968 isolées du lagon de l'atoll de Clipperton (Presence of benzo 3,4 pyrene in the
isolated waters of the lagoon of the Clipperton Atoll). C.R.Hebd.Séances
Acad.Sci.,Paris (D), (267):1772-4

Nicholson, N.L., The Santa Barbara oil spill in perspective. Rep.CCOFI, (16):130-49
1972

O'Sullivan, A.J., Massive oil spillage in Bantry Bay. Mar.Pollut.Bull., (6):3-4
1975

Ottway, S.M., The THUNTANK 6 spill. Rep.Oil Pollut.Res.Unit, Field Stud.Counc., (1971):
1971 29-38

Oudet, L., In the wake of the TORREY CANYON: reflections on a disaster. London, Royal
1972   Institute of Navigation, 72 p. (Translated from the French)

Payne, F.A., Measurement of the quantity of Bunker C in the ARROW. In Project Oil,
1970   A report on some aspects of Operation Oil. Dartmouth, Nova Scotia, Defence
Research Establishment Atlantic, p. 1-7 (Unpubl.Rep.)

Pearce, J.B. and J. Ogren, Observations on the effects of oil from the tanker OCEAN EAGLE
1968   on marine organisms of San Juan Bay and vicinity. 1. Highlands, New Jersey,
Sandy Hook Marine Laboratory, 27 p. (Unpubl.Rep.)

_____, Observations on the effects of oil from the tanker OCEAN EAGLE on marine
1968   organisms of San Juan Bay and vicinity. 2. Observations one year after the
catastrophe. Highlands, New Jersey, Sandy Hook Marine Laboratory, 28 p.
(Unpubl.Rep.)

Peer, D.L., Project Oil, Sublittoral biological survey team. Dartmouth, Nova Scotia,
1970   Marine Ecology Laboratory. (Unpubl.Rep. 3-5):10 p.

Pelkonen, K. and P. Tulkki, The PALVA oil tanker disaster in the Finnish SW Archipelago.
1972   3. The littoral fauna of the oil polluted area. Aqua Fenn., (1972):129-36

Quayle, D.B., The VANLENE oil spill. Manuscr.Rep.Pacif.Biol.Stn.Nanaimo B.C., (1289):21 p.
1974

Ramamurthy, V.D., Oil tanker disaster in northwest coast of India. Curr.Sci., (43):293-4
1974

Ramseier, R.O., G.S. Gantcheff and L. Colby, Oil spill at Deception Bay, Hudson Strait.
1973   Sci.Ser.Can.Inland Water Dir.Ottawa, (29):60 p.

Rashid, M.A., Degradation of Bunker C oil under different coastal environments of
1974   Chedabucto Bay, Nova Scotia. Estuar.Coast.Mar.Sci., (2):137-44

Raytheon Company, New Haven - Long Island Sound. Oil spill impact study for Environmental
1971   Protection Agency. Portsmouth, Rhode Island, Raytheon Company, Environmental
Systems Center, 63 p.

Resources Technology Corporation, Studies and investigations of the fate and effect of the
1972   Shell oil spill, Platform B, Block 26, South Timbalier Bay (1 December 1970 -
30 November 1971). Houston, Texas, Resources Technology Corporation, Final
report, 117 p.

Roppel, A.Y., Effects of the Santa Barbara oil seep on birds, mammals and other marine
1969   life. Report to Director, Bureau of Commercial Fisheries, 24 p. (Unpubl.)

Sanders, H.L., Some biological effects related to the West Falmouth oil spill. In
1973   Proceedings of a Workshop on inputs, fates and effects of petroleum in the
marine environment, 21-25 May 1973, Airlie, Virginia. Washington, D.C.,
National Academy of Sciences, Vol. 2:766-88

_____, The West Falmouth saga: how an oil expert twisted the facts about a landmark
1974   oil spill study. New Eng., (1974):1-7

Scarrat, D.J., Project Oil. Sublittoral biological survey team. Preliminary underwater
1970   survey, 26 February-3 March 1970. New Brunswick, Fisheries Research Board of
Canada, 1970, Rep. (1):13 p. (Unpubl.)

Scarrat, D.J.,   Project Oil.   Sublittoral biological survey team,   Second underwater survey,
1970   10-12 March 1970.   New Brunswick, Fisheries Research Board of Canada, 1970,
Rep. (2):6 p. (Unpubl.)

Scarrat, D.J. and V. Zitko,   Sublittoral sediment and benthos sampling and littoral
1973   observations in Chedabucto Bay, April 1973.   In   Oil and the Canadian environment, Proceedings of a conference sponsored by the Institute of Environmental
Science and Engineering, 16 May 1973, D. MacKay and W. Harrison (Eds.).
Toronto, University of Toronto, p. 78-9

Smith, P.V.,   The occurrence of hydrocarbons in recent sediments from the Gulf of Mexico.
1952   Science, Wash.,D.C., (116):437-9

Spray, M., L.C. Frost and M.H. Martin,   The effects of detergents on soil-borne seeds and
1974   seedling establishment.   Cornish Stud., (2):15-22

St. Amant, L.S.,   Impacts of oil on the Gulf Coast.   Trans.North Am.Wildl.Nat.Resource Conf.,
1971   (36):206-19

Stander, G.H.,   Stilabii oil spill fumble.   Mar.Pollut.Bull., (2):83-4
1971

Stevens, D.B. and W.W. Bruce,   Hudson River oil pollution.   In   Hudson River ecology.
1969   Proceedings of a symposium, G.P. Howells and G.K. Lauer (Eds.).   New York,
New York University, p. 457-73

Straughan, D.,   Santa Barbara oil pollution project.   Mar.Pollut.Bull., 1(4):61
1970

Sutcliffe, W.H.,   Carbon, nitrogen and adenosine triphosphate in particulate material from
1970   Chedabucto Bay.   Dartmouth, Nova Scotia, Marine Ecology Laboratory, 9 p.
(Unpubl.Rep.)

Tarren, C. and R. Campbell,   Effects on micro-organisms in the soil of detergents used to
1974   combat oil pollution at the Lizard, Cornwall.   Cornish Stud., (2):23-6

Thomas, M.L.H. and M.L. Harley,   Long-term effects of Bunker C oil from the Tanker ARROW
1973   on intertidal and lagoonal biota.   In   Oil and the Canadian environment,
D. MacKay and W. Harrison (Eds.).   Toronto, Ontario, University of Toronto,
p. 84-5

Trites, R.W.,   Summary of physical, biological, socio-economic and other factors relevant
1974   to potential oil spills in the Passamaquoddy Region of the Bay of Fundy.   Tech.
Rep.Fish.Res.Board Can., (428)

United States Environmental Protection Agency,   Oil spill, Long Island Sound, 21 March 1972.
1973   Environmental effects.   Environ.Prot.Ag.Oil Hazardous Mater.Program Ser.,
Wash.,D.C., (OHM 73-06-001):146 p.

Walker, J.D. and R.R. Colwell,   Microbial ecology of petroleum utilization in Chesapeake Bay.
1973   In   American Petroleum Institute, Proceedings of a Joint conference on prevention
and control of oil spills, Washington, D.C., 13-15 March 1973.   Washington, D.C.,
American Petroleum Institute, p. 685-90

Walstad, K. and F.E. Hearth,   Oil pollution in the Santa Barbara Channel: a comprehensive
1969   bibliography with particular emphasis on the oil spill of 28 January 1969.
Santa Barbara, California, University of California, University Library, 195 p.

Wardley-Smith, J., Magellan Straits spill. Mar.Pollut.Bull., (5):163
1974

Watson, J.A. et al., Biological assessment of diesel spill in the vicinity of Anacortes,
1971　Washington, May 1971. Final report. Dallas, Texas, Texas Instruments, 169 p.

Westley, B. and D.H. Brown, Chemical detection of detergent residues in contaminated soils
1974　at the Lizard, Cornwall. Cornish Stud., (2):27-32

Wong, C.S., D.R. Green and W.J. Cretney, Pelagic tar in the North Pacific Ocean. In
1973　Proceedings of a Workshop on inputs, fates and effects of petroleum in the
marine environment, 21-25 May 1973, Airlie, Virginia. Washington, D.C.,
National Academy of Sciences, Vol. 2:400-15

_____, Quantitative tar and plastic waste distributions in the Pacific Ocean.
1974　Nature, Lond., 247(5435):30-2

Zafiriou, O.C., Petroleum hydrocarbons in Narragansett Bay. 2. Chemical and isotopic
1973　analysis. Estuar.Coast.Mar.Sci., 1(1):81-7

Zitko, V. and W.V. Carson, Analysis of water and sediment samples. St. Andrews, New
1970　Brunswick, Fisheries Research Board of Canada, Biological Station, Report No. 2.
Project Oil - Chemistry, 6 p. (Unpubl.Rep.)

_____, Some analytical characteristics of Bunker C oil from the Tanker ARROW.
1970　St. Andrews, New Brunswick, Fisheries Research Board of Canada, Biological
Station, Report No. 1. Project Oil - Chemistry, 6 p. (Unpubl.Rep.)

Anon., Report of the Task Force - Operation Oil. Vol. 2. Ministry of Transport. Ottawa,
1970　Information Canada, Cat. No. T22-2470/2, 104 p.

_____, Field Studies Council, Oil Pollution Research Unit, Orielton Field Centre,
1971　Annual report, 46 p.

_____, Tamano oil spill. Mar.Pollut.Bull., 3(9):133
1972

2.2.7 Other Effects and Properties

Baier, R.E., Organic films on natural waters: their retrieval, identification and modes of
1972　elimination. J.Geophys.Res., 77(27):5062-75

Barger, W.R., W.H. Daniel and W.D. Garrett, Surface chemical properties of banded sea
1974　slicks. Deep-Sea Res., (21):83-9

Betancourt, O.J., Effect of natural weathering on the shore-bound oil after the
1973　Chedabucto Bay spill. In Oil and the Canadian environment. Proceedings of a
conference, 16 May 1973, D. MacKay and W. Harrison (Eds.). Toronto, University
of Toronto, p. 87-9

Boehm, P.D. and J.G. Quinn, Solubilization of hydrocarbons by the dissolved organic matter
1973　in seawater. Geochim.Cosmochim.Acta, (37):2459-77

Boswell, J.L., Experiments to determine the effect of a surface film of crude oil on the
1950　absorption of atmospheric oxygen by water. Texas A&M Research Foundation

Brown, S.O. and B.L. Reid, Experiments to test the diffusion of oxygen through a surface
1951   layer of oil. Texas A&M Research Foundation

Burwood, R. and G.C. Speers, Photo-oxidation as a factor in the environmental dispersal of
1974   crude oil. Estuar.Coast.Mar.Sci., (2):117-35

Cousens, J.D., Vapor-liquid equilibrium of hydrocarbons in sea water. In Oil and the
1973   Canadian environment. Proceedings of a conference, 16 May 1973, D. MacKay and
        W. Harrison (Eds.). Toronto, University of Toronto, p. 92-5

Cowell, E.B., The ecological effects of oil pollution on littoral communities. Symposium
1971   proceedings sponsored by the Institute of Petroleum held in London, England.
        New York, Elsevier Publishing Co. Ltd., 257 p.

Crapp, G.B., Chronic oil pollution. In The biological effects of oil pollution on
1971   littoral communities, E.B. Cowell (Ed.). London, Institute of Petroleum

_____, Monitoring the rocky shore. In The ecological effects of oil pollution on
1971   littoral communities, E.B. Cowell (Ed.). New York, Elsevier Publishing Co. Ltd.,
        p. 102-13

_____, The ecological effects of stranded oil. In The ecological effects of oil
1971   pollution on littoral communities, E.B. Cowell (Ed.). New York, Elsevier
        Publishing Co. Ltd., p. 181-6

Davavin, I.A., On different sensitivity of DNA fractions of Polysiphonia opaca to oil
1974   pollution. Kiev, Naukova Dumka

Davis, C.C., The effects of pollutants on the reproduction of marine organisms. In
1972   Marine pollution and sea life, M. Ruivo (Ed.). West Byfleet, Surrey, U.K.,
        Fishing News (Books) Ltd., p. 308-9

Freegarde, M. and C.G. Hatchard, The ultimate fate of crude oil at sea. Interim report
1970   No. 7. An investigation of the photodecomposition of crude oil. Holton Heath,
        Poole, Dorset, Admiralty Materials Laboratory, Rep.AML (Admir.Mater.Lab.) U.K.,
        (5/70):10 p. (Unpubl.)

Freegarde, M., L.G. Harvey and C.G. Hatchard, The ultimate fate of crude oil at sea.
1971   Interim report No. 8. The dispersion and solubility of crude oil in water.
        Holton Heath, Poole, Dorset, Admiralty Materials Laboratory, Rep.AML (Admir.
        Mater.Lab.) U.K., (17/71):11 p. (Unpubl.)

Gibbs, C.F., A new approach to the measurement of rate of oxidation of crude oil in sea
1972   water systems. Chemosphere, (3):119-24

Glaeser, J.L. and G.P. Vance, A study of the behaviour of oil spills in the Arctic.
1972   Paper presented at the fourth annual Offshore technology conference, Houston,
        Texas, 1-3 May 1972, Paper No. (OTC 1551):14 p.

Gordon, D.C., Jr., P.D. Keizer and N.J. Prouse, Laboratory studies of the accommodation of
1973   some crude and residual fuel oils in sea water. J.Fish.Res.Board Can., (30):
        1611-8

Great Britain, Ministry of Defence, Navy Department, The effects of natural factors on
1971   the movement, dispersal, destruction of oil at sea. Select Committee Report,
        26 July 1968. London, Ministry of Defence, Para. 61, 42 p.

Great Britain, Ministry of Defence, Navy Department, The fate of oil spilt at sea. A
1973   report on the effects of natural factors on the movement, dispersal and
       destruction of oil at sea. Report by a Ministry of Defence Working Party.
       Orpington, Kent, Defence Research Information Centre, 54 p.

Guard, H.E. and A.B. Cobet, Fate of petroleum hydrocarbons in beach sand. Oakland,
1972   California, U.S. Naval Biochemical Research Laboratory, Final Report, 89 p.

_____, The fate of a bunker fuel in beach sand. In American Petroleum Institute,
1973   Proceedings of a Joint conference on prevention and control of oil spills,
       Washington, D.C., 13-15 March 1973. Washington, D.C., American Petroleum
       Institute, p. 827-34

Halstead, B.W., Toxicity of marine organisms caused by pollutants. In Marine pollution
1972   and sea life, M. Ruivo (Ed.). West Byfleet, Surrey, U.K., Fishing News (Books)
       Ltd., p. 584-94

Hearst, P.J., The fate of spilled navy distillate fuel. Port Hueneme, California, U.S.
1974   Naval Civil Engineering Laboratory, 31 p.

Hellmann, H. and H. Zehle, Langfristige Untersuchungen zum Verhalten von Rohölen auf
1972   Gewässern (Long-term investigation into the behaviour of crude oils on water).
       Dtsch.Gewäss.Mitt., (16):46-52

Herfjord, H.J. and H.J. Neumann, Zum Sauerstofftransport durch Mineralölschichten auf
1972   Wasser (On the oxygen transport across a mineral oil layer on water). Erdöl-
       Erdgas-Z., 88(1):2-5

Hornstein, B., The visibility of oil-water discharges. In American Petroleum Institute,
1973   Proceedings of a Joint conference on prevention and control of oil spills,
       Washington, D.C., 13-15 March 1973. Washington, D.C., American Petroleum
       Institute, p. 91-9

Jacobs, W.A., Report of the Bureau of Mines on the action of sea water on fuel oil,
1926   October 1925. In United States Interdepartmental Committee on Oil Pollution
       of Navigable Waters, Oil pollution of navigable waters. Appendix 4. Washington,
       D.C., U.S. Government Printing Office, p. 70-5

Jeffery, P.G., Large-scale experiments on the spreading of oil at sea and its disappearance
1973   by natural factors. In American Petroleum Institute, Proceedings of Joint
       conference on prevention and control of oil spills, Washington, D.C.,
       13-15 March 1973. Washington, D.C., American Petroleum Institute, p. 469-74

Johnson, B.D., Some aspects of the formation, stabilization and behaviour of water-in-
1973   petroleum emulsions. In Oil and the Canadian environment. Proceedings of a
       conference, 16 May 1973, D. MacKay and W. Harrison (Eds.). Toronto, University
       of Toronto, p. 98-103

Kennedy, J.M. and E.G. Wermund, Oil spills, IR and microwave. Photogramm.Eng., (37):1235-42
1971

Kleen, A.E. and N. Pilpel, The effects of artificial sunlight upon floating oils. Water
1974   Res., (8):79-83

Kondo, G. et al., Studies on the change of spilled oil with the time elapsed. 1. Physical
1972   change of floating oil. Kankyo Gijutsu (Environ.Technol.), 1(10):5 p.
       (In Japanese)

Kramer, A., An analytical and integrative approach to sensory evaluation of foods. J.Sci.
1973   Fd.Agric., (24):1407-18

Leinonen, P.J. and D. MacKay, The multicomponent solubility of hydrocarbons in water.
1973   Can.J.Chem.Eng., (51):230-3

Little, P.A., A note on oil pollution. Proc.Swansea Scit.Field Nat.Soc., (1):230-2
1934

Lysyj, I. and E.C. Russell, Dissolution of petroleum-derived products in water. Water Res.,
1974   (8):863-8

MacKay, D. and A.W. Wolkoff, Rate of evaporation of low-solubility contaminants from
1973   water bodies to atmosphere. Environ.Sci.Technol., (7):611-4

MacKay, G.D.M., Stability of Bunker C - seawater emulsion. Halifax, Nova Scotia Technical
1970   College, Department of Chemical Engineering, 7 p. (Unpubl.Rep.)

MacKay, G.D.M. et al., The formation of water-in-oil emulsions subsequent to an oil spill.
1973   J.Inst.Petrol.,Lond., (59):164-72

Mayo, D.W., D.J. Donova and L. Jiang, Long-term weathering characteristics of Iranian
1974   crude oil: the wreck of the NORTHERN GULF. Spec.Publ.U.S.Natl.Bur.Stand.,
        (409):201-8

Mazmanidi, N.D. and G.I. Kovaleva, Experimental data on the effect of oil on some chemical
1972   properties of seawater. Okeanologiya, (12):684-9

McCauley, R.N., The biological effects of oil pollution in a river. Limnol.Oceanogr.,
1966   (11):475-86

McClean, A.Y., The behaviour of oil spilled in a cold water environment. Paper presented
1972   at the fourth annual Offshore technology conference, Houston, Texas,
        1-3 May 1972, (OTC 1522):12 p.

McClean, A.Y. and D.A. Odedra, The properties of Sable Island crude oil in relation to
1974   its behaviour in the event of a spill at sea. Paper presented at the sixth
        annual Offshore technology conference, Houston, Texas, 6-8 May 1974,
        (OTC 1981):12 p.

Meyers, P.A. and J.G. Quinn, Association of hydrocarbons and mineral particles in saline
1973   solution. Nature, Lond., (244):23-4

Milgram, J.H. and R.G. Bradley, The determination of the interfacial tension between two
1971   liquids. J.Fluid Mech., (50):469-80

Nagy, E., Water-in-oil emulsions. In Oil and the Canadian environment. Proceedings of
1973   a conference, 16 May 1973, D. MacKay and W. Harrison (Eds.). Toronto,
        University of Toronto, p. 107-11

_____, Evaporation of oil from slicks. In Oil and the Canadian environment.
1973   Proceedings of a conference, 16 May 1973, D. MacKay and W. Harrison (Eds.).
        Toronto, University of Toronto, p. 112-7

Nichols, J.A., Non-persistent oils in the marine environment. Stevenage, Warren Spring
1973   Laboratory, (CR 703 (OP)):6 p. (Unpubl.Rep.)

Parker, C.A. and M. Freegarde, The ultimate fate of crude oil at sea. Interim report
1969   No. 2. Proposed work to determine the magnitude of losses by dispersion and
        solution in seawater. Holton Heath, Poole, Dorset, Admiralty Materials
        Laboratory, 1969, Rep.AML(Admir.Mater.Lab.)U.K., (B/191 (M)):9 p. (Unpubl.)

Polak, J. and B.C.Y. Lu, Mutual solubilities of hydrocarbons and water at 0 and $25^{\circ}C$.
1973   Can.J.Chem., (51):4018-23

Programmes Analysis Unit, The environmental and financial consequences of oil pollution
1973   from ships. Appendix 2: the fate of oil at sea. Didcot, Programmes Analysis
        Unit, 11 p.

Rashid, M.A., Degradation of Bunker C oil under different coastal environments of
1974   Chedabucto Bay, Nova Scotia. Estuar.Coast.Mar.Sci., (2):137-44

Sano, H. et al., Studies on the change of spilled oil with the time elapsed. 2. Chemical
1972   change of floating oil. Kankyo Gijutsu (Environ.Technolo.), 1(11):6 p. (in Japanese

Shiels, W.E., J.J. Goering and D.W. Hood, Crude oil phytotoxicity studies. In Environ-
1973   mental studies of Port Valdez, D.W. Hood et al. (Eds.). Ocas.Publ.Inst.Mar.Sci.,
        Univ.Alaska, (3):413-46

Stokes, V.K. and A.C. Harvey, Drop size distribution in oil water mixtures. In American
1973   Petroleum Institute, Proceedings of Joint conference on prevention and control
        of oil spills, Washington, D.C., 13-15 March 1973. Washington, D.C., American
        Petroleum Institute, p. 457-65

Suess, M.J., Occurrence of polycyclic aromatic hydrocarbons in coastal waters and their
1970   possible effects on human health. Arch.Hyg.Bakt., (154):1-7

Wolfe, L.S. and D.P. Hoult, Effects of oil under sea ice. Publ.Fluid Mech.Lab.MIT,
1972   (72-10):38 p.

Yentsch, C.S., E.S. Gilfillan and J.R. Sears, The fate and behaviour of crude oil on
1973   marine life. Gloucester Marine Station, University of Massachusetts, 69 p.

Zalosh, R.G., A numerical model of droplet entrainment from a contained oil slick.
1974   Washington, D.C., U.S. Coast Guard, Office of Research and Development, Final
        Rep., (CG.D.65-75):84 p.

2.3 DISPERSANTS/DETERGENTS

Alzieu, C., Choice of products for use against the pollution of the marine environment by
1972   oil spills. 3. Relative toxicity of anti-petroleum products to two marine
        organisms. Rev.Trav.Inst.Pêches Marit.Nantes, 36(1):103-19

American Petroleum Institute, Development of bioassay procedures for oil and oil dispersant
1971   chemicals. Washington, D.C., American Petroleum Institute

Baldini, I. and F. Cugurra, Ichthyotoxic effects of some antipollution products. Water Res.,
1974   (8):323-4

Bardach, J.E., M. Fujiya and A. Holt, Detergents: effects on the chemical senses of the fish
1965   Tetalurus natilis (Le Sueur). Science, Wash.,D.C., (148):1605-7

Becker, C.D. et al., Regional survey of marine biota for bioassay standardization of oil
1973   and oil dispersant chemicals. Publ.Am.Petrol.Inst., (4167):102 p.

Bellan, G.L., Toxicity testing at the Station Marine d'Endoume. In Ecological aspects of
1964   toxicity testing of oils and dispersants. Proceedings of a Workshop,
L.R. Beynon and E.B. Cowell (Eds.). Barking, Essex, Applied Science Publishers,
p. 63-7

Bellan, G.L., D.J. Reish and J.-P. Foret, Action toxique d'un détergent sur le cycle de
1971   développement de la polychète Capitella capitata (Fab.) (Toxic action of a
detergent on the development cycle of the polychaete Capitella capitata).
C.R.Hebd.Séances Acad.Sci.,Paris (D), (272):2476-9

————, The sublethal effects of a detergent on the reproduction, development and
1972   settlement in the polychaetous annelid Capitella capitata. Mar.Biol., (14):
183-8

Bellan, G.L. et al., Action in vitro de détergents sur quelques espèces marines (In vitro
1972   action of detergents on some marine species). In Marine pollution and sea
life, M. Ruivo (Ed.). Fishing News (Books) Ltd., p. 245-8

Berglund, H., Influence of wetting agents on growth of marine multicellular green algae.
1969   Vatten, (2):157-65

Beynon, L.R., Oil spill dispersants. J.Inst.Petrol.,Lond., 57(553):1-17
1971

Blackman, R.A.A., Toxicity of oil-sinking agents. Mar.Pollut.Bull., 5(8):116-8
1974

Bleakley, R.J. and P.J.S. Boaden, Effects of an oil spill remover on beach meiofauna.
1974   Ann.Inst.Océanogr.,Monaco, (50):51-8

Bock, K.J. and H. Mann, Toxikologische Untersuchung von Emulgatoren für die Bekämpfung von
1972   Ölverschmutzungen (Toxicological investigations of oil-spill removers).
Arch.Fischereiwiss., (23):64-7

Braaten, B., A. Granmo and R. Lange, Tissue-swelling in Mytilus edulis L. induced by
1972   exposure to a nonionic surface active agent. Norw.J.Zool., (20):137-40

Briant, J. and C.R. Gatellier, Prévention et lutte contre la pollution au cours des
1971   opérations de forage et de production en mer. 4. Traitement des nappes par
dispersion et dégradation (Prevention and the fight against pollution in the
course of drilling operations and production in the sea. 4. Treatment of
spills by dispersion and degradation). Rev.Inst.Fr.Pétrole, (26):802-11

Brown, D.H., Toxicity studies on the components of an oil-spill emulsifier using Lichina
1973   pygmaea and Xanthoria parietina. Mar.Biol., 18(4):291-7

Bryan, G.W., The effects of oil spill removers (detergents) on the gastropod Nucella lapellus
1969   on a rocky shore and in the laboratory. J.Mar.Biol.Assoc.U.K., (49):1067-92

Canevari, G.P., The case for oil spill dispersants. Paper presented at the 1970
1970   Evangeline section regional meeting of the Society of Petroleum Engineers of
the American Institute of Mining, Metallurgical and Petroleum Engineers
(AIME), Lafayette, Louisiana, 9-10 November 1970, Pap. (SPE 3235):8 p.

Canevari, G.P., Development of the "next generation" chemical dispersants. In American
1973   Petroleum Institute, Proceedings of Joint conference on prevention and control
          of oil spills, Washington, D.C., 13-15 March 1973. Washington, D.C., American
          Petroleum Institute, p. 231-40

Chaplin, A.E., The effects of oil-spill emulsifiers on the metabolism of crustacea. Rep.
1971   Oil Pollut.Res.Unit, Field Stud.Counc., (1971):39-45

Cowell, E.B., J.M. Baker and G.B. Crapp, The biological effects of oil pollution and oil
1972   cleaning materials on littoral communities, including salt marshes. In Marine
          pollution and sea life, M. Ruivo (Ed.). West Byfleet, Surrey, U.K., Fishing
          News (Books) Ltd., p. 359-64

Cracy, H.B. et al., Relative toxicities and dispersing evaluation of eleven oil-dispersing
1969   products. J.Water Pollut.Control Fed., (41):2062-9

Crapp, G.B., The biological consequences of emulsifier cleansing. In The biological
1971   effects of oil pollution on littoral communities, E.B. Cowell (Ed.). London,
          Institute of Petroleum, p. 150-68

_____, Field experiments with oil and emulsifiers. In The ecological effects of
1971   oil pollution on littoral communities, E.B. Cowell (Ed.). London, Institute
          of Petroleum, p. 114-28

_____, Laboratory experiments with emulsifiers. In The ecological effects of oil
1971   pollution on littoral communities, E.B. Cowell (Ed.). London, Institute of
          Petroleum, p. 129-41

Edison Water Quality Laboratory, Oil spill dispersants product data. Edison, New Jersey,
1971   U.S. Environmental Protection Agency, Edison Water Quality Laboratory, 140 p.

Foret-Montardo, P., Evolution dans le temps de la toxicité des détergents issus de la pétroléo-
1971   chimie: étude réalisée sur Scolelepis fuliginosa (Polychète sédentaire)
          (Evolution with time of the toxicity of detergents issued from petroleochemistry:
          a study carried out on Scolelepis fuliginosa (Sedentary polychaeta). Téthys,
          (3):173-82

_____, Notions générales sur les détergents et leur toxicité (General ideas on
1971   detergents and their toxicity). Ann.Soc.Sci.Natl.Toulon, (23):145-52

_____, Problèmes biologiques posés par la dégradation des détergents issus de la
1971   pétroléochimie (Biological problems posed by the degradation of detergents
          issued from petroleochemistry). In Journées d'études sur les pollutions
          marines et l'aménagement du littoral, Nice, 29-30 Septembre 1970. Nice, CERBOM,
          p. 49-56

Foret-Montardo, P. and G.L. Bellan, Données préliminaires sur les problèmes de "biodégra-
1972   dabilité" de détergents issus de la pétroléochimie (Preliminary data on the
          problems of biodegradability of detergents issued from petroleochemistry).
          Rapp.P.-V.Réun.CIESM, (21):253-4

Gatellier, C.R. et al., Experimental ecosystems to measure fate of oil spills dispersed by
1973   surface active products. In American Petroleum Institute, Proceedings of a
          Joint conference on prevention and control of oil spills, Washington, D.C.,
          13-15 March 1973. Washington, D.C., American Petroleum Institute, p. 497-504

Gunkel, W., Toxicity testing at the Biologische Anstalt Helgoland, West Germany. In
1974   Ecological aspects of toxicity testing of oils and dispersants. Proceedings
       of a Workshop, L.R. Beynon and E.B. Cowell (Eds.). Barking, Essex, Applied
       Science Publishers, p. 75-85

Hazel, C.R. et al., Evaluating oil spill clean-up agents: development of testing procedures
1940   and criteria. Publ.Calif.State Water Resour.Control Board, Sacramento, (43)

Hellmann, H. and F.-J. Bruns, Die anwendungstechnische und vergleichende Prüfung von
1970   rohölverteilenden Chemikalien auf Gewässern (Application and comparative
       examination of chemical agents for the dispersion of crude oil on water
       surfaces). Erdöl Kohle Erdgas Petrochem., (23):594-9

Hellmann, H. and H. Zehle, Die Ölviskosität als wirksamkeitsbegrenzender Faktor bei der
1972   Ölbekämpfung in Gewässern mit Hilfe von ölverteilenden Chemikalien (The oil
       viscosity as a limiting efficiency factor in oil control in water with the
       help of chemical oil dispersants). Tenside, (9):61-5

Honda, S., Examination of the treatment efficiency of oil dispersers. Osaka Ind.Tech.Lab.Q.,
1972   (23):225-9 (In Japanese with English summary)

Honda, S. and Y. Murakami, Oil spill dispersers. Mizu Shori Gijutsu (Water Purific.Liquid
1972   Wastes Treatment), 13(1):41-9 (In Japanese)

Intercompany Operating Committee for the Conservation of Lake Maracaibo, Studies on the
1974   effects of two oil collecting agents on aquatic organisms of Lake Maracaibo

Jeffery, P.G., Dispersants for oil spill clean-up operations. Stevenage, Warren Spring
1971   Laboratory, (LR 162(PC)):9 p.

Jeffery, P.G. and J.A. Nichols, Dispersants for oil spill clean-up operations at sea, on
1973   coastal waters and beaches. Stevenage, Warren Spring Laboratory, (LR 193(OP)):14 p

Kaim-Malka, R.A., Action in vitro des détergents non ioniques sur l'isopode valvifère
1972   Idotea balthica basteri Audouin 1827 (In vitro action of non-ionic detergents
       on the valviferous isopod Idotea balthica basteri Audouin 1827). Téthys,
       (4):51-61

_____, Action in vitro des détergents non ioniques sur l'isopode Sphaeroma serratum
1972   (Fabricius) (In vitro action of non-ionic detergents on the isopod Sphaeroma
       serratum (Fabricius)). Téthys, (4):587-96

_____, Action des détergents sur deux espèces de Crustacés (Action of detergents on
1972   two species of crustacea). Rapp.P.-V.Réun.CIESMM, (21):255-8

Kerminen, S. et al., Memorandum on the potential for use of emulsifiers in Finnish
1971   conditions. Finland, National Board of Waters, 22 p. (Unpubl.Rep.)

Kondo, G., About the propriety of dispersing method for spilled oil. Paper presented at
1972   the second International ocean development conference, October 1972, 19 p.

Kutt, E.C. and D.F. Martin, Effect of selected surfactants on the growth characteristics
1974   of Gymnodinium breve. Mar.Biol., (28):253-9

Lacaze, J.-C., Utilisation d'un dispositif expérimental simple pour l'étude de la pollution
1971   des eaux in situ. Effets comparés de trois agents émulsionnants anti-pétrole
        (Use of a simple experimental device for the study of water pollution in situ.
        Comparative effects of three anti-petroleum emulsifying agents). Téthys,
        (3):705-15

_____, Influence de l'éclairement sur la biodégradation d'un tensioactif non
1973   ionique utilisé pour la dispersion des nappes de pétrole en mer (Effect of
        illumination on the biodegradation on non-ionic surface-active agents used for
        the dispersion of spills of petroleum in the sea). C.R.Hebd.Séances Acad.Sci.,
        Paris, (D), (277):409-12

Lallier, R., Changes in the differentiation of the sea urchin larva by action of a detergent
1973   upon the unsegmented egg. Experientia, (29):1022-3

Latiff, S.A., The effect of oil slick dispersant, Esso Corexit 7664 and the joint effect
1971   with crude oil on the survival of marine glass fish, Ambassis sp. and
        palaemonid shrimps. Malay.Agric.J., (48):13-9

Lindén, O., Effects of oil spill dispersants on the early development of Baltic herring.
1974   Ann.Zool.Fenn., (11):141-8

Mackie, A.M., Avoidance reactions of marine invertebrates to either steroid glycosides of
1970   starfish or synthetic surface-active agents. J.Exp.Mar.Biol.Ecol., (5):63-9

Maggi, P., Choix de produits pour lutter contre la pollution du milieu marin par les hydro-
1972   carbures. 4. Toxicité relative de sept produits émulsionnants antipétrole
        (Choice of products for controlling pollution of the marine environment by
        hydrocarbons. 4. Relative toxicity of seven anti-petroleum emulsifying
        products). Rev.Trav.Inst.Pêches Marit.,Nantes, (36):121-4

Maggi, P. and D. Cossa, Nocivité relative de cinq détergents anioniques en milieu marin.
1973   1. Toxicité aiguë à l'égard de quinze organismes (Relative toxicity of five
        anionic detergents in the marine environment. 1. Acute toxicity towards
        fifteen organisms). Rev.Trav.Inst.Pêches Marit.,Nantes, (37):411-7

Massachusetts, Division of Water Pollution Control, Use of chemicals, materials or techniques
1970   to treat oil spills. Boston, Massachusetts, Division of Water Pollution Control,
        Publ., (5394):7 p.

McCarthy, L.T., Jr., I. Wilder and J.S. Dorrler, Standard dispersant effectiveness and
1973   toxicity tests. Environ.Protect.Ag.Environ.Protect.Technol.Ser., (EPA-R2-73-201):
        65 p.

McManus, D.A. and D.W. Connell, Toxicity of the oil dispersant, Corexit 7664, to certain
1972   Australian marine animals. Search, (3):222-4

Michel, P., Choice of products for use against the pollution of the marine environment by
1972   oil spills. Efficiency of antipetroleum products. Rev.Trav.Inst.Pêches Marit.,
        Nantes, 36(1):85-102

Murphy, T.A., Problems with the use of chemical dispersants for handling oil spills. In
1970   American Petroleum Institute, Proceedings of the Industry-government seminar on
        oil spill treating agents, Washington, D.C., 8-9 April 1970. Washington, D.C.,
        American Petroleum Institute, p. 150-61

Nagell, B. M. Notini and O. Grahn, Toxicity of four oil dispersants to some animals from
1974   the Baltic Sea. Mar.Biol., (28):237-43

Okubo, K., Y. Takita and K. Sakuma, Study on the oil dispersant. 1. The effects upon
1972　　aquatic life and the emulsification property of non-ionic oil dispersants.
Bull.Tokai Reg.Fish.Res.Lab., (71):79-86 (In Japanese with English summary)

Orlov, L.N. et al., Action of a demulsifier on a freshly prepared water and petroleum
1972　　emulsion. Neftepererab.Neftekhim.,Mosk., (1):11-2 (In Russian)

Perkins, E.J., E. Gribbon and J.W.M. Logan, Oil dispersant toxicity. Mar.Pollut.Bull.,
1973　　(4):90-3

Portmann, J.E., A summary of the results of toxicity tests with 36 oil-dispersing mixtures.
1969　　Copenhagen, International Council for the Exploration of the Sea, Fisheries
Improvement Committee, (CM 1969/E:9):5 p.

_____, Toxicity-testing with particular reference to oil-removing materials and heavy
1972　　metals. In Marine pollution and sea life, M. Ruivo (Ed.). West Byfleet, Surrey,
Fishing News (Books) Ltd., p. 217-22

Portmann, J.E. and P.M. Connor, The toxicity of several oil spill removers to some species
1968　　of fish and shellfish. Mar.Biol., (1):322-9

Pybus, C., Effects of anionic detergent on the growth of Laminaria. Mar.Pollut.Bull.,
1973　　(4):73-7

Ruel, M. et al., Guidelines on the use and acceptability of oil spill dispersants. Ottawa,
1973　　Ontario, Environment Canada, Environmental Emergency Branch, Rep., (EPS 1-EE-73-1):
54 p.

Shelton, R.G.J., Effects of oil and oil dispersants on the marine environment. Proc.R.Soc.
1971　　Lond., (B), (177):411-22

Shuttleworth, F., Design of an inshore and beach spraying unit. Stevenage, Warren Spring
1971　　Laboratory, (LR 149(ES)):6 p.

_____, A method of testing oil dispersant chemicals at sea. Stevenage, Warren
1971　　Spring Laboratory, (LR 152(ES)):40 p.

_____, Sea truck: Rotork Marine Limited. Stevenage, Warren Spring Laboratory,
1972　　(LR 168(PC)):7 p.

Shuttleworth, F. and P.G. Jeffery, Dispersants for oil-spill cleanup operation: carriage
1974　　and containment at sea. Stevenage, Warren Spring Laboratory, (LR 195(OP)):17 p.

Soudan, F., Choice of products for use against the pollution of the marine environment by
1972　　oil spills: motive of the choice. Rev.Trav.Inst.Pêches Marit.,Nantes, 36(1):81-3

Spooner, M.F., Effects of oil and emulsifiers on marine life. In Water pollution by oil,
1971　　P. Hepple (Ed.). London, Institute of Petroleum, p. 375-6

Stora, G., Contribution à l'étude de la notion de concentration léthale limite moyenne
1972　　(CL50) appliqué à des invertébrés marins. 1. Etude méthodologique (A contribution to the study of the idea of median lethal concentration (CL50) applied
to some marine invertebrates. 1. Methodological study). Téthys, (4):597-644

Sullivan, C.E., A comparative study of the effects of emulsifiers BP 1002 and BP 1100 on
1971　　three mud sand species. Rep.Oil Pollut.Res.Unit, Field Stud.Counc., (1971):14-21

Swedmark, M., Toxicity testing at Kristineberg Zoological Station. In Ecological aspects
1974   of toxicity testing of oils and dispersants. Proceedings of a Workshop,
       L.R. Beynon and E.B. Cowell (Eds.). Barking, Essex, Applied Science Publishers,
       p. 41-51

Swedmark, M. et al., Biological effects of surface active agents on marine animals. Mar.
1971   Biol., (9):183-201

Tarzwell, C.M., Standard methods for determination of relative toxicity of oil dispersants
1969   and mixtures of dispersants and various oils to aquatic organisms. In
       Prevention and control of oil spills. American Petroleum Institute, 1969,
       p. 179-86

_____, Comments on standard methods for the determination of the relative toxicity
1970   of oil dispersants and mixtures of dispersants and various oils to aquatic
       organisms. In American Petroleum Institute, Proceedings of the Industry-
       government seminar on oil spill treating agents, Washington, D.C., 8-9 April 1970.
       Washington, D.C., American Petroleum Institute, p. 80-5

_____, Toxicity of oil and oil dispersant mixtures to aquatic life. In Water
1971   pollution by oil, P. Hepple (Ed.). London, Institute of Petroleum, p. 163-272

Templeton, W.L., P.C. Walkup and J.R. Blacklaw, Chemical dispersants for oil spillage
1970   cleanup. In American Petroleum Institute, Proceedings of the Industry-
       government seminar on oil spill treating agents, Washington, D.C., 8-9 April 1970.
       Washington, D.C., American Petroleum Institute, p. 57-68

Wardley-Smith, J. and F. Shuttleworth, Development of the W.S.L. dispersant spraying
1971   equipment. Stevenage, Warren Spring Laboratory (LR 151(ES)):9 p.

Warren Spring Laboratory, Warren Spring Laboratory dispersant-spraying equipment.
1970   Stevenage, Warren Spring Laboratory, 13 p.

Wildish, D.J., Arrestant effect of polyoxyethylene esters on swimming in the winter
1974   flounder. Water Res., (8):579-83

Wildish, D.J. and D. Beatty, In vitro hydrolysis of polyoxyethylene esters by tissues of
1973   the American eel and Atlantic salmon. Bull.Environ.Contam.Toxicol., (9):212-7

Wildish, D.J. and W.G. Carson, Acute lethality of some nonionic and cationic surfactants
1972   to Salmo salar L. and Gammarus oceanicus Sergestrale. Manuscr.Rep.Ser.(Oceanogr.
       Limnol.)Fish.Res.Board Can., (1212):14 p.

Wildish, D.J. and N.J. Lister, The acute toxicity of the oil dispersant Gulf Agent 1009,
1973   LS-371 to aquatic fauna. Manuscr.Rep.Ser.Fish.Res.Board Can., (976):8 p.

Wilson, K.W., The ability of herring and plaice larvae to avoid concentrations of oil
1974   dispersants. In The early life history of fish: proceedings of an international
       symposium, Oban, Scotland, 17-23 May 1973, J.H.S. Blaxter (Ed.). Berlin,
       Springer-Verlag, p. 589-602

Yuen, K., Project Oil. Toxicity tests for oil dispersants. Dartmouth, Nova Scotia,
1970   Marine Ecology Laboratory, 2 p. (Unpubl.)

Zajic, J.E. and E. Knettig, Microbial emulsifiers for Bunker C fuel oil. Chemosphere,
1972   (1):51-6

Zillich, J., A biological evaluation of six chemicals used to disperse oil spills. Michigan,
1969   Michigan Department of Natural Resources, 12 p.

Zillioux, E.J. et al., Using Artemia to assay oil dispersant toxicities. J.Water Pollut.
1973   Control Fed., (45):2389-96

2.4   BIODEGRADATION

Agosti, J. and T. Agosti, Oxidation of certain Prudhoe Bay hydrocarbons by microorganisms
1972   indigenous to a natural oil seep at Umiat, Alaska. In Proceedings of the
        American Association for the Advancement of Science Alaskan science conference,
        Fairbanks, Alaska, 1972, B.H. McCrown (Ed.)

Ahearn, D.G., Microbial-facilitated degradation of oil: a prospectus. Publ.La.State Univ.,
1973   (LSU-SG-73-01):1-2

_____, Biodegradation of oil pollutants by yeasts and yeast-like fungi. Micro-
1974   biology Program Progress Report. Abstracts. Washington, D.C., Office of
        Naval Research, Department of the Navy,(ONR Report ACR-197):125-6 (Unpubl.)

Ahearn, D.G. and S.P. Meyers (Eds.), The microbial degradation of oil pollutants.
1973   Proceedings of a Workshop held at Georgia State University, Atlanta, Georgia,
        4-6 December 1972. Baton Rouge, Louisiana, Louisiana State University,
        Center for Wetland Resources, (LSU-SG-73-01):322 p.

Ahearn, D.G., S.P. Meyers and P.G. Standard, The role of yeasts in the decomposition of
1971   oils in marine environments. Dev.Ind.Microbiol., (12):126-34

Ahlfeld, T.E. and P.A. LaRock, Alkane degradation in beach sands. Publ.La.State Univ.,
1973   (LSU-SG-73-01):199-203

Anderes, E.A., Distribution of hydrocarbon oxidizing bacteria in some Pacific Ocean water
1973   masses. Publ.La.State Univ., (LSU-SG-73-01):311-2

Atlas, R.M. and R. Bartha, Biodegradation of polluting oil. Naval Res.Rev., (25):17-22
1972

Atlas, R.M. and C.E. Heintz, Ultrastructure of two species of oil-degrading marine
1973   bacteria. Can.J.Microbiol., (19):43-5

Bartha, R. and R.M. Atlas, Stimulated biodegradation of polluting oil at sea. Ocean.
1972   Abstr., 9(6):4-7

_____, Biodegradation of oil in seawater: limiting factors and artificial stimulation.
1973   Publ.La.State Univ., (LSU-SG-73-01):147-52

Beam, H.W. and J.J. Perry, Co-metabolism as a factor in microbial degradation of cyclo-
1973   paraffinic hydrocarbons. Arch.Microbiol., (91):87-90

Blumer, M., Benzpyrene in soil. Science, Wash.,D.C., (134):474-5
1961

Boyland, E., The biological significance of metabolism of polycyclic compounds. Biochem.
1950   Loc.Symposia, No. 5:40-54

Brisou, J., Biosynthesis of 3,4 benzpyrene and anaerobiosis. C.R.Séances Soc.Biol.,Paris,
1969 (163):772-4

Button, D.K., Petroleum - biological effects in the marine environment. In Impingement
1971 of man on the oceans, D.W. Hood (Ed.). New York, Wiley-Interscience, p. 421-9

―――――――, Hydrocarbon biodegradation kinetics. In Proceedings of a Workshop on inputs,
1973 fates and effects of petroleum in the marine environment, 21-25 May 1973, Airlie,
Virginia. Washington, D.C., National Academy of Sciences, Vol. 1:307-22

Byrom, J.A. and S. Beastall, Microbial degradation of crude oil with particular emphasis
1971 on pollution. In Institute of Petroleum, Microbiology 1971 (a symposium),
Amsterdam, Applied Science Publishers, p. 73-86

Carnes, D. and W.R. Finnerty, Characterization of the anapleroti enzyme, phospho-enol-
1971 pyruvate carboxylase, from Micrococcus cerificans. Bact.Proc., (1971):145

Cobet, A.B. and H.E. Guard, Effects of a bunker fuel on the beach bacterial flora. American
1973 Petroleum Institute, Proceedings of a Joint conference on prevention and control
of oil spills, Washington, D.C., 13-15 March 1973. Washington, D.C., American
Petroleum Institute, p. 815-9

Cooney, J.J. and J.D. Walker, Hydrocarbon utilization by Cladosporium resinae. Publ.La.
1973 State Univ., (LSU-SG-73-01):25-32

Crow, S.A., S.P. Meyers and D.G. Ahearn, Microbiological aspects of petroleum degradation
1974 in the aquatic environment. Mer, (12):37-54

Cundell, A.M. and R.W. Traxler, The isolation and characterization of hydrocarbon-utilizing
1973 bacteria from Chedabucto Bay, Nova Scotia. In American Petroleum Institute,
Proceedings of a Joint conference on prevention and control of oil spills,
Washington, D.C., 13-15 March 1973. Washington, D.C., American Petroleum
Institute, p. 421-6

―――――――, Microbial degradation of petroleum at low temperatures. Mar.Pollut.Bull.,
1973 (4):125-7

―――――――, Hydrocarbon-degrading bacteria associated with Arctic oil seeps. Dev.Ind.
1974 Microbiol., (15):250-5

Davis, J.B., Paraffinic hydrocarbons in the sulfate-reducing bacteria Desulfovibrio
1968 desulfuricans. Chem.Geol., (3):155-60

Fasoli, U. and W. Nümann, A proposal for the application of Mond's mathematical model to
1973 the biodegradation of mineral oil in natural waters. Water Res., (7):409-18

Finnerty, W.R. et al., Microbes and petroleum: perspectives and implications. Publ.La.
1973 State Univ., (LSU-SG-73-01):105-25

Floodgate, G.D., A threnody concerning the biodegradation of oil in natural waters. In
1973 The microbial degradation of oil pollutants. Proceedings of a Workshop,
Atlanta, Georgia, 4-6 December 1972, D.G. Ahearn and S.P. Meyers (Eds.).
Baton Rouge, Louisiana, Louisiana State University, Center for Wetland
Resources, p. 17-24

Floodgate, G.D., C.F. Gibbs and K.B. Pugh, The biodegradation of oil. Mar.Pollut.Bull.,
1971 (2):143

Floodgate, G.D., C.F. Gibbs and K.B. Pugh, Microbial interaction with oil in the marine
1973   environment. In Proceedings of a Workshop in inputs, fates and effects of
        petroleum in the marine environment, 21-25 May 1973, Airlie, Virginia.
        Washington, D.C., National Academy of Sciences, Vol. 2:507-15

Friede, J. et al., Assessment of biodegradation potential for controlling oil spills on the
1972   high seas. Washington, D.C., Department of Transportation, U.S. Coast Guard,
        Office of Research and Development Report Project 4110.1.3.1

Fusey, P. and J. Oudot, Note sur l'accélération de la biodégradation d'un pétrole brut
1973   par les bactéries (Note on the acceleration of the degradation of a crude oil
        by bacteria). Mater.Org., (8):157-64

Gatellier, C.R., Les facteurs limitant la biodégradation des hydrocarbures dans l'épuration
1971   des eaux (Factors limiting the biodegradation of hydrocarbons in the purification
        of waters). Génie Chim.,Paris, (104):2283-9

Gibbs, C.F., Quantitative studies on marine biodegradation of oil. I. Nutrient limitation
1975   at $14^{o}C$. Proc.R.Soc.Lond.,(B), (188):61-82

Gibbs, C.F., K.B. Pugh and A.R. Andrews, Quantitative studies on marine biodegradation of
1975   oil. II. Effect of temperature. Proc.R.Soc.Lond.,(B), (188):83-94

Gibson, D.T., Microbial degradation of aromatic compounds. Science, Wash.,D.C., (161):
1968   1093-7

_____, Initial reactions in the degradation of aromatic hydrocarbons. In
1972   Proceedings of the Conference on degradation of synthetic organic molecules in
        the biosphere, natural, pesticidal and various other man-made compounds,
        San Francisco, California, 12-13 June 1971. Washington, D.C., National Academy
        of Sciences, p. 116-36

Gibson, D.T. and W.K. Yeh, Microbial degradation of aromatic hydrocarbons. Publ.La.State
1973   Univ., (LSU-SG-73-01):33-8

Guire, P.E., J.D. Friede and R.K. Gholson, Production and characterization of emulsifying
1973   factors from hydrocarbonoclastic yeast and bacteria. Publ.La.State Univ.,
        (LSU-SG-73-01):229-31

Gunkel, W., Some reflections on the biodegradation of mineral oils in the marine environ-
1973   ment. In Proceedings of a Workshop on inputs, fates and effects of petroleum
        in the marine environment, 21-25 May 1973, Airlie, Virginia. Washington, D.C.,
        National Academy of Sciences, Vol. 2:516-40

_____, Distribution and abundance of oil-oxidizing bacteria in the North Sea.
1973   Publ.La.State Univ., (LSU-SG-73-01):127-39

Hasham, S.R., Microbiology of crude oil in sea water. Cardiff, University College of
1971   South Wales and Monmouthshire, Department of Microbiology, 1971, 133 p.
        M.Sc. Thesis

Hunt, P.G., The microbiology of terrestrial crude oil degradation. Spec.Rep.Cold Reg.Res.
1972   Eng.Lab.,Hanover, N.H., (168)

Ives, H., Catechol 2,3 oxygenase. In Student projects on the oxidation by marine bacteria
1971   of aromatic compounds found in oil, P.W. Robbins (Ed.). Cambridge, Massachusetts,
        Massachusetts Institute of Technology, Sea Grant Project Office, Report
        (MITSG 71-10):3-16

Jobson, A., F.D. Cook and D.W.S. Westlake, Microbial utilization of crude oil. Appl.
1972   Microbiol., (23):1082-9

Kallio, R.E., Microbial transformations of alkanes. In Fermentation advances,
1969   D. Perlman (Ed.). New York, Academic Press, p. 635-48

Kator, H., Utilization of crude oil hydrocarbons by mixed cultures of marine bacteria.
1973   Publ.La.State Univ., (LSU-SG-73-01):47-65

Knorr, M. and D. Schenk, Zur Frage der Synthese polyzyklischer Aromate durch Bakterien.
1968   Arch.Hyg.Bakt., (152):282-4

Krasil'nikov, N.A., A.V. Tsyban and T.V. Koronelli, Assimilation of saturated alkanes by
1973   marine bacteria. Okeanologiya, 13(5):877-82

Kvasnikow, E.E. and I.P. Krivitsky, On the distribution of micro-organisms utilizing
1968   aromatic hydrocarbons of the Borislav oil deposit. Mikrobiol.Zh., (30):291-5
        (In Ukrainian)

LaRock, P.A. and T.E. Ahlfeld, The relative changes in n-alkane composition and surface
1973   water slick. Publ.La.State Univ., (LSU-SG-73-01):233-6

Liu, D.L.S., Microbial degradation of crude oil and the various hydrocarbon derivatives.
1973   Publ.La.State Univ., (LSU-SG-73-01):95-104

Liu, D.L.S. and D.L. Dutka, Biological oxidation of the hydrocarbons in aqueous phase.
1973   J.Water Pollut.Control Fed., 45(2):232-9

Maciejowska, M. and E. Rakowska, Badania nad rozkladem paliw cieklych przez microorganizmy
1973   morskie (Investigations on decomposition of liquid fuels by marine organisms).
        Pr.Morsk.Inst.Ryb.Gdyni (A), (17):181-203 (In Polish with English summary)

Mahmoud, T.A. and W.B. Davis, The effect of salinity on the removal of some aliphatic
1970   ketones. College Station, Texas, Texas A&M University, Sea Grant Program,
        (TAMU-SG-70-216):106 p.

Mallet, L. and M. Tissier, Biosynthèse expérimental des hydrocarbures polybenzéniques du
1969   type benzo 3,4 pyrène aux dépens des terres de forêts. C.R.Soc.Biol.,Paris,
        (163):63-5

Mallet, L., M.L. Priou and M. Leon, Biosynthesis and bioreduction of the polybenzic
1969   hydrocarbon benzo 3,4 pyrene in the muds of St. Malo Bay. C.R.Hebd.Séances
        Acad.Sci.,Paris (D), (268):202-5

Mallet, L., L. Zanghr and L. Brisou, Examination of the possibilities of bio-synthesis of
1967   polybenzic hydrocarbons of the type benzo 3,4 pyrene by a putrid Clostridium
        in the presence of lipids of a marine plankton. C.R.Hebd.Séances Acad.Sci.,
        Paris (D), (264):1534-7

Mazmanidi, N.D., Diasamidze and Zambachidze, Oil effects on some species of molluscs and
1973   crustacea in the Black Sea. Materials of All Union Symposium on the Studies
        of the Black and Mediterranean Seas, utilization and protection of their
        resources. Part 4. Kiev, Naukova Dumka

Mechalas, B.J., T.J. Meyers and R.L. Kolpack, Microbial decomposition patterns using
1973   crude oil. Publ.La.State Univ., (LSU-SG-73-01):67-80

Merten, E.W., A literature review of the biological impact of oil spills in the marine
1973   water. In Proceedings of a Workshop in inputs, fates and effects of petroleum
       in the marine environment, 21-25 May 1973, Airlie, Virginia. Washington, D.C.,
       National Academy of Sciences, Vol. 2:731-44

Meschin, F.L., Histological modifications of organs and tissues of Lebistes reticulatus (P)
1973   under phenol acute intoxication. Phenol influence on hydrobionts. Leningrad,
       Publishing House "Nauka"

Meyers, S.P. et al., The impact of oil on marshland microbial ecosystems. Publ.La.State
1973   Univ., (LSU-SG-73-01):221-8

Miget, R.J., Microbial seeding to accelerate hydrocarbon degradation. Florida State
1971   University, 127 p. Thesis

Mircea, V. and O. Serbanescu, Etudes microbiologiques dans les eaux du littoral roumain de
1972   la Mer Noire (Microbiological studies in the waters of the Roumanian shore of
       the Black Sea). Cerc.Mar.Inst.Rom.Rech.Mar., (4):17-24

Mironov, O.G. and L.A. Lanskaya, The survival of some plankton and benthoplankton algae in
1968   seawater, contaminated by oil products. Botanical Journal, Vol. 53, No. 5

Mironov, O.G. and A.A. Lebed, Hydrocarbon-oxidising microorganisms in the North Atlantic.
1972   Hydrobiol.J., 8(1):71-4

Mulkins-Phillips, G.J. and J.E. Stewart, Survey of northwestern Atlantic waters and coastal
1973   sediments for hydrocarbon utilizing bacteria. Copenhagen, International Council
       for the Exploration of the Sea, Fisheries Improvement Committee. (CM 1973/E:27):
       9 p.

_____, Surveys for hydrocarbon utilizing bacteria in northwestern Atlantic coastal
1973   areas. In Oil and the Canadian environment, Proceedings of a Conference,
       16 May 1973, D. MacKay and W. Harrison (Eds.). Toronto, University of Toronto,
       p. 65-70

_____, Distribution of hydrocarbon-utilizing bacteria in northwestern Atlantic
1974   waters and coastal sediments. Can.J.Microbiol., (20):955-62

Munday, J.C., Jr., W. Harrison and W.G. MacIntyre, Oil slick motion near Chesapeake Bay
1970   entrance. Water Res.Bull., (6):879-84

Nelson, J.D. et al., Biodegradation of phenylmercuric acetate by mercury-resistant bacteria.
1973   Appl.Microbiol., 26(3):321-6

Niaussat, P. and C. Auger, Evidence and distribution of benzo 3,4 pyrene and perylene in
1970   different organisms of the Clipperton lagunar biocoenosis. C.R.Hebd.Séances
       Acad.Sci.,Paris (D), (270):2702-5

Nyns, E.J. and A.L. Wiaux, Biologie des hydrocarbures 1966-1968. Agricultura, 17(2):3-56
1969

O'Malley, M.L., Hydrocarbon-oxidizing ability of bacteria isolated from Long Island waters.
1970   Garden City, New York, Adelphi University, M.S. Thesis No. 1309:60 p.

O'Neill, T.B., The biodegradation of oil in seawater for naval pollution control.
1972   Port Hueneme, California, U.S. Naval Civil Engineering Laboratory, Tech.Note
       (N-1195):15 p.

Perry, J.J. and C.E. Cerniglia, Crude oil degradation by filamentous fungi. J.Gen.Appl.
1973   Microbiol., (19):151-3

_____, Studies on the degradation of petroleum by filamentous fungi. Publ.La.
1973   State Univ., (LSU-SG-73-01):89-94

Petrikevich, S.B., G. YeDanil'seva and M.N. Miesel, Accumulation and chemical transformation
1964   of 3,4 Benzpyrene by microorganisms. Dokl.Biol.Sci., (159):845-7

Poglazova, M.N. et al., Further investigations on the decomposition of benzo(a) pyrene by
1967   soil bacteria. Dokl.Biol.Sci., (172):649-51

_____, Destruction of benzo(a) pyrene by soil bacteria. Life Sci., (6):1053-62
1967

_____, The oxidation of benzo(a) pyrene by microorganisms in relation to its
1968   concentration in the medium. Dokl.Biol.Sci., (179):199-201

Pritchard, P.H. and T.J. Starr, Microbial degradation of oil and hydrocarbons in continuous
1973   culture. Publ.La.State Univ., (LSU-SG-73-01):39-45

_____, The degradation of oil in continuous culture. Brockport, New York, State
1974   University of New York, Department of Biology

Rhoads, D.B. and A. Ornstein, Evidence for the existence of protocatechuate 3,4-oxygenase
1971   in marine yeast isolated with parahydroxibenzic acid. Cambridge, Massachusetts,
          Massachusetts Institute of Technology, Sea Grant Project Office, Rep.
          (MITSG 71-10):39-55

Robbins, P.W., Student projects on the oxidation by marine bacteria of aromatic compounds
1971   found in oil. Cambridge, Massachusetts, Massachusetts Institute of Technology,
          Sea Grant Project Office, Rep. (MITSG 71-10):55 p.

Robichaux, T.J. and H.N. Myrick, Chemical enhancement of the biodegradation of oil pollution.
1971   Paper presented at the third annual Offshore technology conference, Houston,
          Texas, 19-21 April 1971, 10 p.

Rogoff, M.H., Oxidation of aromatic compounds by bacteria. Adv.Appl.Microbiol., (3):193-221
1961

Rogoff, M.H. and I. Wender, The microbiology of coal. 1. Bacterial oxidation of phenan-
1957   threne. J.Bacteriol., (73):264-8

Schwarz, J.R., J.D. Walker and R.R. Colwell, Deep-sea bacteria: growth and utilization of
1975   n-hexadecane at in situ temperature and pressure. Can.J.Microbiol., (21):682-7

Sedita, S.J., Biodegradation of oil. Houston, Texas, Houston Research Inc., 48 p.
1973

Seesman, P.A., J.D. Walker and R.R. Colwell, Biodegradation of oil by marine bacteria at
1976   potential off-shore drilling sites. Amer.Inst.Biolog.Sci.,Wash.,D.C., (17):293-8

Seki, H., Silica gel medium for enumeration of petroleumlytic microorganisms in the marine
1973   environment. Appl.Microbiol., (26):318-20

Seki, H. et al., Bacteria on petroleum globules in the Philippine Sea in January 1973.
1974   J.Oceanogr.Soc.Jap., (30):151-6

Shecket, G., Oxidation of naphthalene by a marine bacterium. Cambridge, Massachusetts,
1971 Massachusetts Institute of Technology, Sea Grant Project Office, Rep.
(MITSG 71-10):17-31

Soli, G., Degradation of petroleum hydrocarbons by marine bacteria. Rev.Int.Océanogr.Méd.
1971 CERBOM, (24):127

_____, Hydrocarbon-oxidizing bacteria and their possible use as controlling agents
1971 of oil pollution in the ocean. China Lake, California, U.S. Naval Weapons
Center, Research Department, Tech.Rep., (1):19 p.

_____, Marine hydrocarbonoclastic bacteria: types and ranges of oil degradation.
1973 Publ.La.State Univ., (LSU-SG-73-01):141-6

Soli, G. and E.M. Bens, Bacteria which attack petroleum hydrocarbons in a saline medium.
1972 Biotechnol.Bioeng., (14):319-30

_____, Selective substrate utilization by marine hydrocarbonoclastic bacteria.
1973 Biotechnol.Bioeng., (15):285-97

Storrs, P.N., Petroleum inputs to the marine environment from land sources. In Proceedings
1973 of a Workshop on inputs, fates and effects of petroleum in the marine environment,
21-25 May 1973, Airlie, Virginia. Washington, D.C., National Academy of Sciences,
Vol. 1:50-8

Stuart, D., The metabolism of quinoline and related compounds by marine bacteria.
1971 Cambridge, Massachusetts, Massachusetts Institute of Technology, Sea Grant
Project Office, Rep. (MITSG-71-10):33-7

Traxler, R.W., Bacterial degradation of petroleum materials in low temperature marine
1973 environments. Publ.La.State Univ., (LSU-SG-73-01):163-70

Treccani, V., Microbial degradation of aromatic hydrocarbons. Z.Allg.Mikrobiol. 5,
1962 Hydrocarbons. Prog.Indust.Microbiol., (4):3-33

Van Der Linden, A.C. and G.J.E. Thijsse, The mechanisms of microbial oxidations of petroleum
1965 hydrocarbons. Adv.Enzymol., (27):469-546

Walker, J.D. and R.R. Colwell, Microbial ecology of petroleum utilization in Chesapeake Bay.
1973 In American Petroleum Institute, Proceedings of a Joint conference on prevention
and control of oil spills, Washington, D.C. 13-15 March 1973. Washington, D.C.,
American Petroleum Institute, p. 685-90

_____, Measuring the potential activity of hydrocarbon-degrading bacteria. Appl.
1976 Microbiol., (31):189-97

_____, Role of autochthonous bacteria in the removal of spilled oil from sediment.
1976 Environ.Pollut. (In press)

_____, Some ecological factors to be considered in microbial degradation of oil.
n.d. BioScience (In press)

Walker, J.D., H.F. Austin and R.R. Colwell, Heterotrophic potential of marine bacteria.
n.d. Deep-Sea Res. (In press)

Walker, J.D., L. Cofone, Jr. and J.J. Cooney, Microbial petroleum degradation: the role
1973   of *Cladosporium resinae*. In American Petroleum Institute, Proceedings of a
        Joint conference on prevention and control of oil spills, Washington, D.C.,
        13-15 March 1973. Washington, D.C., American Petroleum Institute, p. 821-5

Walker, J.D., R.R. Colwell and L. Petrakis, Bacterial degradation of motor oil. J.Water
1974   Pollut.Control Fed., 47(8):2058-66

_____, Evaluation of petroleum-degrading potential of bacteria from water and
1975   sediment. Appl.Microbiol., 30(6):1036-9

_____, Petroleum degradation by bacteria from Chesapeake Bay sediment: fate of
n.d.   fractions of petroleum subject to degradation by fresh, frozen and enriched
        cultures. J.Gen.Microbiol. (In press)

_____, Biodegradation of petroleum by Chesapeake Bay sediment bacteria. Can.J.
n.d.   Microbiol. (In press)

_____, A study of the biodegradation of Louisiana crude oil employing computerized
n.d.   mass spectrometry. Paper presented to the API/EPA/USCG Conference on prevention
        and control of oil pollution. Can.J.Microbiol. (In press)

Walker, J.D., R.R. Colwell and Z. Vaituzis, A petroleum-degrading achlorophyllous alga,
1975   *Prototheca zopfii*. Nature, Lond., (254):223-4

Walker, J.D., L. Petrakis and R.R. Colwell, Comparison of the biodegradability of crude
1976   and fuel oils. Can.J.Microbiol., Vol. 22

Walsh, F. and R. Mitchell, Inhibition of bacterial chemoreception by hydrocarbons. Publ.
1973   La.State Univ., (LSU-SG-73-01):275-8

Zajic, J.E. and B. Supplison, Emulsification and degradation of Bunker C fuel oil by
1972   microorganisms. Biotechnol.Bioeng., (14):331-43

ZoBell, C.E., Action of microorganisms on hydrocarbons. Bact.Rev., (10):1-48
1946

_____, Assimilation of hydrocarbons by microorganisms. Adv.Enzymol., (10):443-86
1950

_____, Microbiology of oil. Mem.N.Z.Oceanogr.Inst., (3):39-47
1959

ZoBell, C.E. and J. Agosti, Bacterial oxidation of mineral oils at sub-zero Celsius.
1972   Paper prepared for presentation at the 72nd annual meeting, American Society
        for Microbiology, Philadelphia, Pennsylvania, 24 April 1972, 10 p.

Anon., Controlling oil pollution of harbors through biological means. Naval Res.Rev.,
1970   23(5):24-6

2.5 LEVELS OF HYDROCARBONS

Barbier, M. et al., Hydrocarbons from seawater. Deep-Sea Res., 20(4):305-14
1973

Brooks, J.M. and W.M. Sackett, Sources, sinks and concentrations of light hydrocarbons in
1973   the Gulf of Mexico. J.Geophys.Res., (78):5248-58

Brooks, J.M. et al., Baseline concentrations of light hydrocarbons in the Gulf of Mexico.
1973   Environ.Sci.Technol., (7):639-42

Brown, R.A. et al., Measurement and interpretation of nonvolatile hydrocarbons in the
1974   ocean. Part 1. Measurements in Atlantic, Mediterranean, Gulf of Mexico and
Persian Gulf. Linden, New Jersey, Exxon Research and Engineering Co., 221 p.

Carlberg, S.R., A three year study of the occurrence of non-polar hydrocarbons (oil) in
1973   Baltic waters 1970-1973. Copenhagen, International Council for the Exploration
of the Sea, Fisheries Improvement Committee, 1973 (CM 1973/E:28):8 p.

Carlberg, S.R. and C.B. Skarstedt, Some results of determination and monitoring of oil-
1970   content in seawater. Copenhagen, International Council for the Exploration of
the Sea, Hydrography Committee, 1970 (CM 1970/C:23):2 p.

Clark, R.C., Jr. and M. Blumer, Distribution of n-paraffins in marine organisms and
1967   sediments. Limnol.Oceanogr., (12):79-98

DiSalvo, L.H. et al., Hydrocarbons of suspected pollutant origin in aquatic organism of
1973   San Francisco Bay. Methods and preliminary results. Microbial degradation oil
pollutants, Workshop, p. 205-20

Farrington, J.W., Benthic lipids of Narrangansett Bay - fatty acids and hydrocarbons.
1972   Kingston, Rhode Island, University of Rhode Island, Thesis, 157 p.

Farrington, J.W. and J.G. Quinn, Biogeochemistry of fatty acids in recent sediments from
1973   Narrangansett Bay, Rhode Island. Geochem.Cosmochim.Acta, (37):259-68

Fossato, V.U., Elimination of hydrocarbons by mussels. Mar.Pollut.Bull., (6):7-10
1975

Gordon, D.C., Jr. and P.A. Michalik, Concentration of Bunker C fuel in the waters of
1971   Chedabucto Bay, April 1971. J.Fish.Res.Board Can., 28(12):1912-4

Gordon, D.C., Jr., P.D. Keizer and P.S. Chamut, Estimation of hydrocarbon concentrations
1974   in the water column of Come-by-Chance Bay, 1971-73. Tech.Rep.Fish.Res.Board Can.,
(442):15 p.

Gortalum, G.M., Content of 3,4 benzpyrene in generator tar from Estonian shales. Vop.Gig.
1958   Truda v Slantsevoi Prom Estonskoi.SSR, (563):159-64

Harington, J.S., Occurrence of oils containing 3,4 benzpyrene and related substances in
1962   asbestos. Nature, (193):43-5

Jardas, I. and I. Munjko, Preliminary observations of oil and phenol distribution in the
1972   central Adriatic. Bilj.Inst.Oceanogr.Ribarst., (29):6 p.

Kinney, P.J., Baseline hydrocarbon concentrations. Occas.Publ.Inst.Mar.Sci.Univ.Alaska,
1973   (3):397-410

Ledet, E.J. and J.L. Laseter, Alkanes at the air-sea interface from offshore Louisiana
1974 and Florida. Science, N.Y., (186):261-3

Levy, E.M., The presence of petroleum residues off the east coast of Nova Scotia, in the
1971 Gulf of St. Lawrence and the St. Lawrence river. Water Res., 5(9):723-33

Levy, E.M. and A. Walton, Dispersed and particulate petroleum residues in the Gulf of
1973 St. Lawrence. J.Fish.Res.Board Can., 30(2):261-7

Majori, L., F. Petronio and G. Nedoclan, L'inquinamento marino da idrocarburi nell'alto
1973 Adriatico. Nota. 3: Risultati di un'indagine nei Golfi di Trieste e di Venezia
negli anni 1971 e 1972 (Seawater pollution from hydrocarbons in the high Adriatic
Sea. Note 3: Results of research in the Gulfs of Trieste and Venice in the years
1971 and 1972). Ig.Mod., 66(2):21 p.

Michalik, P.A. and D.C. Gordon, Jr., Concentration and distribution of oil pollutants in
1971 Halifax Harbor - 10 June to 20 August 1971. Tech.Rep.Fish.Res.Board Can.,
(284):26 p.

Monaghan, P.H., J.H. Seelinger and R.A. Brown, The persistent hydrocarbon content of the
1973 sea along certain tanker routes - a preliminary report. In American
Petroleum Institute, Report on the 18th Tanker conference, Hilton Head Island,
South Carolina, 7-9 May 1973. Washington, D.C., American Petroleum Institute,
p. 232-59

Morris, R.J. and F. Culkin, Lipid chemistry of eastern Mediterranean surface layers.
1974 Nature, Lond., (250):640-2

Parker, P.L., J.K. Winters and J. Morgan, A baseline study of petroleum in the Gulf of
n.d. Mexico. Port Aransas, Texas, University of Texas, Marine Science Institute,
25 p. (Unpubl.)

Sackett, W.M., Significance of low molecular weight hydrocarbons in Eastern Gulf Waters.
1974 Paper presented to the conference on marine environmental implications of
offshore drilling, Eastern Gulf of Mexico, St. Petersburg, Florida,
31 January-2 February 1974

Sackett, W.M. and J.M. Brooks, Use of low molecular-weight-hydrocarbon concentrations as
1974 indicators of marine pollution. Spec.Publ.U.S.Natl.Bur.Stand., (409):171-3

Shaw, D.G., Lipids in shallow bottom sediments. Environ.Sci.Technol., (7):740-2
1973

Whittle, K.J., P.R. Mackie and R. Hardy, Hydrocarbons in the marine ecosystem. S.Afr.
1974 J.Sci., (70):141-4

Wong, C.S. et al., Baseline information on chemical oceanography and petroleum-based
1974 hydrocarbons in the Southern Beaufort Sea. Victoria, B.C., Environment Canada,
Ocean Chemistry Division, Ocean and Aquatic Affairs, Pacific Region, Interim
Report, 51 p.

Youngblood, W.W. and M. Blumer, Alkanes and alkenes in marine benthic algae. Mar.Biol.,
1973 (21):163-72

Zsolnay, A., The relative distribution of saturated and olefinic hydrocarbons in the
1972 Baltic in September 1971. Copenhagen, International Council for the Exploration
of the Sea, Fisheries Improvement Committee, (CM 1972/E:12):14 p.

Zsolnay, A., Measurements made at sea of the saturated and aromatic hydrocarbons in the
1973   Baltic in April 1973. Copenhagen, International Council for the Exploration
of the Sea, Fisheries Improvement Committee, 1973 (CM 1973/E:22):13 p.

─────────, The relative distribution of non-aromatic hydrocarbons in the Baltic in
1973   September 1971. Mar.Chem., (1):127-36

─────────, Hydrocarbons and chlorophyll: a correlation in the upwelling region off
1973   West Africa. Deep-Sea Res., (20):923-5

─────────, Hydrocarbon content and chlorophyll correlation in the waters between
1974   Nova Scotia and the Gulf Stream. Spec.Publ.U.S.Natl.Bur.Stand., (409):255-6

2.6 NATURAL SEEPS/OFFSHORE OPERATIONS

Allen, D.R., Environmental aspects of oil producing operations - Long Beach, California.
1972   J.Petrol.Technol., (24):125-31

Baldwin, A.H. and E.B. Cowell, Improvements in oil and gas recovery techniques with
1974   particular reference to the North Sea and their impact on the environment.
Paper presented at the ninth World Energy Conference, Detroit, Michigan, 1974.
(Ref. 3.1-4):24 p.

Barbier, M. et al., Hydrocarbons from sea water. Deep-Sea Res., 20(4):305-14
1973

Bertrand, A.R.V., Prévention et lutte contre la pollution au cours des opérations de forage
1971   et de production en mer. Annexe A. Principaux accidents de pollution en mer
(Prevention and the fight against pollution in the course of drilling operations
and production at sea. Appendix A. Principal pollution accidents at sea).
Rev.Inst.Fr.Pétrole, (26):830-9

Castela, A. and M. Masson, Prévention et lutte contre la pollution au cours des opérations
1971   de forage et de production en mer. 2. Pollution par les chantiers de forage
et de production en mer (Prevention and the fight against pollution in the
course of drilling operations and production at sea. 2. Pollution by drilling
rigs and production at sea). Rev.Inst.Fr.Pétrole, (26):765-79

Cundell, A.M., Oil pollution and offshore field development. Chem.N.Z., (36):184-7
1972

Edwards, J.H., Safety in offshore oil exploration and production: practices for maintaining
1972   clean seas. J.Environ.Plann.Pollut.Control, 1(1):12-23

Ingram, G.E. and P.A. Dee, Reliability engineering applied to offshore petroleum production
1973   systems. Paper presented at the fifth annual Offshore technology conference,
Dallas, Texas, 1973, (OTC 1756):10 p.

Johannes, R.E. and K.L. Webb, Release of dissolved organic compounds by marine and fresh
1970   water invertebrates. Occas.Pap.Inst.Mar.Sci.Univ.Alaska, (1):257-73

Johnson, T.C., Natural oil seeps in or near the marine environment: a literature survey.
1971   Washington, D.C., U.S. Coast Guard, Office of Research and Development, 30 p.

Landes, K.K., Mother Nature as an oil polluter. Bull.Am.Assoc.Petrol.Geol., (57):637-41
1973

Lester, T.E. and J.R. Beynon, Pollution and the offshore oil industry. Mar.Pollut.Bull.,
1973   (1):23-5

Marshall, A.G., Offshore oil in South-East Asian waters: potential environmental problems.
1974   Environ.Conserv., 1(1):69-70

McKelvey, V.E., Environmental protection in offshore petroleum operations. Ocean Manage.,
1973   1(1):119-27

Mikolaj, P.G. and J.P. Ampaya, Tidal effects on the activity of natural submarine oil
1973   seeps. Mar.Technol.Soc.J., (7):25-8

Mikolaj, P.G., A.A. Allen and R.S. Schlueter, Investigation of the nature, extent and
1972   fate of natural oil seepage off Southern California. Paper presented at the
         fourth annual Offshore technology conference, Houston, Texas, 1-3 May 1972,
         (OTC 1549):16 p.

Miller, R.T. and R.L. Clements, Reservoir engineering techniques used to predict blowout
1972   control during the Bay Marchand fire. J.Petrol.Technol., (24):234-40

Pottier, J., Prévention et lutte contre la pollution au cours des opérations de forage et
1971   de production en mer. 1. Généralités (Prevention and the fight against
         pollution in the course of drilling operations and production at sea. 1.
         Generalities). Rev.Inst.Fr.Pétrole, (26):759-64

Späing, I., Schutz der Meeresküsten gegen Verunreinigung. Technische Vorkehrungen bei
1968   der Ausbeute von Öl und Gas in Küstengebieten (Protection of coasts against
         pollution. Technical precautions in the production of oil and gas in coastal
         regions). In Föderation Europäischer Gewässerschutz (FEG), Schutz der
         Meeresküsten gegen Verunreinigung, Symposium, Hamburg, 5-7 October 1967.
         Zürich, FEG, p. 53-5

United States, National Academy of Engineering, Marine Board, Panel on Operational Safety
1972   in Offshore Resource Development, Outer continental shelf resource development
         safety: a review of technology and regulation for the systematic minimization
         of environmental intrusion from petroleum products. Washington, D.C., U.S.
         Geological Survey, 197 p.

Warner, D.G., Safety and pollution control - offshore petroleum producing facilities. In
1972   Preprints of the eighth annual conference and exposition of the Marine Technology
         Society, 11-13 September 1972. Washington, D.C., Marine Technology Society,
         p. 741-52

_____, Spill prevention in offshore petroleum producing facilities. In American
1973   Petroleum Institute, Proceedings of Joint conference on prevention and control
         of oil spills, Washington, D.C., 13-15 March 1973. Washington, D.C., American
         Petroleum Institute, p. 31-7

Wilson, R.D., Estimate of annual input of petroleum to the marine environment from natural
1973   marine seepage. In Proceedings of a Workshop on inputs, fates and effects of
         petroleum in the marine environment, 21-25 May 1973, Airlie, Virginia.
         Washington, D.C., National Academy of Sciences, Vol. 1:59-96

_____, Estimate of annual input of petroleum to the marine environment from offshore
1973   production operations. In Proceedings of a Workshop on inputs, fates and
         effects of petroleum in the marine environment, 21-25 May 1973, Airlie,
         Virginia. Washington, D.C., National Academy of Sciences, Vol. 1:97-100

Wilson, R.D. and O.J. Shirley, A dual program for prevention of oil spills in offshore
1973   drilling and producing operations. In American Petroleum Institute,
Proceedings of a Joint conference on prevention and control of oil spills,
Washington, D.C., 13-15 March 1973. Washington, D.C., American Petroleum
Institute, p. 61-3

2.7 OIL LUMPS IN THE SEA

Ayappan, N.S. et al., Preliminary observations on tar-like material observed on some
1972   beaches. Curr.Sci., (41):766-76

Burman, I. and O.H. Oren, Tar pollution in the Levant basin and along the Mediterranean
1973   coast of Israel. In CIESMM, Journées d'études sur les pollutions marines,
problèmes posés par les rejets directs en mer d'effluents pollués, Athens,
3-4 November 1972. Monaco, Commission internationale pour l'exploration
scientifique de la mer Méditerranée, p. 49-51

Butler, J.N. and B.F. Morris, Quantitative monitoring and variability of pelagic tar in
1974   the North Atlantic. Spec.Publ.U.S.Natl.Bur.Stand., (409):75-8

Butler, J.N., B.F. Morris and J. Sass, Pelagic tar from Bermuda and the Sargasso Sea.
1973   Spec.Publ.Bermuda Biol.Stn., (10):346 p.

Dwivedi, S.N. and A.H. Parulekar, Oil pollution along the Indian coastline. Spec.Publ.
1974   U.S.Natl.Bur.Stand., (409):101-5

Hildebrand, H.H. and G. Gunter, A report on the deposition of petroleum tars and asphalts
1955   on the beaches of the northern Gulf of Mexico, with notes on the beach
conditions and the associated biota. University of Texas, Institute of Marine
Science, 101 p.

Jeffery, L.M., Preliminary report on floating tar balls in the Gulf of Mexico and Caribbean.
1973   College Station, Texas, Texas A&M University, Department of Oceanography, 6 p.

Jeffery, L.M. et al., Summary report on pelagic, beach and bottom tars of the Gulf of
1973   Mexico and controlled weathering experiments. College Station, Texas, Texas
A&M University, Department of Oceanography, 28 p.

_____, Progress report on pelagic, beach and bottom tars of the Gulf of Mexico
1973   and controlled weathering experiments. College Station, Texas, Texas A&M
University, Department of Oceanography, 90 p.

_____, Pelagic tar in the Gulf of Mexico and Caribbean Sea. Spec.Publ.U.S.Natl.
1974   Bur.Stand., (409):233-5

Levy, E.M. and A. Walton, An evaluation of two neuston samplers. Dartmouth, Nova Scotia,
1971   Atlantic Oceanographic Laboratory, Bedford Institute, AOL Report, (1971-9):10 p.

McGowan, W.E., W.A. Saner and G.L. Hufford, Tar ball sampling in the western north Atlantic.
1974   In U.S. National Bureau of Standards. Marine pollution monitoring (petroleum).
Proceedings of a Symposium and workshop, 13-17 May 1974. Washington, D.C., U.S.
Government Printing Office, p. 83-4

Morris, B.F. and J.N. Butler, Petroleum residues in the Sargasso sea and on Bermuda beaches.
1973 In American Petroleum Institute, Proceedings of Joint conference on prevention and control of oil spills, Washington, D.C., 13-15 March 1973. Washington, D.C., American Petroleum Institute, p. 521-9

Ohya, M., T. Otsuki and M. Saito, Oil pollution in the Izu Islands waters. J.Oceanogr.Soc.
1973 Jap., (29):121-9

Okera, W., Tar pollution of Sierra Leone beaches. Nature, Lond., (252):682
1974

Price, A.R.G., A further look at pollution in the Atlantic Ocean. Biol.Conserv., (5):297-8
1973

Reddy, C.V.G. and S.Y.S. Singbal, Chemical characteristic of tar-like materials found on
1973 the beaches along the east and west coast of India in relation to their source of origin. Curr.Sci., (42):709-11

Saner, W.A. and M. Curtis, Tar ball loadings on Golden Beach, Florida. Spec.Publ.U.S.
1974 Natl.Bur.Stand., (409):79-81

Sherman, K. et al., Distribution of tar balls and neuston sampling in the Gulf Stream
1974 system. Spec.Publ.U.S.Natl.Bur.Stand., (409):243-4

Sleeter, T.D., B.F. Morris and J.N. Butler, Quantitative sampling of pelagic tar in the
1974 North Atlantic, 1973. Deep-Sea Res., (21):773-5

Sweet, W.E., Jr., Tar balls in the sea: a new source concept. Paper presented at the
1974 sixth annual Offshore technology conference, Houston, Texas, 6-8 May 1974, (OTC 2002):7 p.

Traxler, R.W. and R.H. Pierce, Jr., Standard and inter-comparison criteria: tar balls and
1974 particulate matter. Spec.Publ.U.S.Natl.Bur.Stand., (409):161-2

Wong, C.S., D.R. Green and W.J. Cretney, Quantitative tar and plastic waste distribution
1974 in the Pacific Ocean. Nature, Lond., 247(5435):30-2

Wong, C.S., D. MacDonald and R.D. Bellegay, Distribution of tar and other particulate
1974 pollutants along the Beaufort Sea coast. Victoria, B.C., Environment Canada, Ocean Chemistry Division, Ocean and Aquatic Affairs, Pacific Region, 1974. Interim Report, 69 p. (Unpubl.MS)

2.8 PROPERTIES OF OIL

Abbott, M.B., On the spreading of one fluid over another. Part 1. Discharge of oil into
1961 a canal. Houille Blanche, (5):622-8

_____, On the spreading of one fluid over another. Part 2. The wave front.
1961 Part 3. Discharge of oil into still water, radial flow. Houille Blanche, (6):827-46

Abbott, M.B. and T. Hayashi, Unsteady radial flow of oil being discharged from a source
1967 on the ocean. In Proceedings of the fourteenth Coastal engineering conference. Tokyo, Japanese Society of Civil Engineering, p. 226-9

Alofs, D.J. and R.L. Reisbig, The effects of waves on oil spill movements. In Proceedings
1971 of the Symposium on technology for the future to control industrial and urban wastes, Rolla, Missouri, February 1971. Columbia, Missouri, University of Missouri, Continuing Education Series, p. 101-2

———, An experimental evaluation of oil slick movement caused by waves. J.Phys.
1972 Oceanogr., (2):439-43

Atlas, R.M., Fate and effects of oil pollutants in extremely cold marine environments.
n.d. Pasadena, Jet Propulsion Laboratory Calif. Institute of Technology, ONR Government Order NAonr-30-73 (Unpubl.MS)

Ayers, R.C., Jr., H.O. Jahns and J.L. Glaeser, Oil spills in the Arctic ocean: extent of
1974 spreading and possibility of large-scale thermal effects. Science, Wash.,D.C., (186):843-6

Bien, W., Modelling petroleum spills at supertanker ports. Toronto, University of Toronto,
1973 Department of Geography, 122 p. M.S. Thesis

Blumer, M., Dissolved organic compounds in sea water: saturated and olefinic hydrocarbons
1970 and simply branched fatty acids. Occas.Publ.Inst.Mar.Sci.Univ.Alaska, (1):153-67

Buckmaster, J., Viscous-gravity spreading of an oil slick. J.Fluid Mech., (59):481-91
1973

Campbell, W.J. and S. Martin, Oil and ice in the Arctic Ocean: possible large-scale
1973 interactions. Science, Wash.,D.C., (181):56-9

Carruthers, W. and A.G. Douglas, 1-2 Benzanthracene derivatives in a Kuwait mineral oil.
1961 Nature, (192):256-7

Conomos, T.J., Movement of spilled oil in San Francisco Bay as predicted by estuarine
1974 nontidal drift. Spec.Publ.U.S.Natl.Bur.Stand., (409):97-100

Estes, J.E. et al., Volumetric determination of marine oil spills using coordinated air-
1973 borne and surface sampling data. In American Petroleum Institute, Proceedings of a Joint conference on prevention and control of oil spills, Washington, D.C., 13-15 March 1973. Washington, D.C., American Petroleum Institute, p. 117-25

Harrison, W., Environmental impact of crude oil and naphtha spills, South Riding Point,
1973 Grand Bahama Island. Report prepared for the Government of the Bahamas by Environmental Research Associates Inc., Toronto, University of Toronto, Department of Geography, 50 p.

Horton, A.W. et al., Composition versus carcinogenicity of distillate oils. Amer.Chem.Soc.
1963 Div.Petrol.Chem.Preprints, 8(4):C59-C65

McMinn, T.J., Crude oil behaviour on Arctic winter ice. Washington, D.C., U.S. Coast
1972 Guard, 67 p.

Murray, S.P., Turbulent diffusion of oil in the ocean. Limnol.Oceanogr., (17):651-60
1972

Murty, T.S. and F.G. Barber, Rip current patterns in oil slicks. In Proceedings of the
1972 first Canadian symposium on remote sensing, Ottawa, 7-9 February 1972, Vol.1:241-68

Murty, T.S. and M.L. Khandekar, Simulation of movement of oil slicks in the Strait of Georgia
1973   using simple atmosphere and ocean dynamics. In American Petroleum Institute, Proceedings of Joint conference on prevention and control of oil spills, Washington, D.C., 13-15 March 1973. Washington, D.C., American Petroleum Institute, p. 541-6

Murty, T.S., M.L. Khandekar and G.V. Rao, The movement of oil slicks. Rapp.P.-V.Réun.CIEM,
1974   (167):66-74

O'Brien, J.A., Oil spreading on water from a stationary leaking source. Chem.Eng.,
1970   No. (244):CE407-CE409

_____, Wind tunnel experiments on oil slick transport. J.Hydraul.Res., (9):197-215
1971

Orthlieb, F.L., Forecasting oil slick behaviour - a preliminary guide. Washington, D.C.,
1971   U.S. Coast Guard, Office of Research and Development, 9 p.

Otto, L., De oceanografische aspecten van de olievervuiling (Oceanographic aspects of oil
1972   pollution). Naut.Technol.T., (1):151-4

Premack, J. and G.A. Brown, Predictions of oil slick motions in Narragansett Bay. In
1973   American Petroleum Institute, Proceedings of Joint conference on prevention and control of oil spills, Washington, D.C., 13-15 March 1973. Washington, D.C., American Petroleum Institute, p. 531-40

Reisbig, R.L., Oil spill drift caused by the coupled effects of wind and waves. Rolla,
1973   Missouri, Missouri University, Division of Engineering Research, 61 p.

Reisbig, R.L. et al., Measurement of oil spill drift caused by the coupled parallel
1974   effects of wind and waves. Mém.Soc.R.Sci.Liège, (6):67-77

Ridgway, N.M., Direction of drift of surface oil with wind and tide. N.Z.J.Mar.Freshwat.
1972   Res., (6):178-84

Sauer, W.E. and V. Klemas, Oil layer thickness monitor. Paper presented at the sixteenth
1971   annual ISA conference, Chicago, (71-840):3 p.

Schwartzberg, H.G., The spreading and movement of oil spills. Eng.Ext.Ser.Purdue Univ.,
1970   (137):773-82

Sonu, C.J., S.P. Murray and W.G. Smith, Environmental factors controlling the spread of
1971   oil. Naval Res.Rev., (24):11-9

Sorensen, R.M. and E.B. Spencer, Two-dimensional wind setup of oil on water. J.Waterways
1971   Harb.Coast.Eng.Div.Am.Soc.Civil.Eng., 97(WW3):517-30

Spillane, K.T., Movement of oil on the sea surface. Aust.Meteorol.Mag., 19(4):158-77
1971

Stewart, R.J., J.W. Devanney III and W. Briggs, Oil spill trajectory studies for Atlantic
1974   coast and Gulf of Alaska. Report to U.S. Council on Environmental Quality. Cambridge, Massachusetts, Massachusetts Institute of Technology, 1974

Suchon, W., An experimental investigation of oil spreading over water. Cambridge,
1970   Massachusetts, Massachusetts Institute of Technology, Department of Mechanical Engineering, M.S. Thesis

Tayfun, M.A. and H. Wang, Monte Carlo simulation of oil slick movements. J.Waterways
1973   Harb.Coast.Eng.Div.Am.Soc.Civil.Eng., 99(WW3):309-24

Waldman, G.D., T.K. Fannelop and R.A. Johnson, Spreading and transport of oil slicks on the
1972   open ocean. Paper presented at the fourth annual Offshore technology conference,
         Houston, Texas, 1-3 May 1972, (OTC 1548):12 p.

Waldman, G.D., R.A. Johnson and P.C. Smith, The spreading and transport of oil slicks on
1973   the open ocean in the presence of wind, waves and currents. Wilmington,
         Massachusetts, Avco Corporation, (AVSD-0068-73-RR):84 p.

Wang, S. and L.-S. Hwang, A numerical model for simulation of oil spreading and transport
1974   and its application for predicting oil slick movement in bays. Pasadena,
         California, Tetra Tech Inc.

Warner, J.L., J.W. Graham and R.G. Dean, Prediction of the movement of an oil spill on
1972   the surface of the water. Paper presented at the fourth annual Offshore
         technology conference, Houston, Texas, 1-3 May 1972, (OTC 1550):8 p.

2.9 GENERAL/DISCUSSIONS

Alzieu, C., Choice of products for use against the pollution of the marine environment by
1972   oil spills. 3. Relative toxicity of anti-petroleum products to two marine
         organisms. Rev.Trav.Inst.Pêches Marit.,Nantes, 36(1):103-19

American Petroleum Institute, Environmental research: annual report. Washington, D.C.,
1974   American Petroleum Institute, Coordinating Research Council, p. 2-67

Arpino, R., Sea pollution by petroleum products. Recherche, 30(4):71-3
1973

Aubert, M. and D. Pesando, Télémédiateurs chimiques et équilibre biologique océanique. 2.
1971   Nature chimique de l'inhibiteur de la synthèse d'un antibiotique produit par une
         diatomée. Mar.Int.Océanogr.Réd., (21):17-22

Aubert, M., J. Aubert and M. Gauthier, Le milieu marin et les matières organiques. Rev.
1972   Int.Océanogr.Méd.CERBOM, (28):181-93

Badger, G.M.D. and T.M. Spotiswood, The formation of aromatic hydrocarbons at high
1963   temperatures. Part 17. Pyrolysis of a (Premium grade) petroleum (at
         700 degrees C). Aust.J.Chem., (16):392-400

Baker, J.M. and C.B. Crapp, Toxicity tests for predicting the ecological effects of oil
1974   and emulsifier pollution on littoral communities. In Ecological aspects of
         toxicity testing of oils and dispersants. Proceedings of a Workshop,
         L.R. Beynon and E.B. Cowell (Eds.). Barking, Essex, Applied Science Publishers,
         p. 23-40

Becker, C.D. et al., Regional survey of marine biota for bioassay standardization of oil
1973   and oil dispersant chemicals. Publ.Am.Petrol.Inst., (4167):102 p.

Bestougeff, M.A., Petroleum hydrocarbons in fundamental aspects of petroleum geochemistry,
1967   B. Nagy and U. Colombo (Eds.). New York, Elsevier Publ. Co.

Binet, L. and L. Mallet, Diffusion of polybenzene hydrocarbons in the living environment.
1963   Gaz.Hop., (135):1142 (in French)

Blokker, P.C. and H.J. Marcinowski, Refinery effluents - Western Europe. Petrol.Rev.,
1971   25(299):395-9

Boesch, D.F. and C.H. Hershner, The ecological effects of oil pollution in the marine
1974   environment. In Oil spills and the marine environment. A report to the Energy
        Policy Project of the Ford Foundation, D.F. Boesch, C.H. Hershner and
        J.H. Milgram (Eds.). Cambridge, Massachusetts, Ballinger Publishing Company,
        p. 1-55

Bonz, P.D., Fabric boom concept for containment and collection of floating oil. Washington,
1973   D.C., U.S. Environmental Protection Agency, Office of Research and Development,
        (EPA-670/2-73-06)

Borneff, J. et al., Experimental studies on the formation of polycyclic aromatic hydro-
1968   carbons in plants. Environ.Res., 2:22-9

_____, Die Synthese von 3,4-Benzypyren und anderen polyzyklischen, aromatischen
1968   Kohlenwasserstoffen in Pflanzen. Arch.Hyg.Bakt., (152):279-82

Boyland, E., Polycyclic hydrocarbons. Brit.Med.Bull., 20:121-6
1964

Boyland, E. and P. Sims, Metabolism of polycyclic compounds. 24. The metabolism of
1964   Benz(a) anthracene. Biochem.J., (9):439-506

Brockis, G.J., Sources of sea pollution by oil. London, British Petroleum Company, 22 p.
1972   (Unpubl.Rep.)

Brummage, K.G., The sources of oil entering the sea. In Proceedings of a Workshop on
1973   inputs, fates and effects of petroleum in the marine environment, 21-25 May 1973,
        Airlie, Virginia. Washington, D.C., National Academy of Sciences, Vol.1:1-6

Butler, A.R., Oil pollution - carelessness or crime. Proc.U.S.Naval Inst., (1972):52-7
1972

Canada, Ministry of Transport, Report of the Task Force - Operation Oil. Ottawa,
1970   Information Canada, Vol.2:104 p.

Carlson, G.P., Detoxification of foreign organic compounds by the quahog, Mercenaria
1972   mercenaria. Comp.Biochem.Physiol., (43B):295-302

Charter, D.B., R.A. Sutherland and J.D. Porricelli, Quantitative estimates of petroleum
1973   in the oceans. In Proceedings of a Workshop on inputs, fates and effects of
        petroleum in the marine environment, 21-25 May 1973, Airlie, Virginia.
        Washington, D.C., National Academy of Sciences, Vol.1:7-30

Clar, E., Polycyclic hydrocarbons. New York, Academic Press, 487 p.
1964

Clark, R.C., Biological fates of petroleum hydrocarbons in aquatic macroorganisms.
1973   In Background papers for a Workshop on inputs, fates and effects of petroleum
        in the marine environment. Washington, D.C., National Academy of Sciences

Conney, A.H. and J.J. Burns, Metabolic interactions among environmental chemicals and
1972   drugs. Science, (178):576-86

Cowell, E.B., Some biological effects of oil pollution. Your Environ., (1):84,93-4
1970

Cowell, E.B., Chronic oil pollution by refinery effluent water. In Water pollution by oil.
1971   London, The Institute of Petroleum. P. Hepple (Ed.), 380-1

_____, Zoological studies on shore communities. Oil pollution in perspective.
1971   In The ecological effects of oil pollution on littoral communities, E.B. Cowell
(Ed.). Elsevier Publishing Co., N.Y., pp. 224-34

_____, A critical examination of present practice. In Ecological aspects of
1974   toxicity testing of oils and dispersants. Proceedings of a Workshop,
L.R. Beynon and E.B. Cowell (Eds.). Barking, Essex, Applied Science Publishers,
p. 97-104

Crapp, G.B., Zoological studies on shore communities. Chronic oil pollution. In The
1971   ecological effects of oil pollution on littoral communities, E.B. Cowell (Ed.).
Elsevier Publishing Co., N.Y., pp. 187-203

Crapp, G.B., R.G. Withers and C. Sullivan, Zoological studies on shore communities.
1971   Investigations on sandy and muddy shores. In The ecological effects of oil
pollution on littoral communities, E.B. Cowell (Ed.). Elsevier Publishing Co.,
N.Y., pp. 208-16

Crossley Surveys Inc., Oil spill control survey for onshore and offshore facilities.
1970   API (Am.Petrol.Inst.)Rep., (4023):372 p.

Currier, H.B. and S.A. Peoples, Phytotoxicity of hydrocarbons. Hilgardia, (23):155-73
1954

Davis, J.B., Petroleum microbiology. New York, Elsevier, 604 p.
1967

Demayo, A., Identification of petroleum products in water. Tech.Bull.Dep.Energy Mines
1970   Resourc.,Ottawa, (32):22 p.

Environmental Protection Agency, Environmental studies as they relate to offshore petroleum
1975   operations. Washington, D.C., U.S. Environmental Protection Agency, (Unpubl.)

Environmental Protection Service, Petroleum refinery effluent regulations and guidelines.
1974   Report EPS 1-WP-74-1. Part 2. Can.Gaz., 107(21)

Farrington, J.W., Atmospheric input of hydrocarbons to the sea. Mar.Pollut.Bull., (4):96
1973

Feuerstein, D.L., Input of petroleum to the marine environment. In Proceedings of a
1973   Workshop on inputs, fates and effects of petroleum in the marine environment,
21-25 May 1973, Airlie, Virginia. Washington, D.C., National Academy of
Sciences, Vol.1:31-8

Field Studies Council, Oil Pollution Research Unit, Orielton Field Center. Annual report.
1971   London, Field Studies Council, 46 p.

Fortson, R., Jr. et al., Maritime accidental spill risk analysis: Phase 1; Methodology
1973   development and planning. Washington, D.C., U.S. Coast Guard, Office of
Research and Development, 71 p.

Germany, Federal Republic, Bundesverkehrsministerium, How to prevent oil pollution of the
1969   sea. Guidance and suggestions. Hamburg, Bundesverkehrsministerium, 32 p.

Glude, J.B., Observations on the effects of the Santa Barbara oil spill on intertidal
1969   species. Report to the Director, Bureau of Commercial Fisheries, 21 p. (Unpubl.)

Guillard, R.R.L. and J.A. Hellebust, Growth and the production of extracellular substances
1971   by two strains of Phaeocystis. J.Phyco., 7(4):330-8

Hallhagen, A., Survey of present knowledge and discussion of input of petroleum to the
1973   marine environment in Sweden. In Proceedings of a Workshop on inputs, fates
and effects of petroleum in the marine environment, 21-25 May 1973, Airlie,
Virginia. Washington, D.C., National Academy of Sciences, Vol.1:39-49

Hamelink, J.S., R.C. Waybrant and R.C. Ball, A proposal - exchange equilibria control the
1971   degree chlorinated hydrocarbons are biologically magnified in lentic environments.
Trans.Am.Fish.Soc., (100):207-14

Hartung, R. and G.W. Klinger, Concentration of DDT by sediment polluting oils. Environ.
1970   Sci.Technol., 4:407

Harvey, H.W., The chemistry and fertility of sea waters. Cambridge
1963

Hbuck, H.J., Toxicology of marine pollutants. Mar.Pollut.Bull., (1):44
1970

Hedgepeth, J.W., The impact of impact studies. Helgol.Wiss.Meeresunter., 24:436-45
1973

Hosang, W., Vorkehrungen gegen Ölgefahren in Küstengebieten. (Methods of precaution against
1968   pollution by mineral oil in coastal regions). In Föderation Europäischer
Gewässerschutz (FEG), Schutz der Meeresküsten gegen Verunreinigung; Symposium,
Hamburg, 5-7 October 1967. Zürich, FEG, p. 46-52

Hoult, D.P., Oil on the sea. New York, Plenum Press, New York, 114 p.
1969

Hufford, G.L., The biological response to oil in the marine environment: a review.
1971   Washington, D.C., U.S. Coast Guard, Oceanographic Unit, 23 p.

Hyland, J.L., Acute toxicity of No. 6 fuel oil to intertidal organisms in the lower York
1973   river, Virginia. Williamsburg, Virginia, College of William and Mary, 57 p.
M.S. Thesis

International Agency for Research on Cancer, Monographs on the evolution of carcinogenic
1973   risk of the chemical to Man. Vol.3. Certain polycyclic aromatic hydrocarbons
and heterocyclic compounds.

Kaplovsky, A.J., Tidewater's Delaware refinery. 2. Evaluation of treatment methods.
1959   Sewage Ind.Wastes, (31):432-42

Kawahara, F.K., Laboratory guide for the identification of petroleum products. Cincinnati,
1969   Ohio, FWPCA, 40 p.

Klintberg, R., Petroleumbranschens problem med oljeföroreningarna (Petroleum industry
1968   problem with oil pollution). Ecol.Res.Comm.Bull.Swed., (2):31-42

Krebs, C.T. and I. Valiela, Reduction of field populations of fiddler crabs by uptake of
1974   chlorinated hydrocarbons. Mar.Pollut.Bull., 5(9):140-3

Lawton, J.H. and S. McNeill, Pollution and world primary production. Biol.Conserv.,
1972   4(5):329-34

Lee, R.F. et al., Lipids in marine environment. Rep.CCOFI, (16):95-102
1972

Lichatowich, J.A. et al., Development of methodology and apparatus for the bioassay of oil.
1973   In American Petroleum Institute, Proceedings of a Joint conference on prevention
and control of oil spills, Washington, D.C., 13-15 March 1973. Washington, D.C.,
American Petroleum Institute, p. 659-66

Liss, P.S. and P.G. Slater, Flux of gases across the air-sea interface. Nature, (247):181
1974

McAuliffe, C.D., Partitioning of hydrocarbons between the atmosphere and natural waters.
1973   In Proceedings of a Workshop on inputs, fates and effects of petroleum in the
marine environment, 21-25 May 1973, Airlie, Virginia. Washington, D.C.,
National Academy of Sciences, Vol.1:280-90

Mimura, A., T. Kawano and R. Kodaira, Biochemical engineering; Analysis of hydrocarbon
1969   fermentation. I. Oxygen transfer in an oil-water system. Hakko Kogaku Zasshi,
47(3):229-36

Mironova, A.I., Blastomogenic effects of products resulting from the production of liquid
1959   fuel. Probl.Oncol., 5(5):21-7

Moore, S.F. and R.L. Dwyer, Effects of oil on marine organisms: a critical assessment of
1974   published data. Water Res., (8):819-27

Moore, S.F. et al., Potential biological effects of hypothetical oil discharges in the
1974   Atlantic coast and Gulf of Alaska. Cambridge, Massachusetts. Massachusetts
Institute of Technology, Sea Grant Program, Report (MITSG 74-19):126 p.

Nelson-Smith, A., The effects of oil pollution on shore life. In Proceedings of the
1970   Symposium on the effects of industry on the environment, Orielton, Pembroke,
1970, London, Field Studies Council, p. 36-42

_____, The problem of oil pollution of the sea. Adv.Mar.Biol., (8):215-306
1970

New England Interstate Water Pollution Control Commission, Technical Advisory Board,
1971   Uniform guidelines for prevention and control of oil spills and for oil
terminal and vessel handling of petroleum and petroleum products. Boston,
Massachusetts, New England Interstate Water Pollution Control Commission
Technical Advisory Board, (TR-17):20 p.

New York Ocean Science Laboratory, Proceedings of Offshore Drilling Conference. Montauk,
1971   New York, 13 September 1971

Nunuparov, S.M., Preventing oil pollution of the sea. Moscow, Transport, 167 p.
1971   (In Russian)

Oceanographic Institute of Washington, Risk analysis of the oil transportation system.
1972   A report to the 43rd Legislature State of Washington by the Oceanographic
Institute of Washington. Washington, Oceanographic Institute of Washington,
642 p.

Osterling, J.F. and L.O. Spano, Waste paper used for the cleanup of oil spills. Science,
1973   Wash.,D.C., (181):775

Pérès, J.M. and G. Bellan, Aperçu sur l'influence des pollutions sur les peuplements
1972 benthiques. In Marine pollution and sea life, M. Ruivo (Ed.). West Byfleet, Surrey, Fishing News (Books) Ltd., p. 382

Programmes Analysis Unit, The environmental and financial consequences of oil pollution
1973 from ships. Appendix 1. Discharges of oil into the oceans. Didcot, Programmes Analysis Unit, 77 p.

_____, The environmental and financial consequences of oil pollution from ships.
1973 Appendix 3. The biological effects of oil pollution of the oceans. Didcot, Programmes Analysis Unit, 276 p.

Puce, R.A., Atmosphere hydrocarbons and their relation to the marine pollution. In
1973 Proceedings of a Workshop on inputs, fates and effects of petroleum in the marine environment, 21-25 May 1973, Airlie, Virginia. Washington, D.C., National Academy of Sciences, Vol.2:416-30

Reid, G.W. et al., Evaluation of waste waters from petroleum and coal processing. Environ.
1972 Protect.Technol.Ser.W., (EPA-R2-72-001):21

Ruivo, M., (Ed.), Marine pollution and sea life. West Byfleet, Surrey, Fishing News (Books)
1972 Ltd., 624 p.

Scarratt, D.J., Impact of spills and clean-up technology on living natural resources, and
1974 resource-based industry. Tech.Rep.Fish.Res.Board Can., (428):141-58

Shabad, L.M., Studies in the U.S.S.R. on the distribution and circulation of carcinogenic
1967 hydrocarbons in the human environment and the role of their deposition in tissues in carcinogenesis. Cancer Res., 37(1):132

Sharpley, J.M., Elementary petroleum microbiology. Houston, Texas, Gulf Publishing Co.,
1966 256 p.

Sheldon, R.W., T.P.T. Evelyn and T.R. Parsons, On the occurrence and formation of small
1967 particles in sea water. Limnol.Oceanogr., (12):367-75

Shelton, R.G.J., Oil pollution and commercial fisheries in Britain. In Water pollution
1971 by oil, London, the Institute of Petroleum. P. Hepple (Ed.), 377-9

Spears, R.W., Evaluation of the effects of oil, oilfield brine and oil removing compounds.
1971 In AIME, Environmental quality conference, Washington, D.C., 7-9 June 1971. New York, American Institute of Mining, Metallurgical and Petroleum Engineers, p. 199-216

St. Amant, L.S., The petroleum industry as it affects marine and estuarine ecology.
1972 J.Petrol.Technol., (24)385-92

Strachan, A., Santa Barbara oil spill - intertidal and subtidal surveys. Rep.CCOFI,
1972 (16):122-4

Straughan, D., Biological effects of oil pollution in the Santa Barbara channel. Coastal
1971 Res.Notes, 3(5):11-7

_____, Ecological effects of the Santa Barbara oil spill. In Proceedings of the
1972 Santa Barbara oil Symposium: offshore petroleum production, an environmental inquiry, Santa Barbara, California, 16-18 December 1970. Washington, D.C., National Science Foundation, p. 173-82

Straughan, D., Biological studies of the Santa Barbara oil spill. In 1973 annual meeting,
1973   American Association of Petroleum Geologists, Santa Barbara channel revisited,
       AAPG Trip 3. sepm Seg. American Association of Petroleum Geologists, p. 4-16

Suess, M.J., Polynuclear aromatic hydrocarbon pollution of the marine environment. In
1973   Proceedings of a Workshop on inputs, fates and effects of petroleum in the marine
       environment, 21-25 May 1973, Airlie, Virginia. Washington, D.C., National
       Academy of Sciences, Vol.2:789-96

United States, Coast Guard, Marine transportation system of the Trans-Alaskan Pipeline
1971   system (TAPS). Washington, D.C., U.S. Coast Guard, 156 p.

Warner, R.E., Marine oil pollution - its biological effects and implications. In
1969   Proceedings of a Conference on pollution. Chemical Institute of Canada,
       p. 25-9

Whittle, K., P.R. Mackie and R. Hardy, Hydrocarbons in the marine ecosystem. S.Afr.J.Sci.,
1974   (70):141-4

Wilson, K.W., Toxicity testing for ranking oils and oil dispersants. In Ecological
1974   aspects of toxicity testing of oils and dispersants. Proceedings of a Workshop,
       L.R. Beynon and E.B. Cowell (Eds.). Barking, Essex, Applied Science Publishers,
       p. 11-22

Wilson, K.W., E.B. Cowell and L.R. Beynon, The toxicity testing of oils and dispersants:
1973   a European view. In American Petroleum Institute, Proceedings of a Joint
       conference on prevention and control of oil spills, Washington, D.C.,
       13-15 March 1973. Washington, D.C., American Petroleum Institute, p. 255-61

Yamamoto, S., Oil pollutants in the marine environment. J. Electrochem.Soc., 119, p. 6247
1972

Young, L., The detoxication of carbocyclic compounds. Physiol.Rev., (19):323-5
1939

_____, The oxidation of polycyclic hydrocarbons in the animal body. Biochem.Soc.
1950   Symposia, (5):27-39

Young, L.Y. and R. Mitchell, Negative chemotaxis of marine bacteria to toxic chemicals.
1973   Appl.Microbiol., 25:972-5

Zafiriou, O., M. Blumer and J. Myers, Correlation of oils and oil products by gas
1972   chromatography. Tech.Rep.Woods Hole Oceanogr.Inst., (72-55):110 p.

Zitko, V. and W.V. Carson, The characterization of petroleum oils and their determination
1970   in the aquatic environment. Tech.Rep.Fish.Res.Board Can., (217):29 p.

ZoBell, C.E., The occurrence, effects and fate of oil polluting the sea. Adv.Water Pollut.
1964   Res., 1:85-118

Zoutendyk, P., Oil pollution of the Cape Infanta coastline. Zool.Afr., (7):533-6
1972

2.10 SYMPOSIA/CONFERENCES

Ahearn, D.G. and S.P. Meyers, (Eds.), The microbial degradation of oil pollutants.
1973   Proceedings of a Workshop held at Georgia State University, Atlanta, Georgia, 4-6 December 1972. Baton Rouge, Louisiana, Louisiana State University, Center for Wetland Resources, Publ.La.State Univ., (LSU-SG-73-01):322 p.

Allen, A.A., R.S. Schlueter and L.E. Fausak, Investigation of the behavior and effects
1974   of oil utilizing a manned underwater habitat. Paper presented at the sixth annual Offshore technology conference, Houston, Texas, 6-8 May 1974, (OTC 1979):8 p.

American Petroleum Institute/Federal Water Pollution Control Administration, Proceedings
1969   of the Joint conference on prevention and control of oil spills. Washington, D.C., American Petroleum Institute, 345 p.

American Petroleum Institute, Proceedings of the Industry-government seminar on oil spill
1970   treating agents, sponsored by the American Petroleum Institute and the U.S. Department of the Interior, held in Washington, D.C., 8-9 April 1970. Am. Petrol.Inst.Comm.Air Water Conserv.Publ.,Wash.,D.C., (4055):168 p.

American Petroleum Institute/Environmental Protection Agency/U.S. Coast Guard, Proceedings
1971   of a Joint conference on prevention and control of oil spills. Washington, D.C., American Petroleum Institute, 544 p.

American Petroleum Institute, Proceedings of Joint conference on prevention and control of
1973   oil spills, sponsored by the American Petroleum Institute, U.S. Environmental Protection Agency and U.S. Coast Guard, held in Washington, D.C., 13-15 March 1973. Washington, D.C., American Petroleum Institute, 834 p.

Atlas, R.M. and R. Bartha, Fate and effects of polluting petroleum in the marine environment.
1973   Residue Rev., (49):49-85

Baaij, P.K., Fysische, chemische en biologische aspecten van olieverontreiniging op zee
1972   en de bestrijding ervan (Physical, chemical and biological aspects of oil pollution of the sea and its control). Ing.Grav., 84(7):A136-A140

Bertrand, A.R.V. et al., Prévention et lutte contre la pollution au cours des opérations
1971   de forage et de production en mer (Prevention and the fight against pollution in the course of drilling operations and production in the sea). Rev.Inst.Fr. Pétrole, (26):757-848

Beynon, L.R. and E.B. Cowell, (Eds.), Ecological aspects of toxicity testing of oils and
1974   dispersants. Proceedings of a Workshop on the toxicity testing of oils and dispersants, held at the Institute of Petroleum, London. Barking, Essex, Applied Science Publishers, 149 p.

Boesch, D.F., C.H. Hershner and J.H. Milgram, Oil spills and the marine environment.
1974   A report to the Energy Policy Project of the Ford Foundation. Cambridge, Massachusetts, Ballinger Publishing Company, 114 p.

Briant, J. and C.R. Gatellier, Prévention et lutte contre la pollution au cours des
1971   opérations de forage et de production en mer. Annexe B. Effets généraux de la pollution pétrolière (Prevention and the fight against pollution in the course of drilling operations and production in the sea. Appendix B. General effects of petroleum pollution). Rev.Inst.Fr.Pétrole, (26):839-48

Butler, M.J.A., F. Berkes and H. Powles, Biological aspects of oil pollution in the marine
1972   environment. A review. Manuscr.Rep.Mar.Sci.Cent.McGill Univ., No. (22):118 p.

_____, Biological aspects of oil pollution in the marine environment: a review.
1974   (Enlarged edition.) Manuscr.Rep.Mar.Sci.Cent., (22A):133 p.

Bybee, R.W., Compatibility of petroleum exploitation with marsh and estuarine ecosystems
1973   in Louisiana. In Proceedings of second coastal marsh and estuary management
        Symposium, Baton Rouge, Louisiana, 17-18 July 1972, R.H. Chabreck (Ed.).
        Baton Rouge, Louisiana, Louisiana State University, Division of Continuing
        Education, p. 193-209

Canada, Fisheries Research Board of Canada, Summary of physical, biological, socio-economic
1974   and other factors relevant to potential oil spills in the Passamoquoddy region
        of the Bay of Fundy. Tech.Rep.Fish.Res.Board Can., (428):231 p.

Cowell, E.B., Oil pollution of the sea. Nat.Wales, (13):12-20
1972

Crapp, G.B. and J.M. Baker, Toxicity tests for predicting the ecological effects of oils
1973   and emulsifiers on littoral communities. Rep.Oil Pollut.Res.Unit, Field Stud.
        Counc., (1972):35-9

Culbertson, P.T., Report of the Department of State on the problem and efforts to deal
1926   with it in foreign countries. In United States, Interdepartmental Committee
        on oil pollution of navigable waters. Oil pollution of navigable waters.
        Appendix 8. Washington, D.C., U.S. Government Printing Office, p. 104-19

Denike, E.E., Petroleum hydrocarbons and the sea. Washington, D.C., Office of Technical
1971   Services, Department of Ecology, 20 p.

Devanney, J.W., III, and R.J. Stewart, Bayesian analysis of oil spill statistics. Mar.
1974   Technol., (11):365-82

Dixhoorn, J. van, Detectie en bestrijding van verontreiniging van de Noordzee door olie
1972   (Detection and combatting of pollution of the North Sea by oil). Ing.Grav.,
        84(7):A130-A135

Dodd, E.N., Oils and dispersants: chemical considerations. In Ecological aspects of
1974   toxicity of oils and dispersants, Proceedings of a Workshop, R.R. Beynon and
        E.B. Cowell (Eds.). Barking, Essex, Applied Science Publishers, p. 3-9

Ellis, L.E., The results of a seminar on water pollution by oil. Aviemore, 1970. Bull.
1972   Inf.Féd.Eur.Prot.Eaux, (18):9-17

Engdahl, R., Swedish means and measures of combating oil pollution in the Baltic. In
1972   Proceedings of the international Symposium on the identification and measurement
        of environmental pollutants, Ottawa, 14-17 June 1971, p. 259-64

European Petroleum Organisations, European model code of safe practice for dealing with
1974   oil spills at sea and on shore. London, Applied Science Publishers, 97 p.

Ferguson, G.E., Combating water pollution from large oil spills. J.Am.Water Works Assoc.,
1969   (61):678-80

Finkelstein, M., The problem of oil pollution of the sea. Jerusalem, Israel Institute of
1969   Petroleum, 47 p.

Fryksmark, V., Oljeutsläpp längs västkusten (Oil spills along the west coast). Ecol.Res.
1968   Comm.Bull.Swed., (2):12-6

Gage, S.D., The control of oil pollution in Rhode Island. J.Boston Soc.Civ.Eng., (11):
1924   237-80

Galtsoff, P.S., Biological effects of oil pollution of the sea. Paper presented at the
1969   Symposium on biological effects of oil pollution of the sea, Canadian Society
         of Zoologists and American Society of Zoologists, Burlington, Vermont, U.S.A.,
         49 p.

George Washington University, Legal, economic and technical aspects of liability and
1970   financial responsibility as related to oil pollution. Washington, D.C.,
         George Washington University, National Technical Information Service,
         (PB-198-776):347 p.

Graham, F., Jr., Oil and the Maine coast: is it worth it? Augusta, Maine, Natural
1970   Resources Council of Maine, 43 p.

Great Britain, Department of Trade and Industry, The battle against oil pollution at sea.
1973   Trade Ind., Suppl., 26 July 1973

Henager, C.H., Mechanical and physical aspects of water pollution by oil spillage. J.
1972   Water Pollut.Control Fed., (44):1123-8

Herlinveaux, R.H., The impact of oil spillage from tankers moving from Alaska southward
n.d.   along the British Columbia coast. Part 1. Review of significant environmental
         factors. West Vancouver, British Columbia, Pacific Environment Institute,
         48 p. (Unpubl.Rep.)

Holmes, R.W. and F.A. DeWitt, Jr., (Eds.), Offshore petroleum production, an environmental
1972   inquiry. Proceedings of the Santa Barbara oil symposium, sponsored by the
         National Science Foundation, Division of Graduate Education in Science and the
         Marine Science Institute, University of California, Santa Barbara, held at
         Santa Barbara, California, 16-18 December 1970. Washington, D.C., National
         Science Foundation, 377 p.

Hood, D.W., W.E. Shiels and E.J. Kelley, (Eds.), Environmental studies of Port Valdez.
1973   Occas.Publ.Inst.Mar.Sci.Univ.Alaska, (3):495 p., (3A):800 p.

Horowitz, S.A., Economic principles of liability and financial responsibility for oil
1971   pollution. Prof.Pap.Va.Cent.Naval Analysis, Arlington, (56):25 p.

Ireland, Ministry for Transport and Power, Report of the Working Group on Oil Pollution
1970   to the Minister for Transport and Power. Dublin, Stationery Office,
         (Prl. 1431):76 p.

Jagger, H., The environment aspect: the risks - the weapons. Impact of offshore oil
1974   operations, A.F. Peters (Ed.). Barking, England, Applied Science Publishers,
         p. 121-44

Karrick, N.L., R.C. Clark, Jr. and R.R. Mitsouka, (Eds.), Symposium on oil pollution, the
1972   environment, and Puget Sound, held in Seattle, Washington, 23-24 February 1972.
         Seattle, Washington, National Marine Fisheries Service, Northwest Fisheries
         Center, 36 p.

MacKay, D. and W. Harrison, (Eds.), Oil and the Canadian environment. Proceedings of a
1973   Conference, sponsored by the Institute of Environmental Sciences and Engineering,
         University of Toronto, 16 May 1973. Toronto, University of Toronto, 142 p.

Maginnis, W.S., Report submitted by the United States shipping board to the Interdepart-
1926   mental Committee on Oil Pollution of Navigable Waters, December 1925. In
United States, Interdepartmental Committee on Oil Pollution of Navigable Waters.
Oil pollution of navigable waters. Appendix 7. Washington, D.C., U.S.
Government Printing Office, p. 93-103

Majori, L., F. Petronio and G. Nedoclan, Marine pollution by hydrocarbons in the northern
1972   Adriatic. In CIESMM, Journées d'études sur les pollutions marines, Problèmes
posés par les rejets directs en mer d'effluents pollués, Athens, 3-4 November 1972.
Monaco, Commission internationale pour l'exploration scientifique de la mer
Méditerranée, 1973, p. 31-2

Massachusetts Institute of Technology, Offshore Oil Task Group, The Georges Bank Petroleum
1973   Study. Summary. Cambridge, Massachusetts, Massachusetts Institute of
Technology, (MITSG-73-5):84 p.

_____, The Georges Bank Petroleum Study. Vol. 1. Impact on New England real
1973   income of hypothetical regional petroleum developments. Cambridge, Massachusetts,
Massachusetts Institute of Technology, (MITSG-73-5):284 p.

_____, The Georges Bank Petroleum Study. Vol. 2. Impact on New England environ-
1973   mental quality of hypothetical regional petroleum developments. Cambridge,
Massachusetts, Massachusetts Institute of Technology, (MITSG-73-5):311 p.

_____, Primary physical impacts of offshore petroleum developments. Cambridge,
1974   Massachusetts, Massachusetts Institute of Technology

McIntyre, A.D. and K. Whittle, (Eds.), Petroleum hydrocarbons in the marine environment.
Rapp.P.-V.Réun.Cons.Int.Explor.Mer, 171 (In press)

Mendoza, R.D., Contaminación marina (Marine pollution). IIT Technologia, (73):8-27
1971

Michanek, G., Oljeskador bekämpning och biologisk jämvikt (Oil pollution, controlling
1968   and biological balance). Ecol.Res.Comm.Bull.Swed., (2):1-5

Milgram, J.H., Technological aspects of the prevention, control, and cleanup of oil spills.
1974   In Oil spills and the marine environment. A report to the Energy Policy
Project of the Ford Foundation, D.F. Boesch, C.H. Hershner and J.H. Milgram
(Eds.). Cambridge, Massachusetts, Ballinger Publishing Company, p. 57-110

Moore, S.F., R.L. Dwyer and A.M. Katz, A preliminary assessment of the environmental
1973   vulnerability of Machias Bay, Maine to oil supertankers. Cambridge,
Massachusetts, Massachusetts Institute of Technology, (MITSG-73-6):162 p.

Moss, J.E., Pollution of the coasts of Europe by oil. Paper presented before the U.S.
1962   National Committee for the Prevention of Pollution of the Seas by Oil,
11 December 1962. Washington, D.C., American Petroleum Institute, Division
of Transportation, 10 p.

_____, Petroleum - the problem. In Impingement of man on the oceans, D.W. Hood (Ed.).
1971   New York, Wiley-Interscience, p. 381-419

Nicholls, C.P.L., Oil pollution of Southern California beaches. Shore Beach, 4:128-38
1936

Ottway, S.M., Some effects of oil pollution on the life of rocky shores. Cardiff,
1972   University of Wales, 136 p. M.S. Thesis

Overfield, J.L., Development of a non-toxic simulated oil for marine testing programs.
1972   Paper presented at the fourth annual Offshore technology conference, Houston,
       Texas, 1-3 May 1972, (OTC 1524):8 p.

Paish, H. and Associates, Ltd., The west coast oil threat in perspective - an assessment
1972   of the natural resource, social and economic impacts of marine oil transport
       in Southwest B.C. coastal waters. Vol. 1. Summary. Vol. 2. Main report.
       Vol. 3. Maps. Ottawa, Environment Canada, 338 p.

Peters, A.F., (Ed.), Impact of offshore oil operations. Proceedings of the Institute
1974   of Petroleum summer meeting - The impact of offshore oil operations, held at
       Aviemore, 20-23 May 1974. Barking, Applied Science Publishers, 205 p.

Pottier, J., Prévention et lutte contre la pollution au cours des opérations de forage
1971   et de production en mer. 7. Conclusions (Prevention and the fight against
       pollution in the course of drilling operations and production in the sea. 7.
       Conclusions). Rev.Inst.Fr.Pétrole, (26):828-30

Programmes Analysis Unit, The environmental and financial consequences of oil pollution
1973   from ships. Didcot, Programmes Analysis Unit, 80 p.

————————, The environmental and financial consequences of oil pollution from ships.
1973   Appendix 4: the economic consequences of oil pollution. Didcot, Programmes
       Analysis Unit, 93 p.

Ramadan, F.M. et al., Oil pollution studies in Egyptian territorial waters. Paper
1972   presented at the Arab Educational, Cultural and Scientific Organization
       symposium on pollution, its effects, dangers, and means of protection against
       in the Arab world, Cairo, 22-25 April 1972, 9 p.

Ramseier, R.O., Oil pollution in ice-infested waters. In Proceedings of the International
1972   symposium on the identification and measurement of environmental pollutants,
       Ottawa, 14-17 June 1971, p. 271-6

Rechnitzer, A.B. and C. Limbaugh, An oceanographic and ecological investigation of the
n.d.   area surrounding the Union Oil Company Santa Maria refinery outfall, Oso Flaco,
       California. La Jolla, California, University of California, Institute of
       Marine Resources, (IMR Ref. 56-5):66 p.

Revelle, R. et al., Ocean pollution by petroleum hydrocarbons. In Man's impact on
1971   terrestrial and oceanic ecosystems, W.H. Matthews, F.E. Smith and E.D. Goldberg
       (Eds.). Cambridge, Massachusetts, Massachusetts Institute of Technology Press,
       p. 297-318

Sage, B.L., Oil in the North Sea, an outline of geological exploration, production, and
1973   environmental aspects. Paper presented at a Symposium on oil in Scottish
       waters, Stirling, 18 April 1973, sponsored by the Scottish Marine Biological
       Association, Oban, 16 p.

St. Amant, L.S., Impacts of oil on the Gulf coast. Trans.North Am.Wildl.Nat.Resourc.Conf.,
1971   (36):206-19

Stewart, R.J. and J.W. Devanney, III, Bayesian analysis of large oil spills: an application
1973   of the Erlang sampling distribution. Cambridge, Massachusetts, Massachusetts
       Institute of Technology, Commodity Transport Laboratory, (MITCTL 73-16)

Swedish Natural Science Research Council, Oljeskador och miljövård (Oil pollution and
1968   environmental value). Ecol.Res.Comm.Bull.Swed., (2):46 p.

Swift, W.H. et al., Geographical analysis of oil spill potential associated with Alaskan
1974 oil production and transportation systems. Washington, D.C., U.S. Coast Guard, Office of Research and Development, Final Rep., (CG-D-79-74):267 p.

Tanner, R.I., Effect of long-chained polymers on the size distribution of oil-in-water
1972 emulsions. Providence, Rhode Island, Brown University, Division of Engineering, M.S. Thesis

Tarzwell, C.M., Water quality requirements for aquatic life. Contin.Educ.Ser.Univ.Mich.,
1966 No. (161):185-97

Templeton, W.L., Ecological effects of oil pollution. J.Water Pollut.Control Fed., (44):
1972 1128-34

United States, Bureau of Mines, General report of the Bureau of Mines on pollution by oil
1926 of the coast waters of the United States, 6 January 1926. In United States, Interdepartmental committee on oil pollution of navigable waters. Oil pollution of navigable waters. Appendix 3. Washington, D.C., U.S. Government Printing Office, p. 35-69

United States, Council on Environmental Quality, OCS oil and gas - an environmental
1974 assessment. Washington, D.C., U.S. Council on Environmental Quality, 388 p.

United States, Interdepartmental Committee on Oil Pollution of Navigable Waters, Oil
1926 pollution of navigable waters. Report to the Secretary of State. Washington, D.C., U.S. Government Printing Office, 119 p.

United States, National Academy of Sciences, National Research Council, Environmental
1974 Studies Board, Issues in the assessment of environmental impacts of oil and gas production on the outer continental shelf. In U.S. Council on Environmental Quality, OCS oil and gas - an environmental assessment. Washington, D.C., U.S. Council on Environmental Quality, p. NAS-1 to NAS-43

Vagners, J. and P. Mar, Oil on Puget Sound: an interdisciplinary study in systems
1972 engineering. Seattle, University of Washington Press, 629 p.

Walkup, P.C., Water pollution by oil spillage. J.Water Pollut.Control Fed., (43):1069-80
1971

Wardley Smith, J., Oil pollution - causes and cures. Eng.Lond., (211):644-8
1971

Wardley Smith, J. and R.G.J. Shelton, Disposal of oil spills from water surfaces. In
1971 Proceedings of the eighth World petroleum congress, Moscow, 13-19 June 1971. Amsterdam, Elsevier, p. 67-76

Warner, R.E., Environmental effects of oil pollution in Canada: an evaluation of problems
1969 and research needs. Ottawa, Canadian Wildlife Service, 39 p. (Unpubl.Rep.)

Wayne, T.J. and A.J. Perna, Effects, recovery, and reuse of oil from aqueous environments.
1971 In Proceedings third annual North-east regional anti-pollution conference. Reuse and cycle of wastes, University of Rhode Island, College of Engineering, 21-23 July 1970. Stanford, Connecticut, Technonic Publishing Co., p. 232-43

Wilson, J.E., The naval engineer and operational type oil spills. Naval Engs.J., (83):
1971 96-101

Wolfe, L.S., Some effects of oil spills under sea ice. Cambridge, Massachusetts,
1972   Massachusetts Institute of Technology, Department of Mechanical Engineering,
M.S. Thesis

Wright, H.W., Oil and troubled water. Paper presented at the spring meeting, Pacific
1962   coast district, Division of Production, American Petroleum Institute,
Los Angeles, California, 9-10 May 1962, (801-381):9 p.

Zahka, J.G., Oil spillage monitoring, sampling and recovery systems. In U.S. National
1974   Bureau of Standards. Marine pollution monitoring (petroleum). Proceedings of
a symposium and workshop, 13-17 May 1974. Washington, D.C., U.S. Government
Printing Office, p. 89-90

Zeldin, M., Audubon black paper no. 1: oil pollution. Audubon, (73):99-119
1971

Zinn, D.J., The impacts of oil on the east coast. In Transactions of the 36th North
1971   American wildlife and natural resources conference, Portland, Oregon,
7-10 March 1971. Washington, D.C., Wildlife Management Institute, 1971,
p. 188-206

Anon., Background papers for a Workshop on inputs, fates and effects of petroleum in the
1973   marine environment. Vol.I and II. Prepared under the aegis of Ocean Affairs
Board, National Academy of Sciences, Washington, D.C., 1973. Workshop held at
Airlie, Virginia, 21-25 May 1974: 824 p.

2.11 SELECTED BIBLIOGRAPHIES

Carter, N.M., Index and list of titles, 1965-1972 Canada. Mis.Spec.Publ.Fish.Res.Board Can.,
1972   (18):588 p.

Franklin Institute Research Laboratories, Science Information Services Department,
1974   Bibliography and abstracts of effects of refined oil products on marine
environments. Franklin Institute Research Laboratories, EPA 68-01-2205;
FIRL 80G-C3710-02

Harrison, E.A., Oil spill removal: a bibliography with abstracts. Springfield, Virginia,
1973   National Technical Information Service, (NTIS-WIN-73-002):73 p.

Holmes, R.W., Oil pollution: an index-catalog to the collection of the Oil Spill Information
1972   Center. Vol. 1. Anonymous articles. Santa Barbara, California, University of
California, University Library, 216 p.

_____, Oil pollution: an index-catalog to the collection of the Oil Spill Information
1972   Center. Vol. 2. Authors. Santa Barbara, California, University of California,
University Library, p. 217-468

_____, Oil pollution: an index catalog to the collection of the Oil Spill Information
1972   Center. Vol. 3. Subject Index, A-O. Santa Barbara, California, University of
California, University Library, 274 p.

_____, Oil pollution: an index catalog to the collection of the Oil Spill Information
1972   Center. Vol. 4. Subject Index, O-Z. Santa Barbara, California, University of
California, University Library, p. 275-689

Kondo, G. et al., Bibliography. Reports (concerning the chemical means for treatment of
1970   oil) and studies on the chemical treatment of oil released at sea. Paper
        presented at the meeting of the Joint IMCO/FAO/Unesco/WMO/WHO/IAEA Group of
        Experts on the Scientific Aspects of Marine Pollution, February 1970
        (GESAMP/28):10 p.

Marine Pollution Information Centre, Marine pollution research titles. Plymouth, Devon,
1974   Marine Biological Association of the United Kingdom, Vol. 1

Moulder, D.S. and A. Varley, (Comps.), A bibliography on marine and estuarine oil pollution.
1971   Plymouth, Devon, Laboratory of the Marine Biological Association of the United
        Kingdom, 129 p.

_____, A bibliography on marine and estuarine oil pollution. Supplement 1. Plymouth,
1975   Devon, Marine Biological Association, Marine Pollution Information Centre,
        152 p.

Nadeau, R.J. and T.H. Roush, Biological effects of oil pollution. Selected bibliography.
n.d.   2. Environ.Protect.Ag.Technol.Ser., (EPA-R2-72-055):69 p.

Nelson-Smith, A., A classified bibliography of oil pollution. In The biological effects
1968   of oil pollution on littoral communities, J.D. Carthy and D.R. Arthur (Eds.),
        165-96

Smith, M., Oil spill removal: a bibliography with abstracts. Springfield, Virginia,
1974   National Technical Information Service, (NTIS-WIN-74-037);119 p.

United States, Defense Documentation Center, Oil slicks and films; a DDC bibliography,
1972   September 1961-March 1971. Alexandria, Virginia, U.S. Defense Documentation
        Center, (DDC-TAS-71-64):65 p.

United States, Office of Water Resources Research, Oil spillage. A bibliography.
1973   Washington, D.C., Office of Water Resources Research, 2 Vols., (WRSIC-73-207):
        846 p.

Vaughan, B.E., A bibliography of environmental research: Ecosystems Department 1952-73.
1973   Richland, Washington, Battelle Pacific Northwest Laboratories

Walstad, K. and F.E. Hearth, Oil pollution in the Santa Barbara Channel: a comprehensive
1969   bibliography with particular emphasis on the oil spill of 28 January 1969.
        Santa Barbara, California, University of California, University Library, 195 p.

3. AUTHOR INDEX

This index lists, in alphabetical order, the authors of section 2 "Additional related literature".

Figures in the first column refer to the relevant sub-sections. Figures in the second column are the years of publication. When the latter is put in brackets this indicates that the author's name is cited as co-author.

Author	Sub-section	Years
Abbott, B.C.	2.2.6	1969
Abbott, M.B.	2.8	1961 1967
Ackman, R.G.	2.1.2	1973
	2.1.3	1973
	2.2.3	(1964)
Adlard, E.R.	2.1.3	1972
	2.2.6	1973
Aerojet Electrosystems Co.	2.1.2	1973
Ages, A.B.	2.2.6	1971 1972
Agosti, J.	2.4	1972
Agosti, T.	2.4	(1972)
Ahearn, D.G.	2.4	1971 1973 1974 (1974)
	2.10	1973
Ahlfeld, T.E.	2.4	1973 (1973)
Ahmadjian, M.	2.1.2	1973 (1974)
	2.1.4	(1974)
Aivazova, L.E.	2.2.1	(1974)
Aliev, A.D.	2.2.4	(1973)
Allen, A.A.	2.6	(1972)
	2.10	1974
Allen, D.R.	2.6	1972
Alofs, D.J.	2.8	1971 1972
Alpine Geophys. Assoc. Inc.	2.2.6	1971
Alzieu, C.	2.3	1972
	2.9	1972
American Petrol. Institute	2.3	1971
	2.9	1974
	2.10	1969 1970 1971 1973
Ampaya, J.P.	2.6	(1973)
Andelman, J.B.	2.2.5	1970
Anderes, E.A.	2.4	1973
Anderson, E.K.	2.2.6	1969
Andreeva, S.U.	2.2.4	(1973)
Andrews, A.R.	2.2.4	1974
	2.4	(1975)
Arpino, R.	2.9	1973
Arro, I.K.H.	2.2.5	(1959)
Arthur, D.R.	2.2.4	(1968)
Arvesen, J.C.	2.1.2	(1971)(1972)(1973)
Ashby, E.	2.2.6	1972
Atema, J.	2.2.2	1973 1974
Atlas, R.M	2.4	1972 (1972) 1973
	2.8	–
	2.10	1973
Aubert, J.	2.9	(1972)
Aubert, M.	2.9	1971 1972
Auger, C.	2.2.5	(1970)
	2.4	(1970)
Aukland, J.C.	2.1.2	1971
Austin, H.	2.4	(–)
Avolizi, R.J.	2.2.2	1974
Axelsson, S.	2.1.2	1972 1973
Ayappan, N.S.	2.7	1972
Ayers, R.C., Jr.	2.8	1974
Baaij, P.K.	2.10	1972
Badger, G.M.	2.2.5	1948
	2.9	1963
Baier, R.E.	2.2.7	1972
Baissac, P.deB.	2.2.2	(1974)
Baker, J.M.	2.2.1	1970
	2.2.4	(1972)
	2.3	(1972)
	2.9	1974
	2.10	(1973)
Baldini, I.	2.3	1974
Baldwin, A.H.	2.6	1974
Baldwin, M.F.	2.2.6	1970
Ball, R.C.	2.9	(1971)
Ballinger, D.G.	2.1.4	(1970)
Barber, F.G.	2.8	(1972)
Barbier, M.	2.5	1973
	2.6	1973
Bardach, J.E.	2.3	1965
Barger, W.R.	2.2.7	1974
Barnett, C.J.	2.2.4	(1973) 1975
Bartha, R.	2.4	1972 1973
	2.10	(1973)
Beak, T.W.	2.1.2	–
Beam, H.W.	2.4	1973
Bean, R.M.	2.2.4	1974
Beastall, S.	2.4	(1971)
Beatty, D.	2.3	(1973)
Bechtel, T.J.	2.2.6	(1971)
Becker, C.D.	2.3	1973
	2.9	1973
Bell, W.	2.2.1	1972
Bellamy, D.L.	2.2.6	1967
Bellan, G.L.	2.3	1971 1972 (1972) 1974
	2.9	(1972)
Bellanca, S.C.	2.1.4	(1969)
Bellegay, R.D.	2.7	(1974)
Bender, J.E.	2.2.4	1974
Bennett, H.J.	2.2.3	(1972)

Bens, E.M.	2.4	(1972)(1973)	
Berenblum, I.	2.2.5	1943	
Berglund, H.	2.3	1969	
Berkes, F.	2.10	(1972)(1974)	
Berry, W.L.	2.2.6	1972	
Bertrand, A.R.V.	2.6	1971	
	2.10	1971	
Bestougeff, M.A.	2.9	1967	
Betancourt, O.J.	2.2.7	1973	
Beynon, L.R.	2.3	1971	
	2.6	(1973)	
	2.9	(1973)	
	2.10	1974	
Bien, W.	2.8	1973	
Billing, U.	2.2.1	(1974)	
Binet, D.	2.2.6	1974	
Binet, L.	2.9	1963	
Bingham, E.	2.2.5	1965	
Bittel, R.	2.1.2	1971	
Blacklaw, J.R.	2.3	(1970)	
Blackman, R.A.A.	2.2.3	1973	1974
	2.2.6	1973	
	2.3	1974	
Blanton, W.G.	2.2.3	1973	
Blaylock, J.W.	2.1.3	1973	
Bleakley, R.J.	2.3	1974	
Blokker, P.C.	2.9	1971	
Blumer, M.	2.1.3	(1972)	
	2.4	1961	
	2.5	(1967)(1973)	
	2.8	1970	
	2.9	(1972)	
	2.10	(1972)	
Boaden, P.J.S.	2.3	(1974)	
Bock, K.J.	2.3	1972	
Boehm, P.D.	2.2.7	1973	
Boesch, D.F.	2.9	1974	
	2.10	1974	
Bonz, P.D.	2.9	1973	
Borneff, J.	2.2.5	1963	
	2.9	1968	
Boswell, J.L.	2.2.7	1950	
Bourcart, J.	2.2.6	1965	
Bourne, W.R.P.	2.2.4	1968	
	2.2.6	1972	
Boyce, F.	2.2.6	1969	
Boyd, R.N.	2.1.5	(1974)	
Boylan, D.B.	2.2.4	(1973)	
Boyland, E.	2.4	1950	
	2.9	1964	
Braaten, B.	2.3	1972	
Bradley, R.G.	2.2.7	(1971)	
Briant, J.	2.3	1971	
	2.10	1971	
Briggs, W.	2.8	(1974)	
Brisou, J.	2.4	1969	
Brisou, L.	2.4	(1967)	
Bristow, M.	2.1.2	(1971)	
Broberg, S.	2.2.3	(1973)	

Brockis, G.J.	2.9	1972		
Brooks, J.M.	2.5	1973 (1974)		
Brown, A.C.	2.2.2	1974		
Brown, C.W.	2.1.2	(1973) 1974		
	2.1.4	(1973) 1974		
Brown, D.E.	2.1.2	1972		
Brown, D.H.	2.2.1	1972		
	2.2.6	1974		
	2.3	1973		
Brown, G.A.	2.8	(1973)		
Brown, R.A.	2.1.2	1973		
	2.5	(1973) 1974		
Brown, R.L.	2.1.5	(1973)		
Brown, S.O.	2.2.7	1951		
Bruce, H.E.	2.1.1	1974		
Bruce, W.W.	2.2.6	(1969)		
Brummage, K.G.	2.9	1973		
Brundall, L.	2.2.6	1972		
Brunies, A.	2.2.2	1971		
Bruns, F.-J.	2.3	(1970)		
Bryan, D.E.	2.1.4	1970 (1971)		
Bryan, G.W.	2.3	1969		
Buckmaster, J.	2.8	1973		
Buckmeier, F.J.	2.1.2	(1971)		
Bugbee, S.L.	2.2.4	1973		
Burman, I.	2.7	1973		
Burns, J.J.	2.9	(1972)		
Burns, K.A.	2.2.6	1971		
Burwood, R.	2.2.7	1974		
Bury, R.B.	2.2.4	1972		
Butler, A.R.	2.9	1972		
Butler, J.N.	2.7	1973 (1973) 1974 (1974)		
Butler, M.J.A.	2.10	1972	1974	
Button, D.K.	2.4	1971	1973	
Bybee, R.W.	2.10	1973		
Byrom, J.A.	2.4	1971		
Cahnmann, H.J.	2.1.2	1955		
Calomiris, J.J.	2.4	(-)		
Calvin, M.	2.2.5	(1971)		
Campbell, C.E.	2.1.2	1972		
Campbell, R.	2.2.6	(1974)		
Campbell, W.J.	2.2.6	1973		
	2.8	1973		
Canada, Ministry of Transport	2.9	1970		
Canada, Fish.Res. Board Can.	2.10	1974		
Canevari, G.P.	2.3	1970	1973	
Cardwell, R.D.	2.2.3	1973		
	2.2.4	1973		
Carlberg, S.R.	2.1.4	1972		
	2.1.5	1972		
	2.2.6	1973		
	2.5	1970	1973	
Carlson, G.P.	2.9	1972		
Carnes, D.	2.4	1971		
Carruthers, W.	2.8	1961		

Carson, W.G.	2.3	(1972)
Carson, W.V.	2.1.5	(1970)
	2.2.6	(1970)
	2.9	(1970)
Carter, N.M.	2.11	1972
Carthy, J.D.	2.2.4	1968
Carvacho, A.B.	2.2.4	1971
Castela, A.	2.6	1971
Catoe, C.E.	2.1.2	1970 1972 1973
Cavelieri, E.	2.2.5	1971
Cawlfield, D.E.	2.1.4	(1974)
Cerniglia, C.E.	2.4	(1973)
Chamut, P.S.	2.1.5	(1974)
	2.2.6	(1974)
	2.5	(1974)
Chandler, P.B.	2.1.2	1971
Chang, W.J.	2.1.1	1974
Chaplin, A.E.	2.3	1971
Charter, D.B.	2.9	1973
Charters, A.C.	2.2.6	(1971)
Chet, I.	2.2.4	1971 (1971)
Chia, F.S.	2.2.4	1973
Chiavari, G.	2.1.3	(1971)
Cicatelli, M.S.	2.2.5	1966
Clar, E.	2.9	–
Clark, R.B.	2.2.6	1971
Clark, R.C.,Jr.	2.1.5	1974
	2.5	1967
	2.9	1973
	2.10	(1972)
Clayson, D.B.	2.2.5	1962
Clements, R.L.	2.6	(1972)
Climberg, R.	2.2.6	1973
Coakley, W.A.	2.1.4	1973
Cobet, A.B.	2.2.7	(1972)(1973)
	2.4	1973
Cofone, L.,Jr.	2.4	(1973)
Cohen, Y.	2.2.4	1973 (1974)
Colby, L.	2.2.6	(1973)
Cole, R.D.	2.1.3	1971
Colwell, R.R.	2.1.5	(1973)
	2.2.1	(1968)
	2.2.6	1972 (1973)
	2.4	(1973)(1974)(1975) (1976)(-)
Connell, D.W.	2.3	(1972)
Connell, J.H.	2.2.6	1973
Conney, A.H.	2.2.5	1958
	2.9	1972
Connor, P.M.	2.2.2	1972
	2.3	(1968)
Conomos, T.J.	2.8	1974
Conway, W.D.	2.2.5	(1964)
Cook, F.D.	2.4	(1972)
Coomber, R.S.	2.1.4	1971/2
Cooney, J.J.	2.4	1973
Copeland, B.J.	2.2.6	1971
Corkery, A.	2.1.3	(1960)

Corkett, C.J.	2.2.4	(1974)
Corner, E.D.S.	2.2.2	1973
Cossa, D.	2.3	(1973)
Cousens, J.D.	2.2.7	1973
Cowell, E.B.	2.2.4	1972
	2.2.6	1969
	2.2.7	1971
	2.3	1972
	2.6	(1974)
	2.9	1970 1971 (1973) 1974
	2.10	1972 (1974)
Cox, G.V.	2.2.6	1972
Cox, W.G.	2.2.6	(1972)
Cracy, H.B.	2.3	1969
Craigie, J.S.	2.2.6	1970
Cram, S.P.	2.1.1	(1974)
Crapp, G.B.	2.2.4	(1972)
	2.2.7	1971
	2.3	1971 (1972)
	2.9	1971 (1974)
	2.10	1973
Creaser, L.F.	2.1.3	(1972)
Cretney, W.J.	2.1.4	1974
	2.2.6	(1973)(1974)
	2.7	(1974)
Croker, R.A.	2.2.6	1969
Cromwell, N.H.	2.2.5	1965
Crossley Surveys Inc.	2.9	1970
Crow, S.A.	2.4	1974
Cubit J.	2.2.4	1972
Cugurra, F.	2.3	(1974)
Culbertson, P.T.	2.10	1926
Culkin, F.	2.5	(1974)
Culley, D.D.,Jr.	2.2.2	(1972)
Cundell, A.M.	2.2.6	1973
	2.4	1973 1974
	2.6	1972
Currier, H.B.	2.9	1954
Curtis, D.L.	2.1.1	1974
Curtis, M.	2.7	(1974)
Daniel, W.H.	2.2.7	(1974)
Daudel, P.	2.2.5	1966 (1966)
Davavin, I.A.	2.2.2	1973
	2.2.7	1974
Davenport, J.	2.2.2	1973
	2.2.4	1973
Davis, C.C.	2.2.7	1972
Davis, J.B.	2.4	1968
	2.9	1967
Davis, W.B.	2.4	(1970)
Day, J.H.	2.2.6	1971
Dean, R.G.	2.8	(1972)
DeCoursey, P.J.	2.2.1	1974
Dee, P.A.	2.6	(1973)
Demayo, A.	2.9	1970

Denike, E.E.	2.10	1971		Fabian, B.	2.2.5	1968	
Devanney III, J.W.	2.8	(1974)		Falk, H.D.	2.2.5	1963	
	2.10	(1973)	1974	Falk, M.R.	2.2.3	1973	
DeWitt, F.A.,Jr.	2.2.6	(1970)		Fannelop, T.K.	2.8	(1972)	
	2.10	(1972)		Fantasia, J.F.	2.1.2	1971 1972 1973	
Diasamidze	2.2.2	(1973)		Farrington, J.W.	2.1.1	1974	
	2.4	(–)			2.1.5	1973 1974 (1974)	
Dickman, G.H.	2.2.6	1971				1975	
Dicks, B.	2.2.1	1973			2.5	1972 1973	
	2.2.2	1973			2.9	1973	
	2.2.3	1973		Fasoli, U.	2.4	1973	
Diehl, H.	2.2.5	(1966)		Fauchald, K.	2.2.4	1972	
Dikun, P.R.	2.2.5	(1966)		Fausak, L.E.	2.10	(1974)	
DiSalvo, L.H.	2.1.3	(1974)		Feldman, M.H.	2.1.4	1974	
	2.1.5	1973			2.2.6	1970	
	2.5	1973		Ferguson, G.E.	2.10	1969	
Dixhoorn, J.van	2.10	1972		Feuerstein, D.L.	2.9	1973	
Dodd, E.N.	2.10	1974		Field Studies Council	2.9	1971	
Done, J.N.	2.1.3	1970		Filby, R.H.	2.1.4	1971	
Donova, D.J.	2.2.7	(1974)		Findlay, E.R.	2.2.3	1957	
Dorrings, P.	2.2.6	1973		Finkelstein, M.	2.10	1969	
Dorrler, J.S.	2.3	(1973)		Finley, J.S.	2.1.5	(1974)	
Douglas, A.G.	2.8	(1961)		Finnerty, W.R.	2.4	(1971) 1973	
Drury, D.E.	2.2.3	(1969)		Fisher, R.	2.2.5	(1963)	
Dubois, L.	2.1.3	1960		Floodgate, G.D.	2.2.4	(1974)	
Duce, R.A.	2.2.6	1972			2.4	1971 1973	
Duncan, T.K.	2.2.4	(1974)		Fogel, S.	2.2.4	1971	
Dutka, D.L.	2.4	(1973)		Foret, J.-P.	2.3	(1971)(1972)	
Dwivedi, S.N.	2.7	1974		Foret-Montardo, P.	2.3	1971 1972	
Dwyer, R.L.	2.9	(1974)		Fortson, R.,Jr.	2.9	1973	
	2.10	(1973)		Fortune, P.R.	2.1.2	(1972)	
				Fossato, V.U.	2.2.2	1975	
Ebeling, A.W.	2.2.3	1972			2.5	1975	
	2.2.6	(1969)		Foster, M.	2.2.6	1971	
Eckardt, R.E.	2.2.3	1959		Fournier, G.	2.1.2	(1974)	
	2.2.5	1959		Franklin Inst. Res. Lab.	2.11	1974	
Edgerton, A.T.	2.1.2	1971 1973 (1973)		Freegarde, M.	2.1.1	1970	
Edison Water Quality Lab.	2.3	1971			2.2.7	(1969) 1970 1971	
Edwards, J.H.	2.6	1972		Freestone, F.	2.1.3	(1973)	
Ehrhardt, J.P.	2.2.5	1972		Friede, J.	2.4	1972 (1973)	
	2.2.6	(1968)		Frost, L.G.	2.2.6	1974	
Eisler, R.	2.2.2	1973		Fryksmark, V.	2.10	1968	
	2.2.4	1974		Fujiya, M.	2.3	(1965)	
Eizen, E.G.	2.2.5	1959		Fujiyama, T.	2.2.4	(1957)	
Ellis, L.E.	2.10	1972		Furon, R.	2.2.6	1968	
Engdahl, R.	2.10	1972		Fusey, P.	2.4	1973	
Environmental Prot. Agency	2.9	1974 1975		Gage, S.D.	2.10	1924	
Estes, J.E.	2.1.2	1972		Gaines, T.H.	2.2.6	1970	
	2.8	1973		Galtsoff, P.S.	2.2.1	1935	
European Petrol. Organisations	2.10	1974			2.2.2	1935	
Evans, W.G.	2.2.4	1970			2.10	1969	
				Ganning, B.	2.2.1	1974	
Evelyn, T.P.T.	2.9	(1967)		Gantcheff, G.S.	2.2.6	(1973)	
				Garrett, W.D.	2.2.7	(1974)	

Gatellier, C.R.	2.3	(1971) 1973		Guillard, R.R.L.	2.9	1971	
	2.4	1971		Guinard, N.W.	2.1.2	1971	
	2.10	(1971)		Guinn, V.P.	2.1.4	1969	1971
Gauthier, M.	2.9	(1972)		Guire, P.E.	2.4	1973	
Gelboin, H.V.	2.2.5	1969 1971		Gularte, R.C.	2.1.2	1970	
George Washington University	2.10	1970		Gunkel, W.	2.1.5	1967	
					2.2.6	1973	
Germany, F.R.	2.9	1969			2.3	1974	
Gholson, R.K.	2.4	(1973)			2.4	1973	
Giaccio, M.	2.2.5	1971		Gunter, G.	2.7	(1955)	
Gibbs, C.F.	2.2.7	1971					
	2.4	(1971)(1973) 1975		Haddow, A.	2.2.5	1958	
Gibson, D.T.	2.2.4	1975		Häkkilä, K.	2.2.3	1973	
	2.4	1968 1972 1973		Haila, Y.	2.2.6	1970	
Gilfillan, E.S.	2.2.2	1975		Hall, W.F.	2.1.2	(1973)	
	2.2.7	(1973)		Hallhagen, A.	2.9	1973	
Gilmartin, W.G.	2.2.4	(1971)		Halstead, B.W.	2.2.7	1972	
Glaeser, J.L.	2.2.7	1972		Hamelink, J.S.	2.9	1971	
	2.8	(1974)		Hann, R.W.,Jr.	2.2.6	1974	
Glude, J.B.	2.9	1969		Hard, T.M.	2.1.2	(1971)(1972)	
Goering, J.J.	2.2.1	(1973)		Hardy, R.	2.2.3	(1972)	
	2.2.7	(1973)			2.5	(1974)	
Golomb, B.	2.1.2	(1970)			2.9	(1974)	
Goolsby, A.D.	2.1.2	1971		Harger, J.R.E.	2.2.2	1972	
Gordon, D.C.,Jr.	2.1.2	(1973)		Harington, J.S.	2.5	1962	
	2.1.4	1974		Harley, M.L.	2.2.6	(1973)	
	2.1.5	(1973) 1974		Harrison, E.A.	2.11	1973	
	2.2.1	1972 (1974)		Harrison, W.	2.4	(1970)	
	2.2.6	1971 (1971) 1974			2.8	1973	
	2.2.7	1973			2.10	(1973)	
	2.5	1971 (1971) 1974		Hartung, R.	2.9	1970	
Gortalum, G.M.	2.5	1958		Hartwell, J.L.	2.2.5	1951	
Graf, W.	2.2.5	1966		Hartwell, K.L.	2.2.5	(1969)	
Graham, F.,Jr.	2.10	1970		Harvey, A.C.	2.2.7	(1973)	
Graham, J.W.	2.8	(1972)		Harvey, H.W.	2.9	1963	
Grahn, O.	2.3	(1974)		Harvey, L.G.	2.2.7	(1971)	
Gram, H.G.	2.1.2	1974		Hasham, S.R.	2.4	1971	
Granmo, A.	2.3	(1972)		Hatchard, C.G.	2.2.7	(1970)(1971)	
Grassle, J.F.	2.1.5	1973		Hayashi, T.	2.8	(1967)	
Gray, J.S.	2.2.4	1971		Hayre, H.S.	2.1.2	1971	
Great Britain, Dept. of Trade and Industry	2.10	1973		Hazel, C.R.	2.3	1940	
				Hbuck, H.J.	2.9	1970	
				Hearst, P.J.	2.2.7	1974	
Great Britain, Medical Res. Council	2.2.4	1968		Hearth, F.E.	2.2.6	(1969)	
	2.2.5	1968			2.11	(1969)	
				Hedgepeth, J.W.	2.9	1973	
Great Britain, Ministry of Defence, Navy	2.2.7	1971 1973		Heidelberger, C.	2.2.5	1964 (1971)	
				Heino, A.	2.2.6	1972	
				Heintz, C.E.	2.4	(1973)	
Green, D.R.	2.2.6	(1973) 1974		Hellebust, J.A.	2.9	(1971)	
	2.7	(1974)		Hellmann, H.	2.1.4	1973	
Greenham, M.S.	2.2.6	1973			2.1.5	1971	
Gribbon, E.	2.3	(1973)			2.2.7	1972	
Griffiths, M.	2.2.1	(1972)			2.3	1970 1972	
Grover, P.L.	2.2.5	1971		Henager, C.H.	2.10	1972	
Gruenfeld, M.	2.1.4	1973		Herfjord, H.J.	2.2.7	1972	
Guard, H.E.	2.1.3	(1974)		Herlinveaux, R.H.	2.2.6	1972	
	2.2.7	1972 1973			2.10	-	
	2.4	(1973)		Heros, M.	2.2.6	(1962)	
				Hershner, C.H.	2.9	(1974)	
					2.10	(1974)	

Hertz, H.S.	2.1.5	1974			Jardas, I.	2.2.6	1972	
Hickman, G.D.	2.1.2	(1973)				2.5	1972	
Hildebrand, H.H.	2.7	1955			Jeffery, L.M.	2.7	1973	1974
Hinds, P.	2.1.2	(1971)			Jeffery, P.G.	2.1.1	1973	
HMSO	2.1.5	1971				2.2.7	1973	
	2.2.6	1964				2.3	1971 1973 (1974)	
Hogle, R.D.	2.2.6	1971			Jeffries, H.P.	2.2.2	1972	
Holmes, R.W.	2.2.6	1970			Jeltes, R.	2.1.3	1969	
	2.10	1972			Jiang, L.	2.2.7	(1974)	
	2.11	1972			Jobson, A.	2.4	1972	
Holt, A.	2.3	(1965)			Johannes, R.E.	2.6	1970	
Honda, S.	2.3	1972			Johnson, B.D.	2.2.7	1973	
Hood, D.W.	2.2.1	(1973)			Johnson, G.L.	2.1.2	1971	
	2.2.7	(1973)			Johnson, R.A.	2.8	(1972)(1973)	
	2.10	1973			Johnson, T.C.	2.6	1971	
Hornstein, B.	2.2.7	1973						
Horowitz, S.A.	2.10	1971			Kaim-Malka, R.A.	2.3	1972	
Horton, A.W.	2.2.5	1963			Kallio, R.E.	2.4	1969	
	2.8	1963			Kaneko, T.	2.2.6	(1972)	
Horton, W.	2.2.5	(1965)			Kanter, R.	2.2.2	1974	
Horvath, R.	2.1.2	1970 1971 1974			Kaplovsky, A.J.	2.9	1959	
Hosang, W.	2.9	1968			Karinen, J.F.	2.2.2	1974	
Hoult, D.P.	2.2.7	(1972)				2.2.4	-	
	2.9	1969			Karrick, N.L.	2.10	1972	
Howard, H.W.,Jr.	2.1.4	1972			Kasymov, A.G.	2.2.4	1973	
Howe, C.	2.2.2	1971			Katayama, T.	2.2.4	1957	
	2.2.4	1971			Kator, H.	2.4	1973	
Huberman, E.	2.2.5	(1971)			Katz, A.M.	2.10	(1973)	
Hueper, W.C.	2.2.5	1961 1962 1964			Katz, M.	2.2.4	1971	
Hufford, G.L.	2.7	(1974)			Kauss, P.R.	2.2.1	1972	
	2.9	1971			Kawahara, F.K.	2.1.3	1971	
Hughes, D.R.	2.1.4	1972				2.1.4	1970	
Hunt, P.G.	2.4	1972				2.9	1969	
Hunter, L.	2.1.3	1974			Kawano, T.	2.9	(1969)	
Hutchinson, T.C.	2.2.1	(1972)			Keizer, P.D.	2.1.2	1973	
Hwang, L.-S.	2.8	(1974)				2.1.4	(1974)	
Hyland, J.L.	2.2.4	(1974)				2.1.5	1973 (1974)	
	2.9	1973				2.2.6	(1974)	
						2.2.7	(1973)	
						2.5	(1974)	
Ingram, G.E.	2.6	1973			Kelley, E.J.	2.10	(1973)	
Ingrao, H.C.	2.1.2	(1971) 1972 (1973)			Kennedy, J.M.	2.2.7	1971	
Inoue, N.	2.2.3	(1973)			Kerminen, S.	2.3	1971	
International Ag. for Research on Cancer	2.9	1973			Ketchel, R.J.	2.1.2	1973	
					Khandekar, M.L.	2.8	(1973)(1974)	
Ireland, Ministry for Transport and Power	2.10	1970			Kikuchi, R.	2.1.5	1973	
					Kilvington, C.C.	2.2.2	(1973)	
Irons, D.E.	2.2.6	1973			Kim, H.H.	2.1.2	1973 (1974)	
Ives, H.	2.4	1971			Kinney, P.J.	2.2.6	1970	
						2.5	1973	
Jackson, D.F.	2.1.1	(1973)			Kissill, G.W.	2.2.4	(1974)	
Jackson, P.	2.1.2	1973			Kittredge, J.S.	2.2.2	1971 (1973)	
Jacobs, W.A.	2.2.7	1926			Kleen, A.E.	2.2.7	1974	
Jacobson, S.M.	2.2.4	1973			Klemas, V.	2.1.2	1972	
Jadamec, J.R.	2.1.1	(1974)				2.8	(1971)	
	2.1.4	1974			Klinger, G.W.	2.9	(1970)	
Jagger, H.	2.10	1974			Klintberg, R.	2.9	1968	
Jahns, H.O.	2.8	(1974)			Kloth, T.C.	2.2.3	1972	
					Knettig, E.	2.3	(1972)	
					Knight, W.	2.1.2	(1974)	

Knorr, M.	2.4	1968			Levell, D.	2.2.4	1973	
Kodaira, R.	2.9	(1969)			Levy, E.M.	2.1.3	1973	
Koe, B.K.	2.2.5	(1952)				2.1.4	1972	
Koehring, V.	2.2.1	(1935)				2.2.6	1971 1972 1973	
Kolpack, R.L.	2.4	(1973)				2.5	1971 1973	
Kondo, G.	2.2.7	1972				2.7	1971	
	2.3	1972			Lichatowich,			
	2.11	1970			J.A.	2.1.5	1973	
Kotogiannis, J.E.	2.2.4	1973 (1975)				2.2.3	1972	
Koronelli, T.V.	2.4	(1973)				2.9	1973	
Kosama, K.	2.2.3	(1967)			Limbaugh, C.	2.10	(-)	
Kotin, P.	2.2.3	(1957)			Lindén, O.	2.3	1974	
	2.2.5	(1963)			Lirette, E.F.	2.1.2	(1974)	
Kovaleva, G.I.	2.2.7	(1972)			Liss, P.S.	2.9	1974	
Kramer, A.	2.2.7	1973			Lister, N.J.	2.3	(1973)	
Krasil'nikov,					Little, P.A.	2.2.7	1934	
N.A.	2.4	1973			Liu, D.L.S.	2.4	1973	
Krebs, C.T.	2.9	1974			Lockhead, M.S.	2.2.5	(1950)	
Kriebel, A.R.	2.1.2	1973			Logan, J.W.M.	2.3	(1973)	
Krishen, K.	2.1.2	1973			Long, E.R.	2.2.6	1972	
Krishnaswami,					Loof, S.	2.1.4	1974	
S.K.	2.2.3	1969 1973			Loucks, R.H.	2.2.6	1971	
Kristoffersson,					Lu, B.C.-Y.	2.2.7	(1973)	
R.	2.2.3	1973			Lukens, H.R.	2.1.4	1971 (1971) 1972	
Krivitsky, I.P.	2.4	(1968)			Lynch, P.F.	2.1.2	(1974)	
Kühnhold, W.W.	2.2.3	1974				2.1.4	1973 (1974)	
Kupchanko, E.E.	2.2.3	(1969)(1973)			Lysyj, I.	2.2.7	1974	
Kutt, E.C.	2.3	1974						
Kvasnikow, E.E.	2.4	1968			Macdonald, D.	2.7	(1974)	
					Maciejowska, M.	2.4	1973	
Lacaze, J.-C.	2.1.5	1971			MacIntyre, W.G.	2.4	(1970)	
	2.2.1	1969 1973 1974			MacKay, D.	2.2.7	1973	
	2.3	1971 1973				2.10	1973	
Lacourly, G.	2.1.2	(1971)			MacKay, G.D.M.	2.2.7	1970 1973	
Lallier, R.	2.3	1973			Mackie, A.M.	2.3	1970	
Lalou, C.	2.2.6	1962 1963			Mackie, P.R.	2.2.3	1972 (1973)	
Lami, R.	2.2.1	(1964)				2.5	(1974)	
Landes, K.K.	2.6	1973				2.9	(1974)	
Lange, E.	2.2.5	1973			Maggi, P.	2.3	1972 1973	
Lange, R.	2.3	(1972)			Maginnis, W.S.	2.10	1926	
Lanskaya, L.A.	2.2.1	(1966)(1968)(1969)			Mahmoud, T.A.	2.4	1970	
	2.4	(1968)			Majori, L.	2.1.5	1973	
LaRoche, G.	2.1.5	1973				2.5	1973	
LaRock, P.A.	2.4	1973 (1973)				2.10	1972	
Laseter, J.L.	2.5	(1974)			Makolaj, P.G.	2.1.5	1972	
Latiff, S.A.	2.3	1971			Mallet, L.	2.2.1	1964 (1969)	
Lawrence, D.J.	2.2.6	(1971)				2.2.5	1960 1964 1967	
Lawrence, M.J.	2.2.3	(1973)					(1970)	
Lawton, J.H.	2.9	1972				2.2.6	1961 1962 1963	
Lebed, A.A.	2.4	(1972)					1964 (1965) 1967	
Ledet, E.J.	2.5	1974				2.4	1967 1969	
Lee, R.F.	2.9	1972				2.9	(1963)	
Leenhardt, H.	2.2.2	1925			Mankki, J.	2.2.3	1974	
Leinonen, P.J.	2.2.7	1973			Mann, H.	2.3	(1972)	
Leon, B.	2.2.2	(1974)			Mann, S.	2.2.6	(1973)	
Leon, M.	2.4	(1969)			Mar, P.	2.10	(1972)	
Lester, T.E.	2.6	1973			Marchal, E.	2.2.6	(1974)	
Le Theule, M.T.	2.2.6	(1961)			Marcinowski, H.J.	2.9	(1971)	
LeVantine, A.D.	2.1.1	(1974)			Marine Pollution			
					Info. Centre	2.11	1974	

Mark, H.B.,Jr.	2.1.4	1972			Michalik, P.S.	2.2.6	1971 (1971)	
	2.2.6	1971				2.5	1971 (1971)	
					Michanek, G.	2.10	1968	
Markham, W.E.	2.2.6	1970			Michel, P.	2.3	1972	
Marmelstein, A.D.	2.1.2	(1973)			Miesel, M.N.	2.4	(1964)	
Marshall, A.G.	2.6	1974			Miget, R.	2.1.1	1974	
Martin, D.F.	2.3	(1974)			Miget, R.J.	2.4	1971	
Martin, M.H.	2.2.6	(1974)			Mikolaj, P.G.	2.6	1972	1973
Martin, S.	2.2.6	(1973)			Milgram, J.H.	2.2.7	1971	
	2.8	(1973)				2.10	1974	
Massachusetts, Div. of Water Pollut. Cont.	2.3	1970			Millard, J.P.	2.1.2	1971	1972 1973
					Miller, E.C.	2.2.5	(1958)	
					Miller, J.A.	2.2.5	(1958)	
Massachusetts, Inst. of Tech.	2.10	1973	1974		Miller, R.T.	2.6	1972	
					Mills, E.R.	2.2.2	1972	
Masson, M.	2.6	(1971)			Mimura, A.	2.9	1969	
Matthews, J.E.	2.2.3	1973			Mircea, V.	2.4	1972	
Matthews, P.H.D.	2.1.3	(1972)			Mironov, O.G.	2.2.1	1966	1968 1969
Mattson, J.S.	2.1.4	1970	1971			2.2.4	1973	
Maughan, P.M.	2.1.2	(1973)				2.4	1968	1972
Mayo, D.W.	2.2.7	1974			Mironova, A.I.	2.9	1959	
Mazmanidi, N.D.	2.2.2	1973			Mitchell, B.L.	2.1.3	1973	
	2.2.7	1972			Mitchell, D.M.	2.2.3	1972	
	2.4	1972			Mitchell, P.H.	2.2.2	1912	
McAuliffe, C.D.	2.9	1973			Mitchell, R.	2.2.1	(1972)	
McCarthy, L.T., Jr.	2.3	1973				2.2.4	(1971)	
						2.4	(1973)	
McCauley, R.N.	2.2.7	1966				2.9	(1973)	
McClean, A.Y.	2.2.7	1972	1974		Mitsouka, R.R.	2.10	(1972)	
McColl, W.D.	2.1.2	(1972)			Miyake, Y.	2.2.3	1967	
McCormack, K.	2.1.2	1974			Moffatt, J.D.	2.1.3	(1973)	
McGill, A.S.	2.2.3	(1972)			Mohr, D.	2.1.2	1973	
McGowan, W.E.	2.7	1974			Monaghan, P.H.	2.5	1973	
McIntyre, A.D.	2.10	–			Monkman, J.L.	2.1.3	(1960)	
McKelvey, V.E.	2.6	1973			Moore, H.J.	2.2.3	1975	
McLachlan, J.	2.2.6	(1970)				2.2.4	1975	
McManus, D.A.	2.3	1972				2.2.6	1974	
McMinn, T.J.	2.8	1972			Moore, S.F.	2.1.5	1973	
McNeill, S.	2.9	(1972)				2.9	1974	
Measures, R.M.	2.1.2	1971				2.10	1973	
Mechalas, B.J.	2.4	1973			Morgan, J.	2.5	(–)	
Medeiros, G.C.	2.1.5	1974			Morgan, W.L.	2.1.2	(1970)(1971)	
Meehan, W.R.	2.2.3	1974			Morris, B.F.	2.2.6	1971	
Meek, R.	2.2.6	1969				2.7	1973 (1974)	
Meeks, D.	2.1.2	1971	(1973)		Morris, R.J.	2.5	1974	
Mendoza, R.D.	2.10	1971			Morrow, J.E.	2.2.3	1974	
Mersey Estuary Oil Pollut. Sci. SG	2.2.6	1969			Moss, J.E.	2.10	1962	1971
					Motohiro, T.	2.2.3	1973	
					Moulder, D.S.	2.11	1971	1975
Merten, E.W.	2.4	1973			Mulkins-Phillips, G.J.	2.2.6	1974	
Meschin, F.L.	2.2.3	1973				2.4	1973	1974
	2.4	1973			Munday, J.C.,Jr.	2.4	1970	
Meyer, P.A.	2.1.5	(1975)			Munjko, I.	2.2.6	(1972)	
Meyers, P.A.	2.2.7	1973				2.5	(1972)	
Meyers, S.P.	2.4	(1971) 1973 (1973)			Murakami, Y.	2.3	(1972)	
		1974			Murphy, T.A.	2.3	1970	
	2.10	(1973)						

Murray, S.P.	2.8	(1971) 1972		Okubo, K.	2.3	1972	
Murty, T.S.	2.8	1972 1973 1974		Oliver, J.D.	2.1.5	1973	
Mustonen, M.	2.2.6	1972		Olson, D.G.	2.1.2	1973	
Myers, J.	2.1.3	(1972)(1973)		O'Malley, M.L.	2.4	1970	
	2.9	(1972)		O'Neil, R.A.	2.1.2	1973	
Myers, L.H.	2.2.3	(1973)		O'Neill, T.B.	2.4	1972	
Myrick, H.N.	2.4	(1971)		Orlov, L.N.	2.3	1972	
				Ornstein, A.	2.4	(1971)	
Nadeau, R.J.	2.11	(-)		Orthlieb, F.L.	2.8	1971	
Nagell, B.	2.3	1974		Osterling, J.F.	2.9	1973	
Nagy, E.	2.2.7	1973		O'Sullivan, A.J.	2.2.6	1975	
Neal, J.	2.2.4	(1963)		Oren, O.H.	2.7	(1973)	
Nedoclan, G.	2.1.5	(1973)		Otsuki, T.	2.7	(1973)	
	2.5	(1973)		Ottenwalder, J.	2.2.1	(1969)	
	2.10	(1972)			2.2.5	(1969)	
Nelson, J.D.,Jr.	2.2.6	1973			2.2.6	(1968)	
	2.4	1973		Otto, L.	2.8	1972	
Nelson, R.F.	2.2.6	1972		Ottway, S.M.	2.2.2	(1971) 1973	
Nelson-Smith, A.	2.2.1	1971			2.2.4	1971 (1971)	
	2.9	1970			2.2.6	1971	
	2.11	1968			2.10	1972	
Neumann, H.J.	2.2.7	(1972)		Oudet, L.	2.2.6	1972	
Neushul, M.	2.2.6	(1971)		Oudin, J.L.	2.1.2	1971	
New England				Oudot, J.	2.4	(1973)	
Interstate				Overfield, J.L.	2.10	1972	
Wat. Poll.							
Cont. Comm.	2.9	1971		Paish, H.	2.10	1972	
New York Ocean				Parker, C.A.	2.2.2	1969	
Sci. Lab.	2.9	1971			2.2.7	1969	
Niaussat, P.	2.2.1	1969		Parker, P.L.	2.1.2	1971	
	2.2.5	1969 1970			2.5	-	
	2.2.6	1968		Parsons, R.	2.2.2	1973	
	2.4	1970		Parsons, T.R.	2.9	(1967)	
Nicholls, C.P.L.	2.10	1936		Parulekar, A.H.	2.7	(1974)	
Nichols, J.A.	2.2.7	1973		Pastorelli, L.	2.1.3	1971	
	2.3	(1973)		Payne, F.A.	2.2.6	1970	
Nicholson, N.L.	2.2.6	1972		Pearce, J.B.	2.2.6	1968	
Niemi, A.	2.2.3	(1973)		Pearson, G.	2.2.6	(1969)	
Nightingale, J.	2.1.1	(1973)		Peer, D.L.	2.2.6	1970	
Nigrelli, R.F.	2.2.5	1952		Pelkonen, K.	2.2.6	1972	
Noble, D.	2.1.2	(1973)		Peoples, S.A.	2.9	(1954)	
	2.1.3	(1973)		Percy, J.A.	2.2.4	1973	
Norris, L.A.	2.2.3	(1974)		Perdriau, J.	2.2.6	(1963)(1964)	
Notini, M.	2.3	(1974)		Perdriau, L.V.	2.2.6	(1963)(1964)	
Nümann, W.	2.4	(1973)		Pérès, J.M.	2.9	1972	
Nunuparov, S.M.	2.9	1971		Perkins, E.J.	2.1.5	1972	
Nuwayhid, M.A.	2.2.2	1973 (1974)			2.3	1973	
Nyms, E.J.	2.4	1969		Perna, A.J.	2.10	(1971)	
				Perry, J.J.	2.4	1973 (1973)	
O'Brien, J.A.	2.8	1970 1971		Pesando, D.	2.9	(1971)	
Oceanographic				Peters, A.F.	2.10	1974	
Inst. of Wash.	2.9	1972		Petrakis, L.	2.4	(1974)(1975)(1976)	
Odedra, D.A.	2.2.7	(1974)				(-)	
Ogren, J.	2.2.6	(1968)		Petrikevich, S.B.	2.4	1964	
O'Hara, L.C.M.	2.2.2	(1973)		Petronio, F.	2.1.5	(1973)	
Ohlsson, E.	2.1.2	(1973)			2.5	(1973)	
Ohya, M.	2.7	1973			2.10	(1972)	
Oikari, A.	2.2.3	(1973)		Piccinetti, C.	2.2.5	1967	
Okera, W.	2.7	1974		Picha, F.	2.2.5	(1966)	

Pierce, R.H., Jr.	2.7	(1974)		Rhoads, D.B.	2.4	1971	
Pilon, R.O.	2.1.2	1971	1973	Rice, S.D.	2.2.2	(1974)	
Pilpel, N.	2.2.7	(1974)		Richards, T.L.	2.2.4	1970	
Pincemin, J.M.	2.2.1	1969		Ridgway, N.M.	2.8	1972	
Pitt, K.	2.1.2	(1972)		Rigdon, R.M.	2.2.4	1963	
Poglazova, M.N.	2.4	1967	1968	Robbins, P.W.	2.4	1971	
Polak, J.	2.2.7	1973		Robichaux, T.J.	2.4	1971	
Poranski, P.F.	2.1.2	(1970)		Robinson, M.C.	2.2.3	(1973)	
Porricelli, J.D.	2.9	(1973)		Roe, J.C.	2.2.5	1966	
Portmann, J.E.	2.3	1968 1969 1972		Rogoff, M.H.	2.4	1957	1961
Pottier, J.	2.6	1971		Roppel, A.Y.	2.2.6	1969	
	2.10	1971		Roubal, W.T.	2.1.5	–	
Powles, H.	2.10	(1972)(1974)		Roush, T.H.	2.11	(–)	
Premack, J.	2.8	1973		Rudder, C.L.	2.1.2	1973	
Price, A.R.G.	2.7	1973		Ruel, M.	2.3	1973	
Priou, M.L.	2.4	(1969)		Ruivo, M.	2.9	1972	
Pritchard, P.H.	2.4	1973 1974		Russell, E.C.	2.2.7	(1974)	
Programmes Anal.							
Unit	2.2.7	1973		Sackett, W.M.	2.5	(1973)	1974
	2.9	1973		Sage, B.L.	2.10	1973	
	2.10	1973		Saito, M.	2.7	(1973)	
Prouse, N.J.	2.2.1	(1972) 1974		Sakuma, K.	2.3	(1972)	
	2.2.7	(1973)		Sanders, H.L.	2.2.6	1973	1974
Prytherch, H.F.	2.2.2	1935		Saner, W.A.	2.1.1	(1974)	
Puce, R.A.	2.9	1973			2.7	1974 (1974)	
Pugh, K.B.	2.4	(1971)(1973)(1975)		Sano, H.	2.2.7	1972	
Pulich, W.M., Jr.	2.2.1	1974		Sardou, J.	2.2.1	(1964)	
Purves, C.G.	2.1.2	(1971)(1973)			2.2.5	(1964)	
Pybus, C.	2.3	1973		Sass, J.	2.7	(1973)	
				Sauer, W.E.	2.8	1971	
Quayle, D.B.	2.2.6	1974		Saward, D.	2.2.3	1972	
Quigley, M.M.	2.2.1	1968		Scaccini, A.	2.2.5	1969	1970
Quinn, J.G.	2.2.3	1974		Scaccini-			
	2.2.7	(1973)		Cicatelli, M.	2.2.5	1965 1966 (1969)	
	2.5	(1973)		Scarrat, D.J.	2.2.6	1970	1973
					2.9	1974	
Rakowska, E.	2.4	(1973)		Scharrer, B.	2.2.5	1950	
Ramadan, F.M.	2.10	1972		Schatzberg, P.	2.1.1	1973	
Ramamurthy, V.D.	2.2.6	1974		Schenk, D.	2.4	(1968)	
Ramseier, R.O.	2.2.6	1973		Schlueter, R.S.	2.6	(1972)	
	2.10	1972			2.10	(1974)	
Rao, G.V.	2.8	(1974)		Schoental, R.	2.2.5	(1943)	
Rashid, M.A.	2.2.6	1974		Schramm, W.	2.2.1	1972	
	2.2.7	1974		Schwartzberg, H.G.	2.8	1970	
Rasmussen, L.E.	2.1.2	(1971)		Schwarz, J.R.	2.4	1975	
Ravanko, O.	2.2.1	1972		Schwemmer, G.K.	2.1.2	1974	
Raytheon Co.	2.2.6	1971		Sears, H.S.	2.2.3	(1974)	
Rechnitzer, A.B.	2.10	–		Sears, J.R.	2.2.7	(1973)	
Reddy, C.V.G.	2.7	1973		Sedita, S.J.	2.4	1973	
Reid, B.L.	2.2.7	(1951)		Seelinger, J.H.	2.5	(1973)	
Reid, G.W.	2.9	1972		Seesman, P.A.	2.4	1976	
Reid, W.K.	2.1.3	(1970)		Seki, H.	2.4	1973	1974
Reinheimer, C.J.	2.1.2	(1973)		Selkirk, J.K.	2.2.5	1971	
Reisbig, R.L.	2.8	(1971)(1972) 1973		Senger, L.W.	2.1.2	(1972)	
		1974		Serbanescu, O.	2.4	(1972)	
Reish, D.J.	2.3	(1971)(1972)		Shabad, L.M.	2.2.5	1966 1967 1968	
Resources Tech.					2.9	1967	
Corporation	2.2.6	1972		Shah, K.R.	2.1.4	(1971)	
Revelle, R.	2.10	1971		Shair, F.H.	2.1.3	(1973)	

Sharpley, J.M.	2.9	1966			Stanley, D.R.	2.2.4	(-)	
Shaw, D.G.	2.5	1973			Starr, T.J.	2.4	(1973)(1974)	
Shecket, G.	2.4	1971			Stegeman, J.J.	2.2.2	(1973)	
Sheldon, R.W.	2.9	1967			Stein, L.S.	2.2.2	(1974)	
Shelton, R.G.J.	2.1.5	1969			Stevens, D.B.	2.2.6	1969	
	2.2.3	1971			Stewart, J.E.	2.2.6	(1973)	
	2.3	1971				2.4	(1973)(1974)	
	2.9	1971			Stewart, R.J.	2.8	1974	
	2.10	(1971)				2.10	1973 (1974)	
Sherman, K.	2.7	1974			Stewart, S.R.	2.1.2	(1971)	
Shiels, W.E.	2.2.1	1973			Stockham, J.	2.1.2	(1971)	
	2.2.7	1973			Stokes, V.K.	2.2.7	1973	
	2.10	(1973)			Stora, G.	2.3	1972	
Shirley, O.J.	2.6	(1973)			Storrs, P.N.	2.4	1973	
Shubik, P.	2.2.5	1969			Strachan, A.	2.9	1972	
Shuttleworth, F.	2.3	1971	1972	1974	Strand, J.A.	2.2.3	(1972)	
Simmonds, P.G.	2.1.3	(1973)			Straughan, D.	2.1.1	1974	
Simpson, J.G.	2.2.4	1971				2.2.2	(1972)	
Sims, P.	2.9	(1964)				2.2.6	(1969) 1970 (1973)	
Sindermann, C.J.	2.1.2	1972				2.9	1971 1972 1973	
Singbal, S.Y.S.	2.7	(1973)			Strong, A.E.	2.1.2	(1973)	
Sipos, J.C.	2.2.3	1964			Stuart, D.	2.4	1971	
Sisler, F.D.	2.2.5	1947			Stumpf, H.G.	2.1.2	1973	
Skarstedt, C.B.	2.1.4	(1972)			Sturdevant, D.C.	2.2.3	1972	
	2.5	(1970)			Suchon, W.	2.8	1970	
Slater, P.G.	2.9	(1974)			Suess, M.J.	2.2.5	1970 (1970)	
Sleeter, T.D.	2.7	1974				2.2.7	1970	
Smith, J.T.,Jr.	2.1.2	1971				2.9	1973	
Smith, M.	2.11	1974			Sullivan, C.	2.9	(1971)	
Smith, P.C.	2.8	(1973)			Sullivan, C.E.	2.3	1971	
Smith, P.V.	2.2.6	1952			Sullivan, J.B.	2.1.5	1973	
Smith, R.O.	2.2.2	1935 (1935)			Supplison, B.	2.4	(1972)	
Smith, W.G.	2.8	(1971)			Sutcliffe, W.H.	2.2.6	1970	
Sohel, M.S.	2.1.2	(1971)			Sutherland, R.A.	2.9	(1973)	
Sohn, J.D.	2.1.2	(1971)			Swedish Natural			
Soli, G.	2.4	1971	1972	1973	Sci.Res.Counc.	2.10	1968	
Sonu, C.J.	2.8	1971			Swedmark, M.	2.3	1971 1974	
Sorensen, R.M.	2.8	1971			Sweet, W.E.,Jr.	2.7	1974	
Soudan, F.	2.3	1972			Swift, W.H.	2.10	1974	
Späing, I.	2.6	1968						
Spano, L.O.	2.9	(1973)			Takahashi, F.T.	2.2.2	1973	
Spears, R.W.	2.9	1971			Takita, Y.	2.3	(1972)	
Speers, G.C.	2.2.7	(1974)			Tanner, R.I.	2.10	1972	
Spellicy, R.	2.1.2	(1970)			Tarren, C.	2.2.6	1974	
Spencer, E.B.	2.8	(1971)			Tarzwell, C.M.	2.3	1969 1970 1971	
Spillane, K.T.	2.8	1971				2.10	1966	
Spooner, M.F.	2.2.4	1974			Tayfun, M.A.	2.8	1973	
	2.3	1971			Teal, J.M.	2.2.6	(1971)	
Spotiswood, T.M.	2.9	(1963)			Teal, J.T.	2.2.2	1973	
Sprague, J.B.	2.2.3	1969			Templeton, W.L.	2.2.3	(1972)	
Spray, M.	2.2.6	1974				2.3	1970	
Staley, T.	2.2.6	(1972)				2.10	1972	
St. Amant, L.S.	2.2.2	1957			Terry, S.A.	2.1.2	1971	
	2.2.4	1972			Thijsse, G.J.E.	2.4	(1965)	
	2.2.6	1971			Thomas, J.P.	2.2.1	1971	
	2.9	1972			Thomas, M.L.H.	2.2.6	1973	
	2.10	1971			Thomson, K.P.B.	2.1.2	1972	
Standard, P.G.	2.4	(1971)			Tissier, M.	2.4	(1969)	
Stander, G.H.	2.2.6	1971			Tkachenko, V.N.	2.2.1	1974	

Townsend, L.	2.1.2	(1971)		Ventilla, R.J.	2.2.4	(1971)
Traxler, R.W.	2.2.6	(1973)		Vernberg, W.B.	2.2.1	(1974)
	2.4	1973 (1974)		Vizy, K.N.	2.1.2	1974
	2.7	1974				
Treccani, V.	2.4	1962		Wade, T.L.	2.2.3	(1974)
Trites, R.W.	2.2.6	1974		Waldman, G.D.	2.8	1972 1973
Tsuruga, H.	2.2.4	–		Walker, J.D.	2.1.5	1977
Tsyban, A.V.	2.4	(1973)			2.2.6	1973
Tulkki, P.	2.2.6	(1972)			2.4	1973 (1973) 1974
Tye, R.	2.2.5	(1965)				1975 (1975) 1976 –
				Walkup, P.C.	2.3	(1970)
Uezumi, N.	2.2.3	(1961) 1967			2.10	1971
United States, Bureau of Mines	2.10	1926		Wallace, A.G.	2.1.2	(1973)
				Walsh, F.	2.4	1973
				Walstad, K.	2.2.6	1969
United States, Coast Guard	2.1.2	1971			2.11	1969
	2.9	1971		Walter, C.M.	2.2.4	(1973)
United States, Counc. on Envir. Qual.	2.10	1974		Walton, A.	2.2.6	(1973)
					2.5	(1973)
					2.7	(1971)
United States, Defense Doc. Centre	2.11	1972		Wang, H.	2.1.2	(1973)
					2.8	(1973)
				Wang, S.	2.8	1974
United States, Envir. Prot. Agency	2.2.6	1973		Wardley-Smith, J.	2.2.6	1974
					2.3	1971
					2.10	1971
United States, Interdept. Committee on Oil Pol. of Nav. Waters.	2.10	1926		Warner, D.G.	2.6	1972 1973
				Warner, J.L.	2.8	1972
				Warner, J.S.	2.1.5	1974
				Warner, R.E.	2.9	1969
					2.10	1969
United States, Nat. Acad. of Engineering	2.6	1972		Warren Spring Laboratory	2.3	1970
				Wash, J.	2.1.2	(1972)
United States, Nat. Acad. of Sciences	2.10	1974		Wasik, S.P.	2.1.5	1973 1974
				Watson, J.A.	2.1.2	(1971)
					2.2.6	1971
United States, Office of Water Res. Research	2.11	1973		Waybrant, R.C.	2.9	(1971)
				Wayne, T.J.	2.10	1971
				Webb, K.L.	2.6	(1970)
				Webber, H.H.	2.2.4	1971
University of California	2.2.4	1971		Weber, L.R.	2.1.3	(1973)
				Weiss, F.T.	2.1.5	1973
				Welch, R.I.	2.1.2	1973
Vagners, J.	2.10	1972		Wender, I.	2.4	(1957)
Vaituzis, Z.	2.4	(1975)		Wermund, E.G.	2.2.7	(1971)
Valiela, I.	2.9	(1974)		Westlake, D.W.S.	2.4	(1972)
Van Baalen, C.	2.2.1	(1974)		Westley, B.	2.2.6	1974
Vance, G.P.	2.2.7	(1972)		Whittle, K.J.	2.5	1974
Vanderhorst, J.R.	2.2.4	(1974)			2.9	1974
					2.10	(–)
Van Der Linden, A.C.	2.4	1965		Wiaux, A.L.	2.4	(1969)
				Wiebel, F.J.	2.2.5	(1971)
Van Melle, M.J.	2.1.2	1973		Wilder, I.	2.3	(1973)
Varley, A.	2.11	(1971)(1975)		Wildish, D.J.	2.3	1972 1973 1974
Vasserot, J.	2.2.5	1962		Wilkinson, P.	2.2.4	(1974)
Vaughan, B.E.	2.11	1973		Williams, D.	2.1.2	(1973)
Vauras, J.	2.2.3	(1974)		Wilson, J.E.	2.10	1971
				Wilson, K.W.	2.3	1974
					2.9	1973 1974

Wilson, R.L.	2.6	1973		Zahka, J.G.	2.10	1974	
Winters, J.K.	2.5	(-)		Zajic, J.E.	2.3	1972	
Winters, K.	2.2.1	(1974)			2.4	1972	
Wobber, F.J.	2.1.2	1971		Zalosh, R.G.	2.2.7	1974	
Wohlschlag, D.E.	2.2.3	(1972)		Zambachidze	2.2.2	(1973)	
Wolfe, L.S.	2.2.7	1972		Zanghr, L.	2.4	(1967)	
	2.10	1972		Zarazil, J.	2.2.5	1966	
Wolkoff, A.W.	2.2.7	(1973)		Zechmeister, L.	2.2.5	1952	
Wong, C.S.	2.1.4	(1974)		Zehle, H.	2.1.4	(1973)	
	2.2.6	1973	1974		2.2.7	(1972)	
	2.5	1974			2.3	(1972)	
	2.7	1974		Zeldin, M.	2.10	1971	
Woodley, D.J.A.	2.1.1	(1973)		Zillich, J.	2.3	1969	
Wright, D.E.	2.1.2	1974		Zillioux, E.J.	2.3	1973	
Wright, G.P.	2.1.2	(1973)		Zinn, D.J.	2.10	1971	
Wright, H.W.	2.10	1962		Zissiz, G.J.	2.1.2	1971	
Wright, J.A.	2.1.2	(1974)		Zitko, V.	2.1.5	1970	
					2.2.6	1970	(1973)
Yamamoto, S.	2.9	1972			2.9	1970	
YeDanil'Seva, G.	2.4	(1964)		ZoBell, C.E.	2.2.5	(1947)	
Yeh, W.K.	2.4	(1973)			2.4	1950	1959 1972
Yentsch, C.S.	2.2.7	1973			2.9	1964	
Yoshida, K.	2.2.3	1961	1967	Zoutendyk, P.	2.9	1972	
Young, L.	2.9	1939	1950	Zsolnay, A.	2.1.3	1973	1974
Young, L.Y.	2.9	1973			2.5	1972	1973 1974
Youngblood, W.W.	2.5	1973		Zuk, D.M.	2.1.2	(1974)	
Yuen, K.	2.3	1970					
				Anon.	2.1.2	1973	
Zafiriou, O.C.	2.1.3	1972	1973		2.2.6	1970	1971 1972
	2.2.6	1973			2.4	1970	
	2.9	1972			2.10	1973	